Systems Analysis for Applications Software Design

Systems Analysis for Applications Software Design

David B. Brown
Auburn University

Jeffrey A. Herbanek
IBM Corporation

 HOLDEN DAY, INC.

SYSTEMS ANALYSIS FOR APPLICATIONS SOFTWARE DESIGN

The opinions expressed in this book are solely the authors' and should not be
attributed to the IBM Corporation.

ISBN: 0-8162-1160-4

Printed in the United States of America

1234567890   HA   84

ju
6-2f-86

Contents

Preface

Systems analysts and software designers are at the cutting edge of applying technology that is undergoing revolutionary advances. They are being forced to cope with the demands of these changes simultaneously with the increasingly complex dictates of their organizations. While the objective of utilizing available computer technology to further the goals of an organization is not new, the need to combine dozens of programs, files, and nontechnical users into one integrated system presents a challenge that cannot be met with old tools. Advances in software design tools and software design methodology must be kept abreast of advances in hardware and operating systems software.

This work has the primary objective of providing a methodology for systematically translating complex user needs into computer software systems which can satisfy those needs. It provides the necessary theory and design methodology to enable individuals with a minimal amount of computer experience to acquire skills sufficient to enable them to formulate a software design and communicate this design to the programmer. It fills a widening gap which currently exists between the software specialist and the applications user, created by the diversity of their backgrounds and the resultant communication barriers. In addition, it addresses one of the most critical needs within data processing today—the generation of adequate system documentation for system development and maintenance. Finally, it takes students who are normally conditioned at this point in their academic careers to think of computer interaction in terms of "programs," and broadens their view to an appreciation of systems design.

The book can be viewed in four components, generally classified as: (1) introductory, (2) technical concepts, (3) methodology, and (4) case study. An attempt has been made to separate the technical concepts from the methodology; and further, to separate these from detailed case-study-type examples. The reason for this is motivated by a didactic philosophy on the part of the authors which separates basic conceptual learning from applications so that the student is freed to make new and creative applications in the future. Thus, concepts are presented in general terms first and examples are specifically chosen to promote, rather than limit, the understanding of these concepts. Further, a large variety of examples are integrated into the methodology sections to prevent the reader from locking onto any one application. Finally, a complete case study is presented in a separate part of the book to illustrate the concepts and to provide a model for new applications.

Chapter 1 falls under the introductory category. It begins with an historic perspective on the current problems and goes on to formalize these problems into specific needs. It presents a picture of the current environment in which software systems evolve, and it explores some of the basic philosophies of system design and development. In this context the overall design methodology is introduced to give the reader a look at the general approach to systems design before delving into details. This chapter is intended to build interest and motivate a desire to study the area of Systems Analysis.

Chapters 2 and 4 present the technical concepts. Chapter 2 concentrates upon analytical and documentation tools, while Chapter 4 deals with file design concepts. The objective of Chapter 2 is to present a consistent set of the most effective tools available. In this regard, the many applications of the Warnier diagram technique, the use of data flow diagrams, and some data dictionary concepts are presented. Similarly, Chapter 4 concentrates upon those file design concepts of most importance to the designer. These include concepts of indexing, selection of file type and organization, some elementary data structures, and file search techniques.

The methodology chapters begin with Chapter 3 where the process of analysis is explored. Beginning with the overall goals of the organization, analysis continues until the current system is documented. This is transformed into a series of new system output requirements, which forms the basis for the new system design. Chapter 5 picks up the methodology with the file design, once the basic concepts of file design are presented in Chapter 4. After files are designed, the methodology continues by presenting a stepwise procedure for synthesizing the system in Chapter 6. Chapter 7 transforms the documented design into a plan for the development of the software.

Part 2 of the book presents a complete set of documentation for one of the actual designs which was developed, programmed, and implemented entirely from the concepts and methodology of this book. In order to present this as a realistic model, the documentation of the design is given as it would appear in practice (ready for the programmer and user). While some advantages might be obtained by integrating the case study throughout the book, this would prevent the student from apprehending a complete target output of the software systems design efforts. Since such documentation presumes some knowledge of the affected organization, a commentary follows the documentation in order to obtain the full instructional benefit of the case study.

Two subthemes bear mentioning in presenting the layout of the book. The first reflects the overall ordering of the book: *it is necessary to define and document system outputs prior to attempting to specify processing or input requirements.* This approach, which might seem backward since it is contrary to the normal flow of data, has been proven to be the most efficient for software design. A second tenet is emphasized throughout: *it is necessary to totally complete the documentation before writing the first line of program code.* While this second is rarely practiced, experience has shown that inefficiency, if not catastrophe, results when this principle is violated.

The book was designed to serve primarily as a textbook and secondarily as a reference. There are two major target audiences: academic and industrial. The academic audience is further subdivided into the more technical engineering and computer science students who would take a course using this book early in their academic careers after mastering one or two language courses. Business and management students would take this course as part of their systems design and computer interaction sequence. The industrial target audience for this book would encompass all facets of the computer software industry, including all large and small companies which engage in software development activities.

Each chapter has a large number of questions and problems. Many of these problems go one step beyond the text presentation in an attempt to put the student into an environment in which creativity is encouraged. This is one of the most important aspects of teaching systems analysis and design: it cannot be accomplished just by the memorization of rules or by mastering many diverse techniques. Rather, students must be put into an environment in which they are stimulated to be creative. Thus, the emphasis in this book is as much upon integrating the tools into a methodology as it is upon learning the techniques themselves. Further, the methodology is presented as a discipline to keep the novice from the pitfalls that are difficult to avoid intuitively.

To further this process of instilling creativity, it is suggested that the material be taught roughly in the order in which it appears. The motivation for doing so is given along with a brief overview of the methodology in Chapter 1. It is suggested that this methodology be mastered at this time, even though the student will not have a full understanding of either the activities that each step entails or the tools to be applied. This is necessary so that frequent reference can be made to these *steps* as the course continues.

Heavy emphasis must be given to the mastery of the tools presented in Chapter 2. This will take a period of several weeks in which it is very possible for a student to "get lost" with respect to the overall objectives of the course. For this reason it is strongly advised that one period per week be dedicated to a laboratory session into which the case study can be integrated, thus keeping the target constantly before the student. To aid in this process, Table P.1 presents a cross-reference between the didactic material in the book and the case study. Note the inconsistency in ordering of the didactic material and the target document; the reasons for this are explained throughout the text.

Chapter 3 gives the student a breather from the tools, although many of them are reviewed in the methodology. This is a good point at which to define a comprehensive term project, or the first of a series of small projects, to be performed by the student. In this way the methodology learned can be put into practice immediately. A similar sequence can be followed for Chapters 4 and 5, the tools being presented first, followed by the methodology. Chapter 6 wraps up the methodology with a synthesis of the design documentation, while Chapter 7 transforms this documentation into a plan for the software development. Both of these chapters are methodological, and are therefore somewhat easier to master than the techniques chapters. This is advantageous in providing the student with ample opportunity for applying the techniques learned earlier by completing the project assignments. Any of these projects will be aided by the case study, which provides examples similar to those assigned, but for a different application. The case also serves to show the part that a student's individual project would play if integrated into a total design.

The authors wish to acknowledge the support of their families in this endeavor. Without continued encouragement from Carolyn and Joyce, this work would never have materialized. Mention must also be given to Mrs. Kathy M. Nix, who provided exceptional editorial assistance in the manuscript preparation, and to Mr. Marshall W. Magee whose help in the design of the MASTER System was invaluable.

Table P.1 Case Study Cross-Reference

Book Section	Case Section
2.1.2	C
2.1.3	MM 4.2
2.1.4	MM 4.2
2.1.5	MM 3.0
2.2.1	MM 4.1.1
2.2.2	MM 4.1.2
2.3.1	O & UG 2.0
2.3.2	MM 3.0
2.3.3	MM 2.0
2.3.4	MM 2.0
2.3.5	MM 4.1.3
3.1	C
3.2	MM 2.0
3.3	—
4.0	MM 3.0
5.0	MM 3.0
6.1	MM 4.1
6.2	O & UG 2
6.3	MM 4.2

O & UG	=	Overview and User Guide
MM	=	Design/Maintenance Manual
C	=	Commentary

TECHNIQUES AND METHODOLOGY

Chapter 1

Introduction

Computer technology is advancing at such a rate that any attempts to document the concepts currently being applied are destined for obsolescence before they are published. However, without occasionally taking a snapshot of current practice, we are in danger of losing many of the recent technological gains. A complete mastery of the basic principles can provide the foundation upon which future developments involving new and different applications can be built. The objective of this book is to clearly present some tools which have been proven to be successful in enabling software engineers and systems analysts to bring applications software into existence. This will be done by first presenting the tools and then integrating them into a software design methodology. It is understood that these tools may need to be modified to accommodate changes in software design technology as they occur.

In order to present the study of applications software design methodology in the proper context of computer systems design, an historical perspective is necessary. Thus, this chapter begins with a brief summary which shows how the current technological gains were attained. An attempt is made to project the innovative momentum of software design into the future, although there is no claim to any prophetical gifts on the part of the authors. Given an understanding of the current state of software design technology, the need for tools becomes more apparent. A discussion of this need, which encompasses the various objectives of this book, is given in Section 1.2. This is followed by two sections which present the concept of system evolution and structured documentation in order to introduce the environment and target of the design effort. The overall design procedure is then introduced, followed by a section related to the personal characteristics of the Systems Analyst. Finally, the overall approach of the book is presented in Section 1.7.

1.1 Historical Perspective

The development of computer technology cannot be divorced from the industrial revolution, although the analogy of the spouse might be more accurately replaced with that of a child. Arguments as to the specific date of birth of this child prove fruitless; however, most agree that for a major technological movement, it is, at most, in its adolescence (with all the problems thereof).

That a machine to process data would be born of the industrial revolution is not at all unreasonable. Machines were devised to help with most of man's physical labors during this period. While calculational devices, such as the abacus, obviously predated this period, the concept of *programmable* machines for calculation and data storage did not. The forerunners of the first true computers were merely mechanical devices. Joseph M. Jacquard is credited with the first programmable machine in 1801. This was a loom which used steel cards that served as a type of removable information storage, quite analogous to the various storage media of today's computers. The primary difference (other than the storage media) is that the programmable machines read and control machine performance by their variable input, whereas computers process the input data itself. The early programmable machines led to the development of the CAD/CAM (Computer Aided Design/Computer Aided Manufacturing) area, one of the fastest-developing applications in computer technology today.

Jacquard's loom was a significant indication that this period of time was ripe for the innovation of a machine to perform routine tasks with materials. The problem of transferring this technology to information processing was not one of conceptualization; rather, the hardware technology available at that time had not yet advanced enough to enable the performance of the desired tasks. In 1812 Charles Babbage put together what was called a difference engine to perform elementary calculations. While recognizing the need for intermediate storage, the hardware technology did not exist to enable the storage and retrieval essential to a practical computer.

It was over two generations later before the first practical application of this continually, although relatively slowly, developing technology could be made. In 1880 Dr. Herman Hollerith assembled a mechanical recording device that was used to tabulate census data. He established the first company dedicated to the production of such equipment, Hollerith's Tabulating Machine Company, which later evolved into International Business Machines Corporation.

Competition bore significant fruit as concurrent development of cardstorage and tabulating machines resulted in the formation of the Powers Accounting Machine Company, which was later purchased by the Remington-Rand Corporation. This equipment was used to tabulate the 1910 census data. Innovations were also progressing on the academic front as exemplified by the development of the Mark I calculator by Professor Howard Aiken at Harvard University between 1937 and 1944. While these were significant hardware gains, the storage devices used were still mechanical, as were most of the components used to perform calculation and tabulation. Thus, a major breakthrough occurred in 1946 with the completion of ENIAC (Electronic Numerical Integrator and Calculator), built at the University of Pennsylvania. This breakthrough was an outgrowth of work done by Atanasoff and Berry of Iowa State University, who are credited with the development of the first digital computer during the late 1930s. The ENIAC was used by the Ballistic Research Laboratories of the Army Ordinance Corps until 1955. However, while possessing storage and computational capability, the ENIAC did not have the capability to execute a stored program. The first computer to possess this capability was the EDSAC (Electronic Delay Storage Automatic Calculator) completed in 1949 at Cambridge University, England.

At this point the stage was set for the birth of the first generation of electronic computers, a period of time which lasted roughly from 1951 through 1960. During the late 1940's the following took place: (1) at Princeton University, the Institute for Advanced Study developed a computer with random access storage and parallel binary arithmetic, (2) the Servomechanisms Laboratory at Massachusetts Institute of Technology (MIT) developed a computer with coincident-current magnetic core memory, (3) the University of Manchester in England implemented the first practical electrostatic storage system, and (4) the United States National Bureau of Standards built the Standards Western Automatic Computer (SWAC), the first large scale computer in the United States that had stored program capability. Significant early first-generation computers included the UNIVAC (Universal Automatic Computer), produced by the Echert-Manchly Corporation which later became a division of Remington-Rand Corporation. Also of significance were the ERA 1103, manufactured by Remington-Rand, and the IBM 701 and 702. Advanced first-generation machines included the UNIVAC II, delivered in 1957. However, the IBM 705 which was delivered in late 1955 enabled IBM to take the lead in the increasingly competitive computer industry.

The mystique that grew up around the first-generation machines was characteristic of the naiveté of the 1950s. Science fiction, propagated through the movie industry, took advantage of the idea of the "electronic brain" to attribute to the computer the capability to take over the universe. While the benefits of free and open imaginative artistry are not disputed, some have seized upon the computer movement to build a power base out of superstition. To a large extent such ideas still persist among the ignorant today, to the detriment not only of the computer industry but also technological progress itself. Some in the industry have done little to dissuade the followers of this myth, perhaps enjoying the status inherent in its mystery. Such is counterproductive to progress, and a subtheme of this book is the need to clarify and define capabilities as they actually exist.

While the early computers seem to be quite primitive compared to those in the industry today, it should be recognized that conceptual design, especially in software methodology, is not that different. The primary breakthrough that enabled first-generation computers to be produced was the development of reliable hardware which had programmable capabilities. The advances since that time have also been hardware based, as we shall see in the second- and third-generation machines.

What has commonly been referenced as the second generation of computer design overlapped the first somewhat, lasting roughly from 1958 to 1965. While applications software and operating systems were taking advantage of expanded memory and increased reliability, this era was defined primarily by the integration of a major new component—the transistor. Replacing the first-generation vacuum tubes, the transistor enabled a major step forward in the number of applications that were made. This built momentum in terms of the psychological impact. Generally, corporate, military, and government leaders now considered the full-scale integration of computers into their operations to be inevitable, whereas it had merely been considered a possibility during the first generation. However, the primary mass storage medium was tape; card and paper tape data entry were most common, and sequential operation (generally referenced as batch processing) reflected the fact that first-generation architecture was still being used.

The third generation could also be characterized by the phrase: more of the same, but smaller. Indeed, once again hardware advances paved the way with the development of the microminiature integrated circuit. These smaller, more reliable, and faster components enabled the size of internal memory to expand; they also made possible the control mechanisms for direct-access (disk) devices for mass external storage. More sophisticated operating system software was developed. In one widely used architecture, interrupt capabilities and virtual addressing powerfully enhanced traditional batch processing. This led to multiprogramming and interactive time sharing, thus allowing several users and/or jobs to be resident in the total machine concurrently. Furthermore, virtual memory addressing removed expensive, wasteful storage fragmentation and again increased the number of concurrent central processing unit (CPU) processes. This resulted in much more efficient use of the CPU, which to this point had been delayed by the relatively slow input-output (I/0) devices, and the squandering of precious main memory.

Whether or not a fourth generation has fully arrived could be the subject of considerable debate. Again we would have to look to hardware leading the way. But the radical change in the industry in the late 1970s can be attributed to the efficiency of the production of computer components rather than any major breakthrough in the component types (such as was true when vacuum tubes were replaced by transistors). The means was developed for mass-producing extremely small silicon chips containing integrated circuits, some of which were as sophisticated as entire second-generation computers. This technology was available in 1970, and it led to smaller, more specialized machines, known as minicomputers. It was not until the mass production of the logic chip, however, that microcomputers became a feasible investment possibility. Starting with the personal computer, and employed to a large extent for games and amusement, microcomputers began growing up into business applications to the point that competition in this new industry became fiercely vigorous among venture startup firms and major well-established computer manufacturers alike.

The difference between a micro and a mini might be debated. In the context of generations, the mini was a product of the late third, while micros are the darling children of the early fourth. Generally micros have been characterized as being 8-bit word-length machines (which generally limits memory to 64K bytes), while minis have at least a 16-bit word length. This distinction is breaking down with many manufacturers coming out with 8-16-bit machines, and IBM's introduction of a 16-bit personal computer. Cost, of course is another factor, with sophisticated business micros (including Winchester direct-access hard disk drives) costing about as much as an expensive sports car, while minis are perhaps greater by a factor of ten.

Thus a major new dimension characterizes the advent of the fourth generation. As opposed to its being heralded by a major breakthrough in component design, it is the increased productivity in manufacturing of chips, stemming from miniaturization, that has enabled smaller machines to become practical in business. But the stimulus which has initiated the large capital investment necessary to produce these components is in large part the immense potential of the personal computer industry. Further, the capital investment necessary to bring a new computer onto the market is not nearly so large, since components are now available at a low cost. Thus, a host of new micro- and minicomputer manufacturers have arisen to compete for what appears to be a virtually unlimited opportunity.

This brings us to the current situation; and it might be asked what relationship this has with applications software design. The answer depends upon the latitude the designer has in choosing hardware. If higher powers mandate that a given piece of hardware will be used, or if time and other resources make this choice obvious, then the designer has no choice of hardware, and the design must be tailored within the constraints of utility imposed by it. It should be recognized, however, that obsolescence has often required a conversion to a new technology, ergo, a new machine. Considerable study in the design phase is essential to avoiding errors which would result from assuming that the hardware is static.

In many cases the designer is responsible for hardware selection, and the project scope expands to include this aspect of design. This is particularly true with the advent of microcomputers. Even if a large computer system (generally referenced as a mainframe) is available, it might be more advantageous to implement a new application on a micro which can be dedicated to the particular task at hand. The scarcity of systems development personnel and the administrative problems in obtaining priority in large organizations have often tipped the scales in favor of acquiring the additional micro or mini hardware.

While the above examples point to the need to understand the historical context of the computer industry, they are by no means exhaustive. A primary need on the part of the designer is to recognize the volatility of both hardware and complementary hardware-dependent software technology. This is most fortunate for the novice who, with concerted effort, can be at the forefront of this technology in a short period of time. At the same time it should alarm those who recognize that the experiences of more than five years ago accrue only on a conceptual basis, since the detailed specifics are in a state of flux. Specialists who believe that their lack of documentation has formed a moat around their castle kingdom may soon find it ineffective against more modern weaponry or, worse yet, that the castle, giving way to the high-rise, is of no further concern to the conqueror.

Looking to the future, based upon an extrapolation of the past, we see an era of continuous and increasingly diverse specialization as computer technology becomes more and more complex. But individuals within these specializations must be flexible enough to move with and, at times, ahead of the industry. Those who rely upon their past experience will quickly be left behind. This has been continuously true with the unskilled segment of the labor force, and it has had disastrous consequences for the individuals involved. At times it has even caused major problems for the entire world economy. However, professionals have been for the most part immune to this expected obsolescence in their skills. But as the computer industry progresses at an ever-accelerating rate, it will do so at the expense of those professionals who are unwilling to keep up with it.

1.2 Need for Planning and Design Tools

The historical context presented above concentrated upon the computer industry. While this industry may be progressing faster than most, others are not standing still. Advances are being made in a wide range of industries, each of which has its own areas of expertise and a need for specialists in those areas. It is unreasonable to expect that an individual who is, for example, an expert in petro-chemical processes is also going to be very knowledgeable in computers, despite the fact that he may work with them on a daily basis. Similarly, most managers probably have more than they can handle just keeping up with the personnel, materials, and machines within their operational realm. In fact, as job demands become more and more specialized we can expect that such "noncomputer" specialists will have less and less opportunity to become even remotely familiar with the inner workings of a computer.

This focuses attention upon the real problem that this book is attempting to address—the need for effective communication between disciplines. The computer does not exist for data-processing professionals. Rather, it is justified only by what it can accomplish in behalf of the productive effort of the rest of the organization. This being the case, it is essential that the communication barrier separating the disciplines be transcended. The effective design process will begin by establishing that a need for a proposed new system (or modification of an old system) does exist. Clear, concise, tangible definitions of end-user requirements will follow. A major part of the project will be dedicated to verifying that the design in fact meets the specifically established need. This verification process will continue through development to include carefully integrated modifications, the necessity for which might be uncovered late in the conversion process during user training or system testing. Analytical tools will be employed early in the design process to minimize unnecessary reiteration in the software development process.

To clarify this discussion, it is important that a distinction be made between the users, the designers, and the developers. While in some organizations two or even all three of these might be the same set of individuals, in the vast majority of systems of the size under consideration here, these would be three distinct groups of individuals. Hence, the following definitions:

> **Users**—individuals skilled in some operational areas of the organization (other than computer software design or development) who will ultimately use the computer (directly or indirectly) to facilitate the functioning of these operations.

> **Designers**—those engaged in translating user-defined needs into specifications from which new or improved applications of the computer can be developed.

> **Developers**—those who take the design specifications and produce the software in order to implement a new or improved application.

With an understanding of these definitions a clear distinction between the design and development processes should emerge. The major con-

Table 1.1   Problems Related to Designer Background

Designer Background		
Computer	Application	Problem
Heavy	Light	Inability to understand true nature of the application; may attempt to impose solution which does not address the problem; inability to get users to understand the design documentation.
Light	Heavy	Inability to understand true nature of the computer and its capabilities; may present a problem that is not within the scope of current computer technology; inability to get software specialists to understand the design documentation.

centration of this book is to present the design tools and to integrate them into a methodology for design.

In order to understand how design tools can help the communication process, consider the background of the system designer. Generally the designer may have a heavy computer background with some applications experience, or, vice versa, heavy applications experience with a smaller amount of computer software development experience. Table 1.1 summarizes these two possibilities and presents the problems that correspond to each. Only in the case where the designer has a heavy background in both applications and computer areas will these problems not create difficulty during design. These same problems will also arise throughout development unless this same designer, or someone with at least equivalent abilities, guides the development process through to implementation. While there are a few superpersons around that do possess a thorough balance of expertise in both areas, their numbers are dwindling and will continue to decline due to the diversity, discussed previously, which is evolving in all areas of technology.

The problems stated in Table 1.1 are further complicated in larger systems which require a design team approach. The lack of communications between design team members and the resulting inconsistent communications with the development team can lead to total system and development costs that are several times what was originally anticipated. Although the problems presented in Table 1.1 are diverse, the solution revolves around an improvement in communications. It is the primary objective of this book to present aids to communication to alleviate, if not eliminate, these problems.

It might be argued that perfectly effective communication between designers, developers, and users will not improve a poor design any

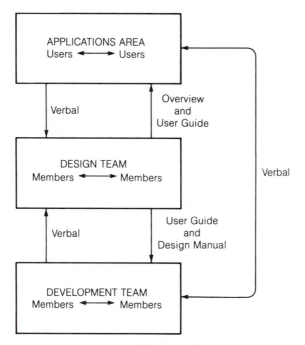

Figure 1.1 Interactions Requiring Effective Communication

more than effective communication between constituents and representatives will necessarily lead to legislation of sound policy. However, if effective communication takes place early enough in the conceptualization of the design and proceeds through the development process, the vast majority of problems will be directly addressed. On the other hand, no design text can compensate for incompetence in the highly technical areas of expertise required to bring a new system into being. A rich background in hardware, software, and the specific application must be present within the design and development teams. This is obtained through intensive technical training and years of experience. The major point is that, even if all of this expertise and competence is available, poor communication techniques might still cause failure. It is this primary problem that is addressed here.

Before closing this section, the difference between *planning* tools and *design* tools should be discussed. Both of these have the common intent of providing effective communication. Figure 1.1 shows the interactions which will necessitate effective communications. Note that these interactions are not only *between* the different major "areas," i.e., users, design, and development. As important are the *internal communications* that are essential to keeping each of these teams functioning. Indeed, many design tools are used to aid a designer in communicating back to himself or herself. All methods of presenting facts, including the many methods of both speaking and writing, fall into the realm of communication. While both planning and design tools are aids to communication, there is a distinct difference between them. Design tools are used to specify something that does not yet exist, while planning tools are generally used to aid the developers in systematically bringing the thing designed into existence. The primary emphasis in this book is upon written communication of a new system *design*, from the design team to the software development team. Considerable secondary benefits in terms of user satisfaction and upper level management understanding are obtained by applying these effective communication methods. Once the design is completed, a plan for development is still required. Thus, Chapter 7 is devoted to the subject of planning. Finally, it should be noted that the methodology given in this book (in Chapters 3, 5, and 6) is a generalized plan for producing a design.

1.3 Concept of System Evolution

Given that good planning and design tools are essential to bringing a software system into existence, we now turn our attention to the approach of the analyst in applying these techniques. It is essential that the analyst recognizes that systems are not developed in a vacuum, nor are they brought into existence as entirely new entities and operated exclusively apart from the internal environment of the organization. As opposed to this, most systems evolve; and the problems inherent in this evolutionary process accrue from a lack of forethought.

This evolutionary process is often called "maintenance"* when it involves the modification of an existing system to produce new or improved outputs. One ramification of this ongoing process to the systems analyst is to provide adequate documentation (i.e., a Maintenance Manual) which will facilitate this inevitable process. A second consideration relates to the analyst's attitude toward the evolutionary process. It must be recognized that the objective is to hit a moving target while shooting from an accelerating platform; and that, while hitting the bullseye is an impossibility, without making allowances for the relative movements, the entire target will be inaccessible.

Two competitive philosophies appear to provide possibilities for overcoming this problem: (1) the *modular* approach, and (2) the *systems* approach. While these two are not mutually exclusive, it assists our study to examine these first as "pure" approaches. The pure modular approach fits well into the concept of system evolution. By designing and implementing the system one module at a time, resources (both design and implementation) are available to allow the system to get initiated and evolve to any level of sophistication. By keeping each

---

*Of course, maintenance is not restricted to system enhancements.

module small, and getting all levels of management involved and participating in the design, greater control is maintained. However, the psychological effects of the modular design might be even more important. When one module is finished, it returns an immediately measurable return. This could be publicized, resulting in an initial momentum to the system. As a system's size, success, and evolution accelerate, so does its momentum. The greater a system's momentum, the harder it is to stop.

Critics of a pure modular approach often overreact in the direction of the systems approach. Obviously the modular approach cannot accommodate all interactions between current and future modules, some of which have not been given any design consideration. Changes in file structure, programming, and procedures may be required to bring currently operating modules up to a point where they can provide the data for a later module. These changes can be very expensive and extremely disruptive of the current system. Thus, they reason, it is better to thoroughly design the entire system before any implementation takes place. The *pure* systems approach would also implement the system as a unit.

Obviously, neither the pure systems approach nor the pure modular approach will be sufficient for most systems. The problems involved in implementing all components of a large system simultaneously are awesome. Many facts are not known at the outset of the design process, thus precluding a total system design from ever being considered complete. In reality there is a trade-off between design effort and ultimate system cost, as illustrated in Figure 1.2. Hence, design cost is shown as

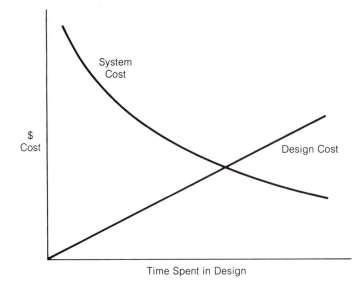

System Cost

$ Cost

Design Cost

Time Spent in Design

Figure 1.2 Design Time-System Cost Trade-off

a straight line, assuming a fixed design team effort. This line also includes the lost value accrued from not having the system operational. The system cost line includes the cost of implementation plus the present worth of operation, redesign, reprogramming, file rebuilding, etc. Note that increased design effort causes this cost to drop sharply at first, indicating a good return for this investment. However, as this effort continues, the cost of delaying implementation becomes excessive and the marginal cost of the design effort exceeds the value obtained by system. (Note: The two lines need not cross to illustrate this point.)

From the discussion above we can see a fallacy in trying to specify all aspects of where the system may evolve, as the pure systems approach would require. However, the logic supporting the systems approach is unaffected; the exceptions merely argue against taking the systems approach too far. Similarly, the high total system cost early on the time scale argues against over-modularization. Some degree of design in view of the total system is warranted, as is some degree of modularization.

The decision, then, is one of the degree of modularization and phased implementation, and not one of selection between the two extremes. It is a decision from a continuum of alternatives rather than from two discrete alternatives. As such, the decision is quite complex. For, although pure design costs can be readily measured, the cost reduction obtained by additional design effort and the cost incurred from delaying the implementation of a module are not easily determined. Understanding the relationship presented in Figure 1.2 enables the systems analyst to make a more intelligent decision in this regard.

In concluding this section, we recommend a modular design for implementation, which has been heavily influenced by total system considerations in terms of overall organizational objectives. It is recognized that design is essential to the success of any software development project. It is also recognized that a point can be reached where prolonged design effort and planning can be detrimental and that implementation must begin. Finally, due to the ease of management, the economic advantage of being able to proceed on a low budget, and the momentum to be attained by generating some quick system benefits, it is recommended that the *implementation* be modular for large systems.

1.4 Concept of Structured Documentation

The documentation of a system is used in every phase of its operation and maintenance. It is no exaggeration to state that the vast majority of data processing disasters today are caused by the lack of adequate documentation. When asked what is their single most damaging problem, most data processing managers will state that it is not a lack of software, but of deficient documentation. The only way for a system to survive inadequate documentation is for the designers and developers to assume the task of operation and maintenance, essentially keeping the documentation on the living tablets of the mind. While this may result in a functioning system, the self-destructive nature of such a situation is analogous to the accurate and reliable clock—in a time-bomb. The better it works now, the greater the long-term consequences. At best, the resultant system becomes statically enslaved to those who know it well.

Documentation is the product of the systems designer, and therefore it is the primary emphasis of this book. Good system documentation must precede the software development, and it must evolve with the system. Managers who allow software development to begin without adequate documentation have no one to blame but themselves when they cannot obtain it after the fact. The problem is, it generally *cannot* be obtained after the fact unless it is done by someone who is unfamiliar with the system. This is true because the "before-the-fact" perspective has been lost. The designers and developers have such intimate familiarity and experience with the system that it is virtually impossible for them to generate the type of documentation needed for system operation and maintenance.

The difference between documentation and good documentation needs emphasis. Quite often when asked for documentation, programmers and analysts will produce volumes of narrative, or a few tables and charts, and call this the system documentation. In most cases, they have made some notes during the course of development on some of the more complex aspects of the system. Calling it documentation does not make it adequate to perform its necessary functions. These functions include the following:

1. To provide an overview so that the highest levels of management can understand the benefits, objectives, approach, and integration of the system into the organization,
2. To provide middle management with an understanding of the impact of the system on their operations,
3. To provide all users with each of their respective user requirements and expected outputs, and
4. To provide all future maintenance personnel with sufficient information to make modifications on the system without having to perform a detailed analysis of existing program code.

The four functions above apply to documentation of existing systems. However, good planning necessitates that all documentation should *precede* the system developments which they detail. The major advantage of this is to provide programmers with clear system specifications, thus greatly facilitating their tasks and eliminating false starts, reprogramming, and other wasted efforts. This leads to a fifth function, which system documentation serves before the fact, that of specifying the design. This is true for both a new system design and a maintenance-type modification. If the documentation does not precede the development the following three problems will arise: (1) the perspective necessary to write detailed documentation will be lost, (2) there will not be adequate design specifications to lead to efficient programming, and (3) the resources necessary for adequate documentation after the fact will be allocated to another project, resulting in procrastination and neglect of the documentation effort.

It is not the intent here to place blame for a lack of good documentation. Usually the problem is with some*thing* rather than some*one*. This *thing* is the lack of proper design resources applied before development. Of course, upper-level management should ensure that these resources are adequate. But if the designers do not recognize that their product is the complete set of documentation, no amount of resources will suffice. In Figure 1.2 this would result in the inefficiency quickly shifting from not enough effort to too much effort without ever passing through the optimal. Also, programmers should not be expected to provide documentation either before or after the fact any more than company lawyers should be responsible for doing accounting (or vice versa). Programmers are specialists and they are best utilized (and generally happiest) when they are allowed to practice their specialty. Thus, programmers can hardly be blamed for poor documentation. We conclude that the problem is one of philosophy rather than personal inadequacy. Considerable time, effort, and money can be saved by upper management in insisting upon, and providing the resources for, the completion of the documentation before the fact. The time and effort for good documentation will have to be employed at some time; why not before the development process begins?

The discussion in this section has concentrated upon the need for before-the-fact documentation. It is important at this point to further define what is meant by structured documentation. Although all good documentation will be structured, not all structured documentation is good. Structuring is the act of building with plan and forethought such that the components fit together to accomplish the intended purpose. Given the five purposes of system documentation, we use the term structured to infer that it is planned by component and constructed such that these purposes are attained. Three objectives of structuring the documentation include: (1) to provide each of the documentation users with a mechanism to quickly and readily obtain all that they need from the documentation, (2) to spare any of the documentation users unnecessary work from being forced to read what they do not need, and (3) to minimize unnecessary redundancy. Note that the third of these objectives necessarily implies the existence of some "necessary" redundancy. The optimal amount and content of redundancy depends upon the satisfaction of the first two objectives.

Table 1.2 Model System Documentation Structure

Major Component	Primary Target	Component Contents
Overview	Highest Management	General Output Input Descriptions
User Guide	All Users, Middle Management	User Requirements Input Specifications
Design/ Maintenance Manual	Programmers and Technical Support Personnel	Output Specifications File Layouts File Structures Program Specifications Flow Diagrams

Some additional concerns regarding the problems of redundancy might be expressed before proceeding. Comparatively speaking, paper is cheap with respect to technical personnel search time, and it might be reasoned that the repetition of certain specifications might be cost-beneficial. Contrary to this approach, we will adopt a policy toward extremely heavy referencing instead of redundancy. This is not done to save paper; rather, experience has shown that effective updating is virtually impossible when there exists a high degree of redundancy within the documentation. If a specification appears only once, and all other uses of it are referenced to it, then a change in that specification will require only one change in the documentation. The alternative is either to totally review the documentation with each update (a good practice in any event, but one which does not guarantee that all such specifications will be found), or to use a cross-reference list of redundancies (which itself must be maintained). Thus, ease of update and the prevention of internal inconsistency is the primary motivation for heavy referencing as opposed to introducing redundancy.

A structure will now be introduced which will provide the model to be followed throughout this book. Table 1.2 presents this overall structure. Note first that there are three components of the documentation: (1) the Overview, (2) the User Guide, and (3) the Design/Maintenance Manual. The Design Manual and the Maintenance Manual are considered simultaneously since the Design Manual will be converted to the Maintenance Manual once the system has become operational. Thus, most of the documentation essential for efficient system development and implementation can be used for maintenance purposes. There are two exceptions: (1) that material which is used purely for design, and (2) the additions made to the Maintenance Manual after implementation. Despite these exceptions, it is a major objective of the approach given here to assure that the Design Manual satisfies the major documentation effort required for effective maintenance.

At this point it is helpful to compare the major components of the proposed documentation structure with the objectives. This leads to the target specification for each of the major components. Each represents a different level of detail, which is targeted at a different audience as shown in Table 1.2. However, it is assumed that the less-detailed documents are also available to and read by those requiring more detail, even though it is not specifically targeted for them. In other words, the Overview will be read by everyone; and the User Guide will be read by the programmers and technical support personnel. This serves to eliminate redundancy, and at the same time higher-level readers are spared unnecessary detail.

To better understand the objective of the components of the documentation, refer back to Figure 1.1. Since we are primarily concerned with design, the design team is the central focal point of this figure. Note how the various components of the documentation flow as output from the design team while the primary inputs to the team are verbal. While the design team should expect some examples and exhibit outputs from the users, these will require considerable upgrading before they are usable. The source of this upgrading, aside from their own creativity, is verbal communication with the users.

Table 1.2 also shows a finer breakdown of each major component in terms of its contents. The Overview concentrates upon the features of the system, which include a brief summary of the output capabilities as well as user compatibility and projected cost savings features. As its name implies, it is to be kept quite general and brief. While it may contain a fair amount of sales-oriented narrative, it should have enough information about the system to give all documentation users a feel for the goals and objectives of the system as well as the overall strategy for meeting those goals and objectives.

The User Guide may be reviewed by higher management, but it is targeted for users and their immediate supervisors (middle management). The user here includes both the data entry personnel and those using the system to obtain information. If these are two distinct groups of people, and if the User Guide is large, it might be desirable to divide it into two volumes, each targeted to its respective user group. Note that a fair amount of programming information, such as menu structure and content, will be obtained from this document, thus eliminating the need for repeating this in the Design/Maintenance Manual. Since this is the case, a small amount of added detail not totally essential to the user might be added to the User Guide. For example, all queries might be spelled out in full instead of just describing the query in narrative form. While either would serve the user's needs, the added detail could save the programmer considerable consternation. Also, mention might be made of the specific file being updated or used. Although this might not be necessary to give the user a feel for the system,

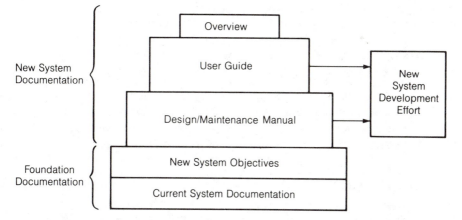

Figure 1.3 Model Documentation Structure

and the basic target of this detail is the programmer, it is justified since very little if any user time is consumed by it.

Finally, the Design/Maintenance Manual is the property of the programmers, analysts, and associated technical personnel. Information contained therein should not be required to operate or use the system, and it is certainly not to be used for system description. As such, these can be excluded from the more global components of the documentation. All components must be developed with the understanding that both their primary targeted audience and their secondary audience must be satisfied by the sum total of the system documentation.

The remaining chapters of this book will detail the contents of the major portions of the structured documentation. Figure 1.3 serves to summarize this section as well as to project the analysis and design process which will be followed. The foundation upon which the new system documentation is based is quite substantive, being the current system documentation and the objectives of the new system. These will be integrated into an idealized logical system documentation by an analytical process discussed in Chapter 3. Based upon this foundation, the new system documentation will be developed as a set of parts interlocked by cross-references between the various subcomponents. When the new system documentation is completed, it provides the input to the development effort.

1.5 Systems Design Procedure

The purpose of this section is to present the stepwise order of the actions necessary to bring a new system into existence. This will provide the background for the detailed presentation to follow in subsequent chap-

Table 1.3   Design Step Relationships to Documentation Components

Step	Documentation
DESIGN ANALYSIS	
1. Analyze the affected management system	Overview and Current System Analysis
2. Analyze system output requirements	Design Manual Output Specifications
3. Analyze input availability	User Manual Input Sources
DESIGN SYNTHESIS	
4. Specify file layouts and structures	Design Manual File Designs
5. Solidify flow diagrams	Design Manual Flow Diagrams
6. Specify input requirements	User Manual User Requirements
7. Develop program specifications	Design Manual Program Specifications
DEVELOPMENT AND IMPLEMENTATION	
8. Develop system software	Use all documentation; update as required
9. Perform conversion; initiate operation and maintenance	Design Manual becomes Maintenance Manual; maintain documentation

ters. As such, the steps presented here will be discussed only to the extent that they can be introduced; a thorough understanding of their objectives and contents will require further study of subsequent chapters.

Table 1.3 presents the steps in the design and development process in terms of the structured documentation component which is produced at each point. It is important to note that the documentation does not grow from one end. Rather, it expands primarily from the middle. With common word processing capability, the updating, augmentation, and rearranging of documentation has become quite routine. Thus, certain parts of the documentation will be created at one point with the intent of their being modified in a subsequent step. Similarly, the steps followed are not perfectly sequential. Rather, they are subject to overlaps, reiteration, and interaction. This will be emphasized, when appropriate, as each of the steps are introduced in a separate subsection below. Beginning in Chapter 3, each of these steps is further discussed in terms of the detailed activities required for its respective accomplishment.

1.5.1 Analyze the Affected Management System

This is performed by a structured approach using a special management system Warnier diagram (MSWD) designed for this purpose (see Chapter 2). The particular part of the total organization affected is defined first in order to determine the scope of the analysis. Then the goals and objectives of this subset of the organization are defined. This leads to a specification of the missions and projects. Projects are further analyzed into activities, each of which is composed of a decision-making process followed by its resultant action or actions. Both the decisions and the actions are further analyzed into their respective procedures. A procedure step will generally either require or generate information, either formally or informally. It is at this point that the analysis is ready to address itself to system output requirements.

The affected management system (AMS) is not the new software; rather, it is the environment within which the new software system must be integrated. In documenting the AMS, some liberties may be taken to modify existing procedures to what they should be (as opposed to what they are). The output of this step provides an understanding of current system needs that is impossible to gain otherwise. This documentation is used primarily as a tool to form the foundation for future steps. If it is thought to be of value to retain this in the new system documentation, it will form an appendix of the Design Manual. If not, it can be filed in the system development archives for future reference.

Another output of this step listed in Table 1.3 is the Overview. The Overview forms the basic document for selling the system. It should be realistic and subject to change as the design evolves. At this point the Overview may be required to obtain funding for the remainder of the design process.

1.5.2 Analyze New System Output Requirements

The analysis performed in the step above crystallizes information needs in the various procedures. It is essential that the structure of step 1 be clarified first, prior to specifying new system output requirements. As mentioned above, there are two systems of concern here: (1) the AMS, and (2) the software system being designed. This is a transitional step. In the first phase of this step consideration is given to the management system. Data Retrieval Transactions (DRTs) and Data Storage Transactions (DSTs) are broadly defined in terms of current (including tentatively projected) data outputs and inputs. These broad definitions need not be detailed; they may reference current documents, such as purchase order forms. These provide a broad definition of the data flows of the AMS, generally only a subset of which will be incorporated into the new design.

The second phase of this step formalizes the data flows and processes. A tentative data flow diagram (See Chapter 2) is developed for the current system to aid all concerned in understanding it. This is an essential communication device since only those directly involved with the day-to-day operations of the affected management system have a thorough understanding of it. This data flow diagram should be totally devoid of technical jargon, since its purpose is to give all levels of management concerned, as well as the systems analysis staff, an understanding of the current system.

During the third phase of this step, attention is turned to the new system and a decision is made as to which portions of the existing system are to be transformed into the new system design. This is thoroughly defined by a documentation of the output requirements of the new system. Further concentration is given to developing the DRTs of the transformed management system Warnier diagram into actual output reports. The output of this step is a complete set of output specifications which go into the Design Manual.

1.5.3 Analyze Input Availability

This step is similar to step 2 in that it uses the output of step 1 as its primary information source. However, there are three primary differences: (1) new system inputs are under consideration rather than outputs, and thus attention is turned to the DSTs originally defined in step 1; (2) the degree of the analysis is not as specific, i.e., only availability is considered, and no specifications are made; and (3) it is recognized that the DSTs analyzed in step 1 may not provide all data requirements and thus other data resources external to the AMS are analyzed. The objective of this step can only be understood in the context of the other steps. Namely, it is to assure that the data required to provide the outputs specified in step 2 are available. This is essential before proceeding to file design.

The output of this step serves to emphasize what it does not do. It does not specify the input; that would be premature at this point. However, it is essential to identify the data sources and any deficiencies at this point such that a reiteration of step 2 can be made if necessary. Thus, the output of this step is a statement of inputs and their sources. This goes into the user manual as a preliminary reference. It will eventually form the basis of the input specification development which is performed in step 6.

It should be re-emphasized at this point that even though the design steps are shown in sequence, they overlap considerably. Allowance must also be made for reiteration, which would be required at this point if, for example, the output requirements documented in step 3 could not be generated from any data that could be made available. While these allowances are to be made, the structure of the stepwise approach serves to provide an orderly method for proceeding.

1.5.4 Specify File Layouts and Structures

This is a very critical step, the first synthesis step and, in a sense, a point of no return. To this point a modification in system output requirements could be accomplished by minor changes performed in short order on a text editor. Once the file structure and layouts are defined, however, changes in output specifications may have ramifications throughout the system. Further, the choice of the file type, its organization, and its structure is by far the most important decision affecting the future operation of the system. For this reason, two chapters are devoted to this step. The first is concerned with file design concepts (Chapter 4), while the second provides a continuation of the design methodology (Chapter 5). The output of this step will be the file design component of the Design Manual.

1.5.5 Solidify Flow Diagrams

Some data flow diagrams for part or all of the existing system were drawn in conjunction with step 1 to aid in understanding the existing management system. These were transitioned into tentative data flow diagrams for the new system. Now that the files have been clearly defined, step 5 concentrates upon the new design with the objective of solidifying the picture of the interaction between system components. Two types of flow diagrams are employed: (1) the data flow diagram, which incorporates standard system symbology, and (2) the Warnier flow diagram which uses the Warnier analytical tools to specify the interactions between the programs and files. Each has its own advantages. The output of this step is the set of flow diagrams which are included in the Design Manual. These provide both an index and a table of contents to all of the remaining system documentation.

1.5.6 Specify Input Requirements

The input requirements proceed directly from the file layouts and the system program structure determined within steps 4 and 5. At this point these inputs are defined to the degree that there is no question in the user's mind as to what the data entry requirements are. This logical structured specification is written in a user-oriented way in such a manner that it can be incorporated into the systems functions section of the User Manual.

The term *input requirements* logically divides into two possibilities: (1) data-entry requirements and (2) user operating requirements. Whether these are organized separately in the User Guide will depend upon whether or not separate users are involved in these two functions. Data entry requirements will be satisfied by data-entry mats and queries (also called *prompts)*, while user operating requirements will be satisfied primarily by responses to menus and queries.

1.5.7 Develop Program Specifications

Given that the system inputs and outputs have been specified, most of the program specification efforts will be straightforward. A major problem encountered at this point involves structuring. The methodology will employ a modular structuring of the system to specify each program component of the system. Techniques will be given such that this will be done by a total enumeration of the system requirements. These program specifications pull all of the system documentation together and thus provide a final validation of the system design.

1.5.8 Develop System Software

The next two steps presented are not part of the design process. Throughout this book it will be maintained that all documentation should be completed before the first line of code is written. However, it will be readily admitted that documentation modifications will be required as software development uncovers flaws in the original design. While these flaws should not occur, we would be naive to think that they will not. Thus consideration will be given to maintaining the documentation through the software development process.

1.5.9 Perform Conversion

This involves the transition of those functions previously performed by the old system to the new system. Here again, no new design considerations will be made at this point; however, considerable reiteration of past steps might be required depending upon how well these past steps were executed. As required changes are made in software or operating procedures, they will be incorporated into the documentation. Finally, once conversion is made, the Design Manual will be transformed into the Maintenance Manual by making any changes which this might require.

1.6 Requirements of the Systems Analyst

The term *systems analyst* is normally applied to a person who analyzes a perceived problem, its environment, and its possible solutions, and then goes on to design a new or improved solution to the problem. Therefore, a systems analyst does synthesis and integration work as well as analysis. In this section the personal traits and talents which seem to characterize the successful systems analyst are discussed. Of course, no checklist of this kind is foolproof, and there are instances of successful analysts who may lack some of these qualities. On the other hand, this discussion should help to bring into focus both the traits and the activities that define the career of the systems analyst.

There are a tremendous variety of jobs that systems analysts perform, both because of the breadth of activities required to design a given software system, and because the systems themselves can differ so markedly. Consequently, analysts are usually generalists with broad backgrounds. This should not be interpreted to mean that an analyst needs to know nothing in depth; rather, it requires that analysts be multitalented and competent in a number of areas.

Accompanying breadth must be flexibility. Systems analysts must be adaptable to the task at hand. This characteristic must be developed to the point that an analyst can acquire new specialties when required to do so. Not only should this ability be present, but it must also be enjoyed. A good analyst does not resist a new learning situation. While fickleness and impatience are to be discouraged, the professional analyst must be willing to leave the comfort of familiar experiences for new challenges.

Flexibility, coupled with the ability to master new specialties, must be tempered with the ability to resist the temptation to become involved in too much detail. In many organizations the analysts formulate and document the design, but they do not complete the development process by coding their own designs. If the analyst is involved in the development process, it will be at a supervisory level since, in most cases, the professional programmers are much more proficient at software development. While analysts are not expected to be expert programmers, neither are they as proficient in the particular application area as are the line managers. In both cases the good analyst will leave the detailed work to the specialists, retaining the generalist's pose.

But the systems analyst does have a specialty. As an intermediary between the ultimate users of the system and its implementors, the analyst plays a crucial role. Written and oral communications skills are vital. In fact, communication is the single most important trait of the successful analyst. This re-emphasizes the need for breadth, since a systems analyst must be able to communicate with people from all of the affected disciplines. This skill requires the ability to speak and write in a language that is understood by the broad array of affected disciplines. But this communication must go beyond the matter of understanding and being understood. The effective analyst will become user-oriented to the degree that an empathy with the user will be established, i.e., feeling what they feel. The art of listening must be thoroughly cultivated.

The analyst's primary function, then, is to provide the communication link between the end users and the programmers. The diversity between these groups is exemplified by the fact that, as a rule, direct communication between these two groups on a technical level is rarely effective. Technical communication with programmers is conducted in a completely different language than that of the end user. The systems analyst must master both.

Although some talents necessary to perform the systems analysis function can be learned through formal education, other characteristics can only be obtained by concerted effort in the field. The analyst will not succeed without an inward sense of purpose. This is not to say that certain characteristics are inborn. Rather, they are developed through experiences, some of which can be quite painful. Thus, if the desire is not present, no amount of formal education will compensate.

As far as education is concerned, two things aid tremendously. The first is a well-rounded cultivation of ideas from many disciplines, with some degree of depth in a few. The necessity for this should be quite evident. The analyst must function to a great extent between disciplines, drawing on experience and intuition in psychological, sociological, management, technical, and communication areas. The types of problems encountered are so large that a narrow education is totally insufficient. Although a narrow academic concentration does not prevent someone from broadening themselves later, quite often those who do specialize lack an appreciation for other disciplines as well as the ability to adapt to the specific problem at hand.

A second educational requirement calls for emphasis upon quantitative and logical problem solving. A technical specialty in some branch of computer science or engineering is essential. Although a person may initially lack the broad background characteristics, the properly motivated, technically trained individual usually can acquire the experience that is needed for the performance of the systems analysis functions. However, the nontechnician, whose mathematical ability breaks down at simple algebra, is not likely to acquire quantitative expertise without great effort.

One might think that the previous two paragraphs contradict each other, and in terms of many traditional academic programs, they do. However, many universities are developing academic programs in systems analysis, which attempt to provide much of the necessary breadth within a technically oriented discipline. Such programs are usually offered at the graduate level, and they assume that the student has both a strong technical undergraduate degree and, possibly, some practical experience.

More than a discipline, systems analysis has been described as almost a way of life. Although it has been described in rather glowing terms, hopefully it is clear that everyone should not strive to be a systems analyst. This is a function to be performed only by a certain type of individual. Although everyone can benefit from systems analysis concepts, and most will use them sometime during their careers, the professional systems analyst life should only be sought by those who believe that their own personal satisfaction will be served by this type of occupation.

Remember that the systems analyst cannot succeed without a fine staff of specialists, each concentrating in a particular technical area. The systems analyst's technical specialties are organization, documentation, and design. These do not bring a system into being by themselves. For each systems analyst, therefore, there must be several systems programmers, software engineers, and others. These are people who grapple with details and have considerably less interaction (i.e., interaction is not their primary occupation).

The purpose of this subsection is not to sell the systems analyst career to the individual. In general the systems analyst is no nobler, no more ethical or moral, and no happier than anyone else. However, there is an obvious need for systems analysts within most organizations. The characteristics of successful systems analysts have been described so that those in upper management, as well as students of computer science, can respond to the need.

1.7 Approach of the Book

Given the need stated above, every effort has been made to organize this book to effectively present those design and planning tools necessary for the efficient generation of applications software. There were a variety of approaches that could be considered, including case study, purely didactic, chronological, and several others. Each of these could range from a purely theoretical treatment to a very practical cookbook approach. The approach taken has been to start with a purely technical treatment of the tools, followed by a detailed application of these tools to real situations illustrated by actual designs with which the authors have been closely involved. This approach has been taken so that the detailed examples will not interfere with the theoretical treatment, and vice versa. The design process is quite complex, requiring the simultaneous consideration of the many tools presented. Thus, it behooves the student to become quite familiar with all of the tools presented in the technical overview of Chapter 2 before proceeding to their applications.

Chapter 3 presents the methodology of the analytical process, roughly in the chronological order in which it would typically be performed. This brings the reader right to the point of file design. It is at this point that additional technical concepts must be presented before proceeding with the methodology. Thus, Chapter 4 presents some basic concepts of indexing, file type and organization, and some basic data structures. After these are presented, the methodology is resumed in

Chapters 5 and 6. Chapter 5 discusses the methodology of the file design, which is a pivotal step between analysis and synthesis. Chapter 6 then outlines the methodology for synthesis of the design, including input and program specification, drawing from the tools presented in Chapter 2. Finally, Chapter 7 is dedicated to planning of both design and development functions. The latter of these translates the final, completed design into an organized plan for the software development.

Part II of the book presents a complete example of one of the actual designs which has been developed, programmed, and implemented entirely from the concepts and methodology described in this book. In order to present this as realistically as possible, the documentation of the design is given as it would appear in practice (ready for the programmer and user). However, it is recognized that the reader will not have the familiarity with the system that would be expected of experienced users and programmers. Thus, a commentary is provided after the system documentation to provide more background into the case. The case itself provides a complete example of the target documentation generated using the tools, techniques, and methodology given in the text.

References

Adams, David R.; Gerald E. Wagner; Terrence J. Boyer. *Computer Information Systems: An Introduction.* Cincinnati: South-Western Publishing Co., 1983.

Booth, Grayce M. *The Distributed System Environment.* New York: McGraw-Hill, 1981.

Eldin, Hamed Kamal, and F. Max Croft. *Information Systems—A Management Science Approach.* London: Mason and Lipscomb Publisher, Inc., 1974.

Edwards, Perry, and Bruce Broadwell. *Data Processing.* Belmont, Calif.: Wadsworth Publishing Co., 1982.

Gore, Marvin, and John Stubbe. *Elements of Systems Analysis.* Dubuque: William C. Brown Company Publishers, 1983.

Kindred, Alton R. *Systems and Management.* Englewood Cliffs: Prentice-Hall, Inc., 1980.

Matthews, Don Q. *The Design of the Management Information System.* New York: Moffat Publishing Company, Inc., 1976.

Questions and Problems

1. Outline the key individuals in the history of computer technology development and briefly state their contribution. Include dates.

2. Repeat problem 1, doing the same with company names.

3. What are the various computer "generations"?

4. What is the difference between a micro, a mini and a mainframe?

5. How does hardware choice affect software, and vice versa?

6. Why does a novice have an advantage in the computer field that he might not have in another? How can this advantage be lost?

7. Discuss the problems associated with the programmer who feels secure because no one else understands the software he has developed.

8. Why are communication tools particularly essential in the area of software design?

9. Counter the argument: Even perfect communication cannot make up for a poor design.

10. State the difference between a planning tool and a design tool. Give an example of each.

11. Define the term *maintenance* in general; then state how it has been applied to software systems.

12. Give the advantages and the disadvantages of: (1) the pure modular approach, and (2) the pure systems approach. How is an optimal balance between these achieved?

13. Show the relationship between planning effort and total system costs. Explain in one sentence the rationale for the decision to stop (or curtail) the design effort and begin development.

14. Name the dangers inherent in poorly documented systems.

15. Name the five functions of structured documentation.

16. Give the reasons for insisting upon a complete set of documentation before software development.

17. What makes the person who knows the most about a system inadequate to produce its documentation after the fact?

18. Give the advantages and disadvantages of redundancy in the documentation.

19. Name the three components of structured documentation and the audiences to whom they are directed.

20. How is redundancy eliminated by structuring the documentation?

21. Is the process of design a purely creative process? Explain in terms of the tangible basis upon which the design rests.

22. What errors would be committed in trying to sit down and write the Overview, then the User Guide, and then the Maintenance Manual, in that order?

23. Define what is meant by the affected management system (AMS).

24. List the nine steps required to bring a new software system into existence.

25. Name 10 characteristics that a systems analyst must possess to be effective. Does anyone possess these traits?

Chapter 2

Software Systems Design Concepts

The difficult part of applications software systems design is in applying the tools, not in understanding the tools themselves. The techniques are, theoretically speaking, quite simple to master. In fact, their simplicity might evoke a false sense of confidence in the novice system designer. The simplicity of the tools results in a great need for creativity in their application, just as the inventiveness required in using simple mechanical tools usually exceeds that required to operate the complex machines upon which they may be employed. To proceed with this analogy, obviously the understanding of the simple screwdriver or wrench is not nearly as important as understanding the machines upon which they are to operate. Here, the machine is analogous to the software design and development process (not the system being computerized or the computer itself). The tools of concern here act directly upon the design process and indirectly in improving the development process. And, just as machines are a means to manufacture products, these processes are a means to producing software.

The objective of this chapter is to present a comprehensive technical review of the tools to be employed in performing software systems design. They are introduced first to prevent their detailed presentation from obscuring the methodology presented later. For this reason, only brief examples will be discussed in this chapter, wherever they are thought to aid the theoretical presentation.

Three types of tools will be examined: (1) analysis and specification tools, (2) flow diagrams, and (3) data dictionary components. These are organized strictly for didactic purposes so that similar techniques can be discussed within the same subsection. As will be shown in the methodology chapters which follow, this is not the order that the techniques are applied in practice to generate a design. Neither is it the order in which results of the various techniques might be found in the system documentation. Either of these orders of presentation, while seemingly logical, would result in a very disorganized study. For this reason we beg the reader to have patience and first master these techniques separate from the methodology into which they will be integrated (i.e., the stepwise procedure given in Table 1.3). Not only will this facilitate the mastery of these tools, it will also provide for an efficient study of the methodology. Further consideration of the integration of tools into methodology is given at the end of this chapter in Section 2.4.

Analyses of the existing and new systems are generally required prior to the actual system design. Thus, a primary tool of analysis, the Warnier diagram, is presented first. However, the use of Warnier diagrams for other applications, including the definition of program logic, will also be presented at this point. Once the analysis is completed, the system design begins to take shape by means of the synthesis of the various components into a functional unit. This synthesis, including the demonstration of the interactions between the components, is accomplished by means of flow diagrams, the second classification of design tools. In turn, the data flow diagram is supported by its accompanying documentation, to which it serves as a master index. A major part of this documentation is sometimes referenced as the data dictionary, which is presented next in terms of the principles applied to its generation and development. The flow diagrams, the data dictionary, and the program specifications should totally define all of the technical details of the new system. This, together with the nontechnical documentation, will be sufficient for bringing the new system into existence.

2.1 Analysis and Specification Tools

Analysis and specification tools fall into several broad categories according to application. This section will begin with an explanation of Warnier diagram symbols and techniques which provide the basic analytical tools used throughout this book. This tool will then be applied to the analysis of management systems and to specifying program logic. Two further variations of these Warnier applications are embedded within the program specifications. These include methods for specifying complex decisions and file organizations. Other techniques which do not use the Warnier symbology are also discussed and compared.

One objective in utilizing the Warnier symbology is to cover as many applications as possible with a single technique. While this promotes a unity in the design approach, we dare not sacrifice the utility that might be obtained from other techniques. Thus, other techniques will be introduced and compared to their Warnier diagram counterparts where necessary. The diversity in the application of the Warnier

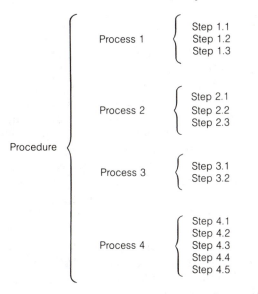

Figure 2.1 General Use of Warnier Diagrams for Analysis

technique can be appreciated in the context of this section in that the application to management systems analysis takes place during the first step, while the synthesis of program specifications is the last step in the design process (see Table 1.3).

2.1.1 Warnier Diagrams

Warnier diagrams* have tremendous advantages over most other methods in communicating system design structure to both technical and administrative personnel. This is due to their simplicity and the ease with which they can be read and interpreted. The purpose of this section is to present the basic principles of Warnier diagrams to facilitate their understanding and use.

Unlike flow charts and data flow diagrams, which tend to synthesize components and state their simultaneous interactions, Warnier diagrams were developed primarily for analysis. They proceed to analyze (decompose into constituent parts) from left to right, using the bracket as the principal symbol. In short, whatever is on the right side of a Warnier bracket is equivalent to and an analysis of the item (or activity) described to the left. Thus, Figure 2.1 shows the analysis of a procedure into several processes, each of which is further analyzed into steps.

---

*The basic concepts of what we call Warnier diagrams were introduced by Warnier and later modified by Orr (see references). Even though all of the applications and variations presented below are not solely attributable to Warnier, we will call these Warnier diagrams in honor of this notable leader in the field of Systems Analysis and Design.

Warnier charts can be used to analyze either actions or things. The significance of the ordering of the analytical components to the right of the bracket is determined by the context. If a procedure (action) is being analyzed into steps, the implied order of performance is from top to bottom unless otherwise qualified, as discussed below. Thus, the ordering is always significant in action analysis. On the other hand, when "things," such as organizations, are analyzed, there is no sequence of performance to consider, and usually the ordering to the right of the bracket is irrelevant. There are certain notable exceptions which are usually quite obvious from their respective contexts. For example, when a Warnier diagram is used to specify the ordering of records within a file, those classifications of records appearing at the top will be found first in the file, irrespective of the actions which are taken to sort the file (see Section 2.1.5).

Basically, the understanding that the Warnier diagram analyzes from left to right (and synthesizes from right to left) is all that is required to generally understand Warnier diagrams. However, there are a number of other symbols which qualify the analysis and make it more precise for system design and, ultimately, program specification. These can be grouped into three categories: (1) frequency qualifiers, (2) logical connectives, and (3) labels and naming entries. These are discussed in the next three sections.

2.1.1.1 Frequency Qualifiers

Figure 2.1 illustrates how the bracket is used to indicate the analysis of a system descriptor into a number of subsystem descriptors, which in turn are further analyzed. It is advantageous to specify how often each descriptor will be repeated or executed. This is done by placing one of the following centered under the descriptor:

Qualifier	Frequency
(0,1)	zero or one time,
(0,n)	zero to n times, where n is any variable,
(n)	exactly n times, where n is any variable,
(m,n)	m to n times, where m and n are variables, or
(1)	exactly one time (default).

Note that 1 is the default value, indicating that if no frequency specification is given, the descriptor will occur or be executed once. If the diagram itself pertains to a given planning cycle, then this would indicate the execution of the descriptor once per cycle. Generally, the presence of parentheses centered below a descriptor infers the use of frequency qualifiers; hence, this notation should not be employed oth-

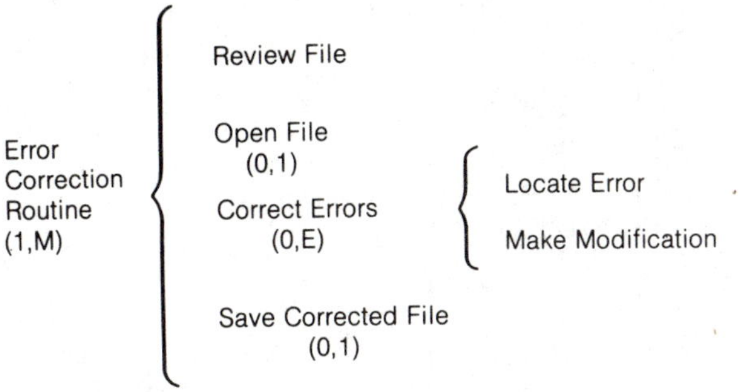

Legend:
M = number of months of operation
E = number of errors detected in a given month

Figure 2.2 Example of the Use of Frequency Qualifiers

erwise unless such use cannot possibly be confused with a frequency qualifier.

Figure 2.2 gives an example of the use of frequency qualifiers. The Error Correction Routine will be performed monthly. No frequency qualifier on the Review File descriptor indicates that it will be performed once for each time the process is repeated. The Open File activity is performed zero or one time depending upon the existence of errors, since no changes will be required if no errors are found. The actual correction (Correct Errors) will be made once for each error detected, and it involves two steps. Finally, the corrected file will be saved.

2.1.1.2 Logical Connectives

Sometimes the ordering of activities requires alteration, with certain activities or steps being conditional. For example, in Figure 2.2, all of the activities below Review File would only be executed if errors were detected during the review. While this might be understandable in the simple sequence of Figure 2.2, it is not definitive enough for system design needs. Thus to eliminate ambiguity a set of logical operators is used to aid in communication. These are as follow:

Symbol	Logical Meaning
(+)	exclusive OR, only one event,
+	inclusive OR, may include several events,
(default)	AND, sequential execution of all events,
$\overline{\text{descriptor}}$	negation of or complement of descriptor.

Figure 2.2 can now be restructured using the logical connectives, as shown in Figure 2.3. Note that the absence of a logical connective

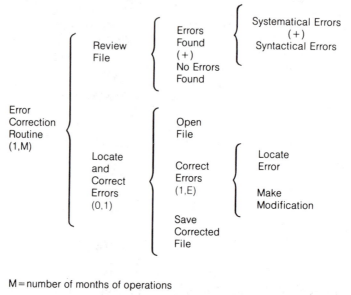

M = number of months of operations

E = number of errors detected in a given month

Figure 2.3 Example of Uses of the OR Connective

implies the AND relationship, and therefore the top-to-bottom sequence of events is in effect. The exclusive OR between two events indicates that only one of them will be performed. The Review File activity can be analyzed into its possible outcomes, which are indicated to be mutually exclusive by means of the exclusive OR symbol. As an example extension, the types of errors can be enumerated; here the inclusive OR symbol is used to indicate that both could occur simultaneously. (This portion of the analysis might be significant if different error types resulted in the use of different correction techniques.) Note the change in the frequency qualifier under the Correct Errors activity, which must be performed at least once. Figure 2.3 is much more definitive than Figure 2.2, thus demonstrating the value of the logical connectives.

2.1.1.3 Warnier Clarification

In addition to facilitating analysis, Warnier diagrams provide an effective means of communicating designs and specifications. In this regard it is essential that no ambiguity exists in the meaning of the diagram being communicated. Most ambiguities can be resolved in one of two ways as follow: (1) by adding descriptor levels to clarify logical ambiguities, or (2) by adding label descriptors which are not part of the logical analysis. Figure 2.4 illustrates how clarification can be brought about by adding a descriptor level. In Figure 2.4a it is ambiguous whether the logical inclusive OR applies to the two subsets of descrip-

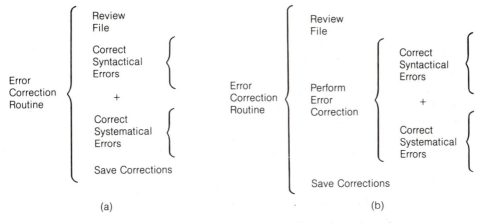

Figure 2.4 Example of Clarification by Adding Descriptor Level

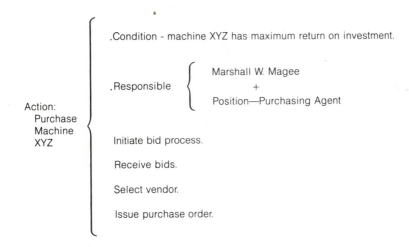

Figure 2.5 Examples of Label Entries

tors or to the two individual descriptors. This is clarified by adding another bracket and a descriptor which generally describes the new subset formed. Thus, brackets can act like mathematical parentheses in clarifying the logical relationships between descriptors.

The second method of clarification involves the addition of descriptive information rather than a logical restructuring. Any descriptor within the Warnier diagram preceded by a decimal point is recognized as being descriptive and not part of the logical analysis. Figure 2.5 presents an action which is analyzed into steps. Two examples of non-analytical descriptors are given, namely: (1) .condition, which is followed by a description of the condition under which the action is to be implemented in practice, and (2) .responsible, which is followed by the actual name or position (indicated by analysis) of the person responsible for implementing the action. For this simplified example, this action is logically analyzed into four steps which by default are linked by the AND operator. In this example the .condition descriptor applies to the entire action, and it could have been positioned under the action descriptor to yield the same meaning. The .condition descriptor can apply to particular steps in this way, and thus it is very powerful. For example, .condition descriptors could be used in Figure 2.2 either in lieu of the frequency descriptors or to give them increased meaning.

The .condition and .responsible labels were presented above strictly as examples. Any useful label might be included within the Warnier diagram. The next section introduces a number of other examples.

2.1.2 Analysis of Management Systems

Figure 2.6 presents a generalized scheme for applying Warnier diagrams to the analysis of a management system. Any management system has general purposes for existence which translate into the mis-

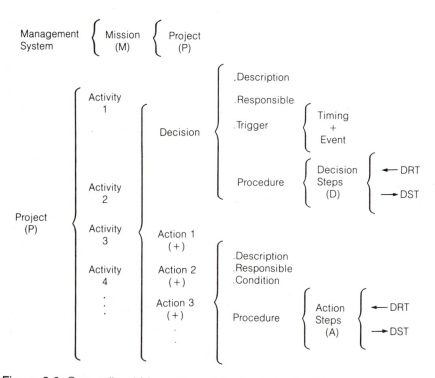

Figure 2.6 Generalized Management System Warnier Diagram (MSWD)

sions which form the first level of the analysis. These are analyzed into several projects for accomplishment, each of which can be further analyzed into their respective activities. Each of the activities is analyzed into the decision and resultant action(s) which are necessary to perform the activity. As indicated, any decision is composed of a process of selection from among alternatives, resulting in the ultimate *action* which results from the execution of the decision.

There is a further analysis for the decision and action descriptors for each activity. Within the analysis of the decision descriptor is an identification of the decision by name and description. Further, there is a label for the individual responsible for making the decision. The descriptor .trigger is followed by a definition of some external event that causes or leads to the necessity for the decision process. This might be due to timing or some event. Finally, a procedure made up of a number of process steps takes place, which results in the ultimate selection from among the alternatives.

Similarly, the action which results from the decision is identified and labeled as the decision descriptor was. However, instead of a trigger there is a condition label which tells which action is implemented depending upon the outcome of the decision made. Technically, the action descriptors should be bound together within another bracket and a general action descriptor assigned to cover all actions. Further, those actions which could be implemented simultaneously might be further demarcated with additional brackets to eliminate ambiguity. However, this is not done within Figure 2.6 to preserve the simplicity of the generalized model.

Each action procedure is analyzed into a number of process steps as was true within the decision descriptor. A final level of analysis is obtained by specifying the Data Retrieval Transactions (DRT) input or the Data Storage Transactions (DST) output within any given step of a decision procedure or an action procedure.

In order to further describe this process of analysis, the following definitions of the terminology employed are given:

Mission—the name of an operation which embodies a very general and broadly defined goal of the management system. Each mission should be broad enough to encompass several projects, but narrow enough to be logically delineated from other missions. The sum total of all missions should totally define all of the goals of the management system related to the design under consideration, and if possible each mission should be mutually exclusive of all others with respect to the projects required for their attainment.

Project—a component of the first level of analysis of a mission into broad classifications of activities necessary for its accomplishment.

Projects generally relate to specific objectives, whereas missions relate to goals.

Activity—the sum total of all processes required in making a decision and carrying out the action which results from it. Note that this is a definition local to our use of Warnier diagrams and that there is a one-to-one correspondence between each decision and the activity associated with it. Further, projects are generally analyzed into a plurality of activities.

Decision—the compilation, organization and study of a set of alternatives followed by the selection of one alternative for implementation.

Action—the implementation which results when a particular decision is made.

Procedure—a group of steps employed to either carry out an action or make a decision.

Step—an element of a procedure which is executed either in making a decision or in the resultant action which follows the decision.

Trigger—the passage of time or the occurrence of an event which leads to the necessity for a decision.

Condition—the output of the decision procedure which defines which alternative action will be implemented.

DRT—Data Retrieval Transaction, the acquiring of data necessary to perform a decision or action step when that data is not made readily available by other steps within the procedure itself.

DST—Data Storage Transaction, the storage of data such that it can be retrieved in another procedure by a DRT.

As a partial example of a management system Warnier diagram, consider a manufacturing organization which would have a *mission* defined for each product produced as well as several missions supportive to the line organization. Consider just one of these supportive missions, the one responsible for acquiring new machinery and equipment. Depending upon the specific application, there might be many *projects*, one for each general area of the plant which requires a consideration of the purchase of new equipment. Here an *activity* would be developed for each of the various pieces of equipment that were under consideration; a *decision* corresponds to each of these activities. Figure 2.7 is an example of development for one activity. Note that the actual

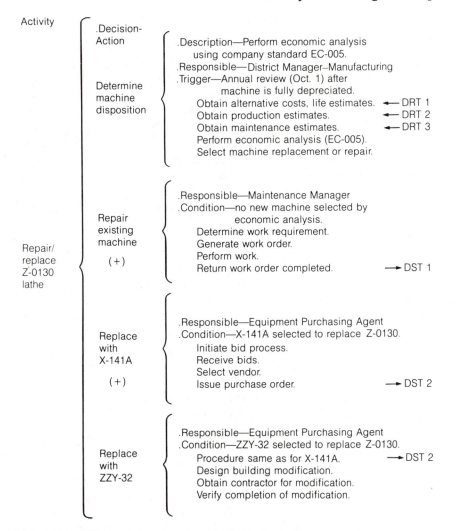

Activity

.Decision-Action
.Description—Perform economic analysis
  using company standard EC-005.
.Responsible—District Manager–Manufacturing
.Trigger—Annual review (Oct. 1) after
  machine is fully depreciated.

Determine machine disposition
  Obtain alternative costs, life estimates. ◄—DRT 1
  Obtain production estimates. ◄—DRT 2
  Obtain maintenance estimates. ◄—DRT 3
  Perform economic analysis (EC-005).
  Select machine replacement or repair.

Repair/replace Z-0130 lathe

Repair existing machine (+)
.Responsible—Maintenance Manager
.Condition—no new machine selected by
  economic analysis.
  Determine work requirement.
  Generate work order.
  Perform work.
  Return work order completed. —►DST 1

Replace with X-141A (+)
.Responsible—Equipment Purchasing Agent
.Condition—X-141A selected to replace Z-0130.
  Initiate bid process.
  Receive bids.
  Select vendor.
  Issue purchase order. —►DST 2

Replace with ZZY-32
.Responsible—Equipment Purchasing Agent
.Condition—ZZY-32 selected to replace Z-0130.
  Procedure same as for X-141A. —►DST 2
  Design building modification.
  Obtain contractor for modification.
  Verify completion of modification.

Figure 2.7 MSWD Example of an Activity Analysis

descriptor replaces its generic name as given in Figure 2.6. In a total analysis of the affected management system (AMS) there might be several missions, dozens of projects, and hundreds of such activities, as analyzed in Figure 2.7.

The generalized model given in Figure 2.6 will provide the basis for the analysis portion of the design process given in Chapter 3, where it will be called a management system Warnier diagram (MSWD). Although many variations and modifications are required in specific applications, this basic model provides the generalized starting point from which these can be made.

2.1.3 Specification of Program Logic

To this point the discussion has been confined to the application of the Warnier diagram technique to management systems analysis. The flexibility of this technique, however, lends itself to several other applications. In this section consideration will be given to the specification of program logic. This application is microscopic compared to the analysis of a management system. Further, program logic specification is the *last* step in the design process. However, it is presented in this technical overview so that free use can be made of this application in subsequent chapters. Two other program specification applications will be presented in Sections 2.1.4 and 2.1.5.

Program specification refers to documentation from which a programmer can directly develop software code. The central tool for the development and communication of program specifications will be called a program specification Warnier diagram (PSWD). The use of Warnier diagrams in specifying procedures and steps necessary to make decisions or accomplish actions was given above in terms of the MSWD. The PSWD is much more specific since it applies exclusively to program code specification.

In some aspects the objective of Warnier diagrams for program logic definition is much the same as that of traditional flowcharting.* Since flowcharting, along with its symbology, is taught in most elementary programming courses, it will not be presented or reviewed here. The following are some advantages of utilizing the Warnier diagram approach to developing and specifying program logic:

1. The PSWD is easier to construct and manipulate, since the developer is not constrained by flowchart symbols;
2. A PSWD may be understood by a wider range of individuals, since the same type of symbology is utilized in several other parts of the analysis and design specification;
3. PSWDs more readily lend themselves to manipulation on text editors; and
4. PSWDs promote a structured (straight-line) programming discipline.

Certainly we would be negligent in implying that Warnier diagrams are always superior to flowcharts for specifying program logic. The pictorial impression transmitted by flowcharts is not totally present in the Warnier diagram; and for certain applications the Warnier diagram may get unwieldy compared to its comparable flow chart. These are

---

*Flowcharts should not be confused with data flow diagrams which will be discussed later in this chapter.

two separate tools, and each should be employed where and when it is most appropriate. Some might further cite the computer routines available which automatically generate flowcharts from program code as an advantage of flowcharts. Rather than showing an advantage, these indeed crystallize a major flowchart shortcoming: they are rarely used by designers to specify program logic *before* coding. Rather, they are usually developed during and after the program is written to aid the programmer or satisfy some documentation standard. However, the objective of program specification is to define the processing requirements so that the programmer can proceed directly to coding. As the PSWD technique is revealed, the simplicity with which this can be accomplished and the ease of specifying integrated systems of programs will become clear.

Figure 2.8 shows the basic straight line approach to utilizing Warnier diagrams for specifying program logic. Several characteristics bear noting. First, this model analyzes the total software system into its component programs, which should be arranged in some logical ordering. Second, for each program there is a specification of a single grouping of inputs and a single grouping of outputs, which may interact with (or be produced by) the program at any time during its execution. For most programs 80–90% of the design work is completed when the inputs and the outputs are totally defined. Generally, referencing will be used to quickly obtain those portions of the documentation already developed which contain complete details of the input and output of concern. Third, the model assumes that a series of statements will all be executed sequentially, since the Warnier rules default to the AND operator and the sequence of performance is top-to-bottom. The "statements" in Figure 2.8 need not be actual programming (e.g., FORTRAN, COBOL, etc.) statements. Rather, they may be expressed in pseudocode

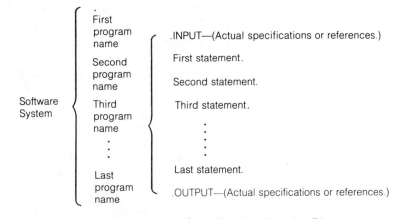

Figure 2.8 Straight Line Program Specification Warnier Diagram

or just plain English. However, each must be clearly translatable into one or more program statements by the programmer. Since one-to-one translation is not essential, the coding procedure as well as the pseudocode might be greatly facilitated by the use of variable names. These may be defined in terms of the input specifications or within one of the statements, and then used subsequently for abbreviation. Of course, any of the Warnier frequency, logical, .condition, or other qualifiers may be applied to more precisely specify the program logic.

The problem with the model given in Figure 2.8 is that very rarely is a program executed in a straight line. Indeed, the major advantage of the computer over the accounting machine is its ability to reiterate patches of code and to do this on a conditional basis. Here a conflict arises between the structured programming concept, which strives for program maintainability, and the efficient use of code, which strives for program brevity and, possibly, minimum execution time. This need not be a problem in the design phase. At this point all program specifications should be structured by a linearly executable set of program statements (compromises in development might be made later). This leads to the model depicted in Figure 2.9.

This second model maintains the structured approach by subdividing the total program into a series of subprograms, each of which might be executed anywhere from zero to a variable (possibly determined by the program) number of times.* Further, the condition for the execution of a given subprogram is spelled out (using the comment: .condition), if in fact a particular subprogram is conditional; otherwise this will be omitted. The condition may be dependent upon the output of a previously executed subprogram, and this would also be explained in the .condition comment.

The extension of Figure 2.9 to a further "sub-subprogram" breakdown should be obvious. Any level of modularization can be attained, depending upon the application. Also there is nothing to prevent one of the statements of one subprogram from conditionally calling another existing subprogram. In this case, when the routine called is completed control would return to the calling statement, a convenient feature characteristic of structured programming. Thus, using the basic principles of Warnier diagrams it is possible to specify program logic of the highest sophistication.

2.1.3.1 Example Using Structured English

Structured English is a general term which is used to indicate a style of program specification which is readable by programmers and non-

---

*Note that this is not specifying that the programmer has to structure his code this way (e.g., he may elect to have only one main program). This decision generally falls within the programmer's area of expertise.

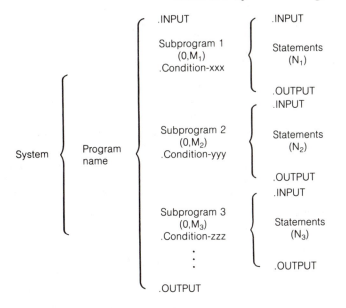

Legend:

M_i = Maximum number of times that subroutine i is repeated.

N_i = Number of statements in subroutine i.

Figure 2.9 Structured Program Specification Warnier Diagram Model

programmers alike because it is written in a form of English. Unlike "plain" English, however, which tends to obscure through variation and emphasize by redundancy, the statements of structured English attempt to eliminate both variation and redundancy. As a result, a linear series of statements using unique and consistent terms is generated. While this makes for boring reading (sometimes leading to the neglect of the proofreading step), it is a delight to the programmer since it greatly facilitates coding. This subsection presents an example of the use of PSWDs where the statements are written in structured English; the next section will present the same example written in pseudocode.

Consider as an example the specification of a program to calculate the amount of compensation due to each distributor of a multilevel distribution company. Such a program would be necessary in order to write monthly checks, which would be part of the program's function. To keep this example manageable, consider the following simplified plan:

Each distributor gets a check at the end of the month for 10% of his/her individual personal sales plus 5% of the sales of any distributor that was sponsored by him/her. In addition, if personal sales exceed $5000, the distributor receives a bonus of 2% of personal sales in excess of $5000 plus 1% of the sales of all those that he/she sponsored. Finally, if the distributor's personal sales are not over $500, the distributor does not receive any compensation from those distributors that he/she sponsored.

While the plain English paragraph above would probably be more than adequate for most programmers, this is a very simple example to illustrate the technique. In reality these types of compensation plans are more complex by several orders of magnitude. To further define this example, assume that a record is kept in a computerized file for each sale. Among other items, each of these records contains the total amount of the sale, the distributor identification number, and the distributor's sponsor's identification number. Thus, at the end of the month the computer can make one pass through these records to determine the amount of compensation to each distributor.

With these basic assumptions stated it is now possible to specify the program using structured English within the PSWD. Figure 2.10 shows one way that this could be done. (As with any other communication media, the specification is not unique.) Note first that the program is logically divided into two subroutines, the first to accumulate the personal and sponsored sales values for each distributor out of the sales record, and the second to calculate the compensation for each distributor. Note that the first is repeated for each sales record while the second is repeated for each distributor. The initialization of registers and the particular programming statements are left totally to the discretion of the programmer; the PSWD tells what to do, not how to do it.

There is an input and an output specified for the program and for each subprogram in Figure 2.10. This is particularly important when subprograms are not contained on the same page. However, when the input or output to a particular program is obvious to the programmer they need not be repeated, which is probably the case in Figure 2.10. Within both the input/output specifications and the statements, the terminology used must be perfectly *consistent*. Unique names are assigned and used consistently throughout the specification. This is a major feature of structured English that separates it from common English narrative. For example, "personal sales amount" is not called "personal sales" or "personal amount" or "sales amount" or just "sales."

Finally, note how easily and concisely the .condition descriptors handle the exceptional cases. Along with the Warnier brackets and frequency qualifiers, they totally structure and define the programmer's task. Structured English outside of the PSWD usually makes heavy use of IF-THEN-ELSE-type statements. While these can be integrated into the PSWD, it is preferable to state the specifications as concisely and completely as possible and leave the choice of program code to the programmer.

.INPUT—Sales records for month, including distributor number, sponsor number, and sales amount for each sale.

Calculate basic sales amounts (S)

.INPUT—same as above.

Read sales record.

Sum sales amount for the appropriate distributor.

Sum sales amount for the appropriate sponsor.

.OUTPUT—Personal sales amount and sponsored sales amount for each distributor.

Distributor Compensation Program .Frequency: monthly

Calculate compensation (D)

.INPUT—Personal sales amount and sponsored sales amount for each distributor.

Amount of check = .1 times the sum of personal sales amount.

Increase amount of check by an amount = .05 times the sum of sponsored sales amount.
.Condition—sum of personal sales amount > 500.

Increase amount of check by amount = .02 times the quantity (sum of personal sales amount − 5000) plus .01 times sum of sponsored sales amount.
.Condition—sum of personal sales amount > 5000.

.OUTPUT—Amount of monthly check for distributor

.OUTPUT—Amount of monthly check for all distributors.

S = number of sales records for month
D = number of distributors

Figure 2.10 Example of a PSWD Using Structured English

2.1.3.2 Example Using Pseudocode

Pseudocode is used to indicate a style of program specification which is to some extent encoded and thus not in typical English. Even in Figure 2.10, the use of equals signs and parentheses for clarification could be considered a first step toward pseudocode. The degree to which the designer should move in the direction of coding is strictly a matter of personal taste. Actually, it is easier for most designers, having had some programming experience, to write specifications in pseudocode rather than structured English. Further, most programmers find that the use of variables rather than names makes their task easier. Finally, the ability to fit more specifications on a single sheet of paper is another plus in favor of pseudocode.

Like any other technique, pseudocode can have its disadvantages when taken to the extreme. The limit, of course, is to totally code the program, which is wasteful of designer time and probably not beneficial in utilizing the expertise of the programmer. And, just as program code is virtually impossible to grasp without intensive effort, so overly developed pseudocode can be cryptic to the programmer. For these reasons the following rules are proposed to help maintain the balance between structured English and pseudocode:

1. Always define a variable by a descriptive label the first time that it is used on any page; after this the variable symbol may be used.
2. Use full explanatory-type descriptors for subroutines and main programs.
3. Integrate English explanations wherever it will aid in the understanding of the processing requirements.

Figure 2.11 presents the example given in Section 2.1.3.1 restated in pseudocode. While this is less readable, all variables are defined for ready reference. Variable names have been chosen in order to facilitate the memory of their meanings. It should be impressed upon the programmers that they are not required to incorporate this nomenclature into their coding. However, within the comments of the program code, a cross reference list to the variable names used in the PSWD will greatly facilitate program maintenance.

A comparison of Figures 2.10 and 2.11 shows how much more compact and definitive pseudocode is over structured English. It is important, however, that an optimum level of coding be sought such that the PSWD conveys what must be done but does not force the programmer to a particular methodology.

It is pertinent to note here that even the example above could not be presented without some assumptions as to the output requirements, file designs, and input specifications of the system of which this program was a small component. This illustrates the point that the development of the program specifications is the last step in the development of the design (see Table 1.3). Indeed, once the other components of the system design documentation are fully developed, the program specification step becomes very straightforward.

2.1.4 Decision Specifications

The model given above for specifying program logic can be greatly facilitated by the integration of special tables or diagrams which greatly abbreviate and clarify the specifications, making them much easier to program. These communication devices are classically called decision tables or decision trees. In this section the concept of a decision table will be presented. This will be followed by the introduction of the

.INPUT—Sales records for month, including distributor
number (DN), sponsor number (SN), and sales
amount (SA) for each sale.

Calculate basic sales amounts (S)

.INPUT—DN, SN, and SA for each
sales record.

.Definitions:
SPS(I) = sum of personal sales for
distributor I.
SSS(J) = sum of sponsored sales for
sponsor J.

Read sales record.

SPS(DN) = SPS(DN) + SA.

SSS(SN) = SSS(SN) + SA.

.OUTPUT—SPS and SSS for all distributors.

Distributor Compensation Program
.Frequency: monthly

Calculate Compensation (D)

.INPUT—SPS and SSS for all distributors.

.Definition:
DI = distributor number for distributor
under consideration.

Check amount (CA) = .1(SPS(DI)).

CA = CA + .05 (SSS(DI)).
.Condition—SPS(DI) > 500.

CA = CA + .02(SPS(DI)) − 5000)
+ .01(SSS(DI)).
.Condition—SPS(DI) > 5000.

.OUTPUT—CA for distributor DI.

.OUTPUT—Amount of monthly check for all distributors.

S = number of sales records for month
D = number of distributors

Figure 2.11 Example of a PSWD Using Pseudocode

CONDITION DESCRIPTION	CONDITION INDICATOR				
i = 1	C_{11}	C_{12}	C_{13}	C_{14}	C_{15}
i = 2	C_{21}	C_{22}	C_{23}	C_{24}	C_{25}
i = 3	C_{31}	C_{32}	C_{33}	C_{34}	C_{35}
i = 4	C_{41}	C_{42}	C_{43}	C_{44}	C_{45}

ACTION DESCRIPTION	PERFORMANCE INDICATOR				
k = 1	P_{11}	P_{12}	P_{13}	P_{14}	P_{15}
k = 2	P_{21}	P_{22}	P_{23}	P_{24}	P_{25}
k = 3	P_{31}	P_{32}	P_{33}	P_{34}	P_{35}
k = 4	P_{41}	P_{42}	P_{43}	P_{44}	P_{45}
k = 5	P_{51}	P_{52}	P_{53}	P_{54}	P_{55}

C_{ij} = numeric, coded, truth value, or blank (if irrelevant).
P_{kj} = indicator of whether the action is to be performed
    (Y if performed, blank otherwise).

ACTION PERFORMANCE RULE: Perform actions k for which P_{kj} = Y
    when all conditions given by C_{ij} are
    in effect (for all i).

Figure 2.12 Generalized 4 Condition, 5 Action Decision Table

decision Warnier diagram (DWD), which itself is a form of decision tree. The following two subsections will present examples which compare and illustrate these powerful techniques.

To introduce the decision table in its generalized form, consider Figure 2.12. The truth table consists of four parts which are generally read in a clockwise order. In this order the components are:

1. Condition description.   A list of all the conditions (inputs) that have any bearing upon the action to be taken (output). Condition descriptions should be as brief as possible while completely describing the meanings of the condition indicators. It may be in plain English or it may use variable names, if they have been clearly defined.

2. Condition indicators.   These are an indication of the state of the conditions given by the condition descriptors. The following are possible types of condition indicators which will be used, depending upon the condition being specified: (a) numeric, where the condition is determined by a unique value or a range of values, (b) coded,

where the conditions are qualitatively described by discrete de-scriptors, each of which is assigned a numeric value, (c) truth, where the condition either occurs or fails to occur (true-false), and (d) blank, where the condition is irrelevant to the action outcome. Ta-ble 2.1 provides examples of the first three types; the fourth is ap-plicable whenever the condition is irrelevant to the action(s) pre-scribed by the performance indicators in the same column (see below).

3. Performance indicator.   This indicates whether or not the ac-tion, given by the action description, is to be performed for a given set of conditions. This set of conditions is given by the condition indicators in the same column immediately above. To simplify the table and eliminate unnecessary entries, only the affirmative is en-tered (Y if the action is to be performed, blank otherwise).

4. Action description.   A list of all possible output actions which could result from the decision under consideration. The description may be in plain English or as a variable definition, if this is clearer to the programmer.

Figure 2.13 presents a common example of a decision table for a student getting started in the morning. Note that, even in this simple problem, there are nine columns required, and this is far from an ex-haustive set of possible conditions. For example, there might be a dif-ferent scheduling requirement for each day of the week, greatly increas-ing the magnitude of the problem.

Some simplification can be obtained, especially when the alterna-tive action set is not large, by using Warnier diagrams as a combination

Table 2.1   Example Condition Indicators

Type	Example Condition Description	Example Condition Indicator
Numeric	Temperature, degrees Fahrenheit	70 GT 100 $50 \leq T \leq 70$
Coded	Day of the week	S - Sunday M - Monday T - Tuesday W - Wednesday H - Thursday F - Friday A - Saturday
Truth	Homework due today	T - True F - False

STUDENT'S "GETTING STARTED" DECISIONS

CONDITION DESCRIPTION	CONDITION INDICATOR								
1. Day of Week	M-F	M-F	M-F	M-F	M-F	M-F	Sa,S	Sa,S	Sa,S
2. Temp. (Degrees F.)	LT60	60-70	GT70	LT60	60-70	GT70	LT60	60-70	GT70
3. Homework due today	T	T	T	F	F	F	-	-	-

ACTION DESCRIPTION	PERFORMANCE INDICATOR								
1. Get up at 6:30 AM	Y	Y	Y	Y	Y	Y			
2. Get up at 9:00 AM							Y	Y	Y
3. Cook hot breakfast	Y	Y		Y	Y				
4. Dress-short sleeves			Y			Y			Y
5. Dress-long sleeves		Y			Y			Y	
6. Dress in sweater	Y			Y			Y		
7. Check over homework	Y	Y	Y						
8. Put on jacket		Y			Y			Y	
9. Put on coat	Y			Y			Y		

Figure 2.13 Example of a Decision Table

decision tree and decision table. In this application we call the resultant diagrams decision Warnier diagrams. The inferred top-to-bottom order of consideration combined with the ease of integration into the program specifications makes the Warnier diagram more powerful than the decision table. For these reasons, and for consistency, use of the DWD is recommended wherever applicable.

Figure 2.14 illustrates the general structure of the DWD. At the extreme left is the reference name or names of variables being defined by the DWD; their values will appear at the extreme right column(s) of the table/diagram. These variables will either be used in subsequent statements of the subroutine or they will be included in the .output list (or both). The first analysis bracket is labeled generically in Figure 2.9 as Defined Variable 1 (DV-1). In practice, this label will assume an actual name or reference, which is understood by the programmer, designating a variable which has been previously defined either by the .input list, former processing, or both. The analysis of this variable is given to the right of the bracket, in terms of the values that it can assume.

Each of the possible outcome values of the first analysis variable is further analyzed by additional variable-value combinations, if any. In

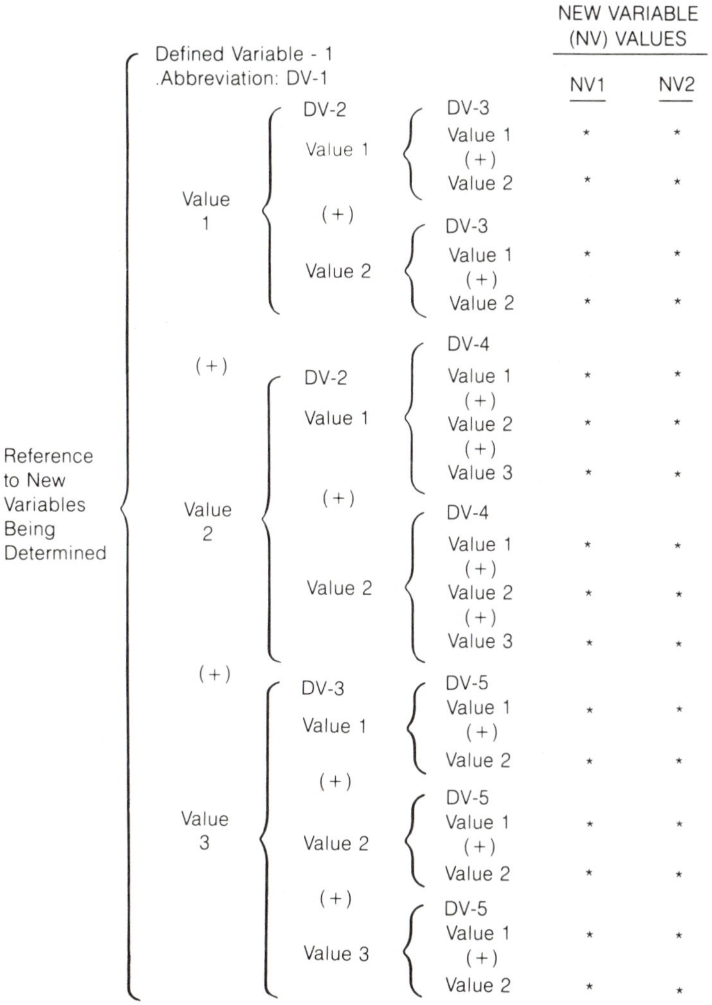

NEW VARIABLE
(NV) VALUES

Defined Variable - 1
.Abbreviation: DV-1

	NV1	NV2

Reference to New Variables Being Determined

DV-2 Value 1
Value 1
(+)
Value 2

DV-3
Value 1
(+)
Value 2   * *
(+)
DV-3
Value 1   * *
(+)
Value 2   * *

(+)

Value 2

DV-2 Value 1
(+)
Value 2   * *
(+)

DV-4
Value 1   * *
(+)
Value 2   * *
(+)
Value 3   * *

DV-4
Value 1   * *
(+)
Value 2   * *
(+)
Value 3   * *

(+)

Value 3

DV-3 Value 1
(+)
Value 2

DV-5
Value 1   * *
(+)
Value 2   * *

DV-5
Value 1   * *
(+)
Value 2   * *

Value 2
(+)

Value 3

DV-5
Value 1   * *
(+)
Value 2   * *

* The entries in this table will be the values of the new variables being defined.

Figure 2.14 Example of a Structure for the DWD

the example structure given in Figure 2.14, value 1 of Defined Variable 1 may occur in combination with the values of Defined Variable 2, each of which, in turn, might occur simultaneously with the values of a third defined variable, DV-3. Notice that the DWD is totally flexible in this regard, requiring only the relevant and feasible combinations of defined variables to be represented. For example, Value 3 of Defined Variable 1 is not analyzed in terms of Defined Variable 2, indicating that its value is irrelevant in determining the values of the output variables.

All feasible combinations of the input variable values must be represented so there are no ambiguities in coding. Continuous variables

ACTIONS

	Time to Awake	Hot Break-fast	Dress	Outer Wear	Check Home-work
Day of Week — Temp — HW Due					
LT 60 — Y (+)	6:30	Y	SW	C	Y
LT 60 — N	6:30	Y	SW	C	
(+) — HW Due					
60-70 — Y (+)	6:30	Y	LS	J	Y
60-70 — N	6:30	Y	LS	J	
(+) — HW Due					
GT 70 — Y (+)	6:30		SS	J	Y
GT 70 — N	6:30		SS	J	
Sa,S — Temp					
LT 60	9:00		SW	C	
(+) 60-70	9:00		LS	J	
(+) GT 70	9:00		SS	J	

Left-side brace structure:

Getting Started Conditions {
  M - F {
    Temp {
      LT 60 (+)
      (+)
      60-70 (+)
      (+)
      GT 70 (+)
    }
  }
  (+)
  Sa,S {
    Temp
    LT 60
    (+)
    60-70
    (+)
    GT 70
  }
}

HW—homework
SW—sweater
LS—long sleeves
SS—short sleeves
J—jacket
C—coat

Figure 2.15 DWD for Example of Figure 2.13

may be subdivided into discrete ranges for representation. The table to the right of the diagrammatic analysis yields the values of the new variables to be defined corresponding to each of the combinations generated by the analysis.

Figure 2.15 presents the example given above in terms of a DWD. The "getting started" alternatives are first analyzed by standard Warnier techniques until the entire set of alternatives is specified. Then the action specifications are added as a table to the right of the analysis. Each line of the action corresponds to one resultant combination in the analysis.

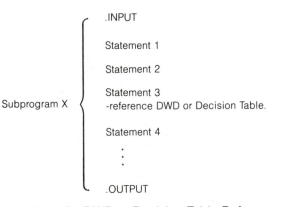

Figure 2.16 Integration of a DWD or Decision Table Reference Into Program Specifications

Finally, consider how a DWD or a decision table would be integrated into the structured program logic model given in Figure 2.9. Any one of the subroutine analyses to the extreme right in Figure 2.9 might be detailed as in Figure 2.16. In this example, Statements 1, 2, and 4 are self-contained, i.e., they can be directly translated into program code without further specification. However, Statement 3 requires a DWD or a decision table to more effectively specify the details of a more complicated decision process. While this could be spelled out using a series of statements, such would be very inefficient in terms of documentation space, as well as designer and programmer time. Note that since there could be several DWD references, a reference number is assigned sequentially to the DWDs, which could refer to a separate section of the documentation. Alternatively, the DWDs and decision tables might be assigned PSWD numbers and be integrated into the PSWDs.

2.1.4.1 Comparison of Techniques

Before going on to illustrate a more complex example that pertains to actual program code specification, it might be useful to discuss the advantages and disadvantages of each of the techniques presented, since seldom are both required for the same specification. Proponents of each technique could easily devise counter-arguments for each of the statements made, and the analyst should feel free to use the technique with which the design and development group is most comfortable.

Figure 2.12 illustrates that the decision table is very structured, and it lends itself to validation by a total enumeration of alternatives. The mathematical specificity of this technique lends itself to direct conversion of the table into program code. It also has the advantage of being able to easily accommodate a large number of actions. Thus, when the number of conditions is fairly small (three or four) and the number of actions is fairly large (over five), the decision table would seem to be preferable.

On the other hand, typically the number of simultaneous output actions under consideration are small; most often actions can be subdivided if this is not the case. For example, the "clothing" decision was totally independent of the "get up time" decision in the example presented above, and these could have been presented equally as well on two separate diagrams or tables. When this is the case, and the number of condition combinations is quite large, the DWD has a distinct advantage. This is especially true when some of the combinations are irrelevant, although this results in greater conciseness to both presentations.

For consistency through the remainder of this book the DWD will be used primarily for the following reasons:

1. It is developed from left to right consistently with the other Warnier diagram applications in this book.
2. It is considerably easier to read, since it is read from left to right in the same sequence as its development (decision tables are read in a clockwise ordering).
3. For the majority of complex cases it provides a more concise specification for the programmer.

This third point is made, to some extent, at the expense of the validation that comes from total enumeration. For this reason we encourage software engineers and programmers to utilize the decision table concept in their validation. However, for a concise specification of complex designs, the DWD tool is hard to beat. Those who doubt this are invited to develop the decision table for the problem presented in the next section.

2.1.4.2 Example of Complex Decision Specification

In order to illustrate DWDs, a detailed decision specification will be presented. This example is taken directly from a real world problem, and it is complex. A simple example would not provide an appreciation of the value of this technique. Therefore we urge the reader to be patient in assimilating this example. Consider the traffic accident records variable determination problem given below.

The output variables to be determined are called the intersection code and the milepost code. The intersection code can take on the following values:

1. Intersection accident, or
2. Nonintersection accident.

The milepost code can take on the following values:

1. Rural mileposted,
2. Rural nonmileposted, or
3. Urban nonmileposted.

Note that there are no urban mileposted roadways considered in this example.

The following variables have already been defined by input on the raw data traffic accident report form:

1. Intersection Code—the recording officer's opinion as to whether the accident was related to an intersection or not.
2. City Code—the code of the city (if any) in which the accident occurred.
3. ON Code—the street code upon which the accident occurred.
4. AT Code—the cross street for intersectional type accidents.
5. Milepost—the milemarker number for mileposted roadways measured from the nearest milemarker to the nearest hundredth of a mile. All urban and some rural streets are not mileposted, and this will be zero for those location types.

The rules for defining the output variables as a function of the input variables might appear as follows. If the accident is indicated to be at an intersection and the city code is a zero, then this is a rural intersectional accident. If the ON code begins with an I or S and the AT code is zero, the intersection code will be 2 and the milepost code will be 1, since this will be treated as a nonintersection mileposted accident. However, if the AT code is nonzero, this will be treated as an intersectional accident; thus, the intersection code will be 1 while the milepost code will be 2. The exact same conditions will hold if the ON code does not begin with an I or an S, with the exception that, if the AT code is zero, the milepost code will be 2. If the city code of intersectional accidents is not equal to zero, this indicates an urban accident. The milepost will generally be zero, with certain exceptions due to inconsistency of reporting. When the milepost code is zero and the AT code is zero, the intersection code will be 2 and the milepost code will be 3. However, if the AT code is not zero, the intersection code will be 1 (milepost code still 3). In those cases where a milepost is entered and the ON code begins with an I or an S, if the AT code equals zero the intersection code will be 2 and the milepost code will be 1 (i.e. it will be processed as a mileposted accident); however, if the AT code does not equal zero, then the intersection code will be 1 and the milepost code will be 3. If the ON code does not begin with an I or S the same output will result; however, the milepost code will always be 3.

The example given above was deliberately complex (although not atypical of real world problems) in order to illustrate the DWD technique. Actually, only about half of the total problem is spelled out to spare the reader unnecessary labor. The solution, in terms of the DWD, is given in Figure 2.17. Note that this diagram states considerably more than the paragraph above, but in an unambiguous manner that requires no rereading or reiteration. For conciseness, the default logical operator has been changed from the AND to the exclusive OR, which is generally the only operator used in this application. The programmer can code

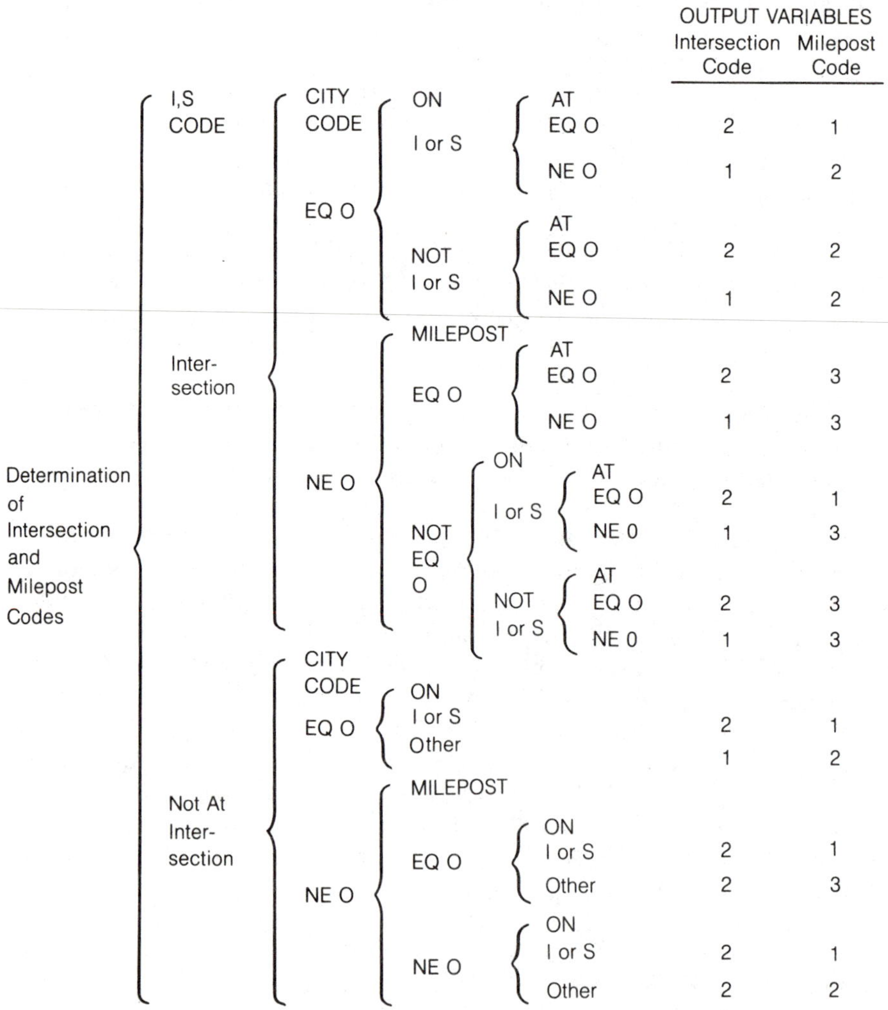

Determination of Intersection and Milepost Codes

Condition					OUTPUT VARIABLES Intersection Code	Milepost Code
I,S CODE	CITY CODE	ON	AT EQ O		2	1
		I or S	NE O		1	2
	EQ O	NOT	AT EQ O		2	2
		I or S	NE O		1	2
Inter-section		MILEPOST	AT EQ O		2	3
		EQ O	NE O		1	3
	NE O	ON	AT EQ O		2	1
		I or S	NE 0		1	3
		NOT EQ O	AT EQ O		2	3
		NOT I or S	NE 0		1	3
Not At Inter-section	CITY CODE EQ O	ON I or S			2	1
		Other			1	2
	NE O	MILEPOST EQ O	ON I or S		2	1
			Other		2	3
		NE O	ON I or S		2	1
			Other		2	2

NOTE: Default operator is exclusive OR, (+), in all cases.

Figure 2.17 Example of Variable Assignments Using a DWD

directly from Figure 2.17, and once the program is written, the figure can be referenced directly in the system maintenance manual to facilitate all future maintenance required on this particular routine. If a decision table is required for programming or validation, it can be developed directly from the DWD.

Finally, it might be noted that in "real life" a statement of the problem as definitive as that given in the paragraph immediately above is somewhat of a luxury. It is usually the job of the analyst to formalize the rules, and narratives are not recommended for this process. Thus, it is recommended that the analyst utilize the DWD directly during fact finding and interviewing, as a tool for crystallizing the fuzzy thoughts expressed.

2.1.5 File Organization Specifications

Warnier diagrams can be used to specify sorting procedures to be employed in order to organize a given file. When this application is made, the resultant diagram will be referenced as an organizational Warnier diagram (OWD). If there are a large number of OWDs required for system specification, then a separate section of the system documentation might be established, or these might be integrated into the PSWDs as was true of the DWDs. Several different subprograms may reference one OWD, and several OWDs may be used by the same subprogram (this is similarly true for DWDs). The reference to an OWD is made analogously to the DWD reference given in Figure 2.16.

In describing the general structure of an OWD, its similarity with the DWD comes to an abrupt end. The analysis is now performed in terms of the variables or fields which determine the sequential organization of the records, and the positioning of the analytical descriptors to the right of the bracket indicates those values which will appear first in the file. Also, the only logical operator in effect here is the AND operator, the normal default. Figure 2.18 gives an example of a generalized OWD.

Figure 2.18 can be interpreted to indicate that the major sort will be performed on variable 1. Within the value 1 subset of variable 1, the records will be sorted on variable 2 in ascending order. Similarly, within the value 2 subset, the records will be arranged in ascending order of variable j. This illustrates the flexibility and clarity communicated by the OWD. Each subset may be sorted by any other combination of variables. The OWD may be extended to the right to specify any number of sort fields. In a situation where every subset is to be sorted identically, the nature of the sort is given in the descriptor to the right of the bracket to indicate commonality, as is given in Figure 2.19. In this case the values do not need to be enumerated and the diagram is greatly simplified.

Figure 2.20 presents an example use of the OWD which extends the example of the DWD application given above. In effect, this OWD takes the records with variables defined as in the example in Section 2.1.4.2 (as well as other variables), and sorts them according to the values assigned by the DWD and/or previously defined variables. For example, all of the milepost code value 1 records come first. Within them, the intersection code value 1 records come first, and within that sub-subset the ON and AT codes are sorted in ascending order. This is followed by the intersection code value 2 sub-subset, which is sorted by milepost within the ON-code variable, both in ascending order. The objective of such an ordering is to group accidents at the same location together so that high accident locations can be found. In this example each roadway type requires a different sort in order to accomplish this purpose.

Note the clarity with which the organization of the data is pre-

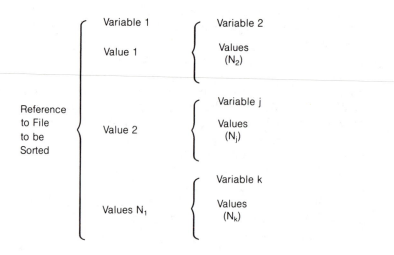

N_x = number of subsets within variable x used in sort specification.

Figure 2.18 Generalized Organizational Warnier Diagram

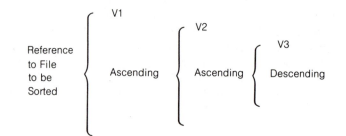

Figure 2.19 OWD With Uniformly Common Sort Fields

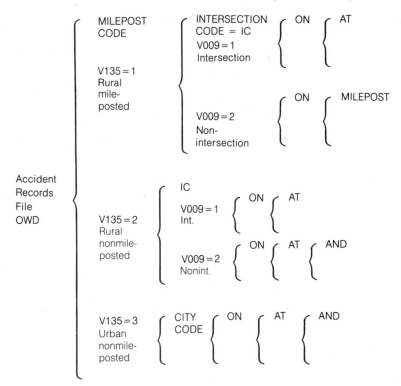

NOTE: All sorts ascending unless otherwise indicated.

Figure 2.20 Example OWD for Accident Records File

sented. The programmer has, on one sheet of paper, a complete, un-ambiguous picture of the resulting output. No methodology need be specified; the programmer is given full latitude to use his experience and expertise to select the utility or write the code which will best produce the organization specified.

2.2 Flow Diagrams

Since readers are expected to have been exposed to some programming language, and since flowcharting symbols and techniques are commonly taught in conjunction with most computer languages, the reader probably has some knowledge of the symbols and techniques of flowcharting. Flowcharts are to programs as flow diagrams are to systems. Since the system designer is primarily interested in system design as opposed to program specification, flow diagrams will be used for representing logical system component interaction. Flowcharts may be used by the implementing programmers to aid in their task; however, the use of flowcharts by the system designer may be counterproductive in that it may stifle the programmers' creativity and prevent them from utilizing the full potential of their training and experience. Further, the ability of flowcharts to effectively communicate the program requirements is questionable. The recommended approach to the designer's specification of programs utilizes the PSWD discussed in Section 2.1.3.

It is quite important that the symbology and objective of flow diagrams and flowcharts not be confused; hence, the discussion above. Flow diagrams are developed very early in the analytical process while the development of program specifications is the last step in the design process. Often the largest impediments to learning a new discipline are the habits acquired in mastering an old one. Therefore, as the symbols and techniques of flow diagramming are uncovered, a concerted effort will be made to distinguish between these two diagrammatic techniques.

Two types of flow diagrams will be discussed below. The first is the traditional data flow diagram, which has the advantage of presenting all external entities, processes, files, and data flows in one picture with labeled symbols, enabling the designer to communicate the full structure of a system design. The second is the Warnier flow diagram which, while sacrificing some informational content, presents the design in a more logical ordering, facilitating quick reference to the supplementary documentation.

2.2.1 Data Flow Diagrams

A data flow diagram (DFD) is a completely labeled symbolic model of a system software design. As a model it will not be perfect in representing all of the facets of the system. However, as a tool for conceptualizing the interactions between the various components of the new system being designed, the DFD is invaluable if consistently developed and read. This is a big "if," since many inexperienced designers do not appreciate how imprecisely developed DFDs can lead to gross misinterpretations on the part of the reader. For this reason, the following subsections stress a discipline in applying consistent rules for both the development and the reading of a DFD. As this discipline is being presented, bear in mind that DFDs can be applied to purely manual, computer-integrated, or totally automated systems. Thus, the development of an accurate DFD depends more upon a knowledge of the process under consideration than it does upon the technical aspects of software development.

2.2.1.1 Data Flow Diagram Symbols

In system design work there are four primary components of data flow specification: (1) the ultimate source or destination of data, (2) data in motion, being transmitted by any means from one point to another, (3)

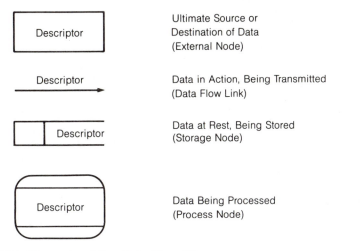

Figure 2.21 Elements of the Data Flow Diagram

data at rest, being stored for future reference or processing, and (4) data being processed, or transitioned from one form to another. Since these are, or can be made, mutually exclusive and collectively exhaustive, symbols can be given to each of these components, and any system can be synthesized using these building blocks. Figure 2.21 presents the symbols for each of the elements of the DFD. Note that each is accompanied by a descriptor which will be used in conjunction with reference numbers to locate the backup documentation for that element of the DFD. The primary purpose for the descriptors, however, is to give the reader an immediate impression of what the designer intends to occur within the system.

There are certain rules for the use of the DFD symbols which must be observed if the DFD is to be an effective communication tool. For example, all descriptors should be present and unique. If certain descriptors are not present, no means will exist for referencing the meaning of the symbol to the backup documentation. Further, a properly worded descriptor provides the reader with an immediate impression of the symbol's meaning and intent. If descriptors within two of the same type symbols are not unique, they will reference the same backup documentation, a situation which is only rarely desirable. Repetitions of the identically same entity are allowed and will be desirable for clarity. Other rules will be stated in conjunction with the specific symbols to which they apply.

In the context of graph theory, the data flow symbol can be viewed as a directed *link*, while the other symbols can be viewed as three types of *nodes*. Thus, there is a uniqueness to the data flow symbol, and a commonality among the other three, which can be studied and exploited. In order to take advantage of this and further our goal of brevity, the three node types will be referenced as: (1) external node, (2) process node, and (3) storage node. The discussion below will begin with the external node, and proceed to explain the data flow link, the process node, and the storage node.

Confusion results from the misuse of the external node, which, by definition, is a *source* or *sink* of data. It is called external because it is out of the scope or purview of the system being designed. External nodes should never be applied to departments of the organization which are actively processing or storing data in conjunction with the system. At times the scope of a system may be such that it totally serves another department of a company; in this context the external node may be used to reference a department. There is a clear inference, however, when the external node is used, that the system itself will not have any capabilities to process, modify, or assume custodianship over the data once it is forwarded to an external node. Further, the generation of the data received from an external node is totally out of the control of the system under design consideration, although it may specify format, range, and other acceptance parameters.

Examples of possible external node assignments include the customers, bank, employees, and the user at a terminal. The determination of the external entities is a logical first step in constructing the DFD. The answers to two questions define the external entities: (1) Who, outside of the system, will receive information from the system? and (2) Who, outside of the system, will be contributing information to the system? A definition of the external nodes begins to crystallize the objectives of, and resources available to, the system (i.e., the system outputs and inputs). While not yet specifying the content of this data, the involved individuals and organizations are defined, which is the the first step in system data definition.

The data flow link descriptor briefly indicates the subset of data which is being transferred between nodes. More importantly, it provides a reference to the specifications of this data subset, including content, format, volume, and other information in the backup documentation. With the exception of queries, which will be discussed below, data only flows in one direction for any data flow link. The arrowhead indicates this direction. Data flow links without arrowheads, or data flow links with two arrowheads, do not appear in carefully drawn DFDs.

The process node is the most complex node, since it conveys more information than any of the other DFD symbols. Figure 2.22 illustrates the three parts of the process node. In the top portion of the process node, a reference number is assigned, usually preceded by a P. This may be assigned to the nodes sequentially as they are developed, but the final assignment is usually made after the DFD is fully drawn. The ordering of the assigned numbers is important, as will be seen when methods for reading DFDs are discussed below. Basically, numbers are assigned in sequences through the DFD to guide the reader in reading and understanding one path at a time. Reference numbers are used to

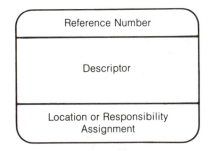

Figure 2.22 Parts of the Process Node

obtain the backup documentation for a given node, which might be quite extensive since it must thoroughly define all aspects of the process. Initially they are assigned as integers. As the development of the DFD continues, process nodes may be subdivided using the DFD symbols, including additional process nodes, for clarity. When this is done a decimal number is formed to identify the parent node, and this may be continued to any extent necessary to thoroughly detail the processes of the system. For example, node P2 may be analyzed into P2.1, P2.2, and P2.3; in turn, node P2.2 might be further exploded into P2.2.1, P2.2.2, etc. Accompanying the process nodes of a given analysis of a parent node would be the other nodes and links necessary to thoroughly explain the process.

Because referencing for process nodes is handled by reference numbers, there is not as much need to create terse and unique descriptors, as was the case with the data flow links. Rather, the descriptor may be a bit more lengthy in describing the process. These descriptors should either be verbs or action nouns to emphasize the action that is being performed. For example, "Calculate Unit Cost" is an acceptable verb phrase descriptor, while "Calculation of Unit Cost" is an equivalent action noun phrase. In either case, the descriptor should clearly but briefly title the action that is being performed in the process under consideration.

The final node to be discussed is the file* or storage node, which is detailed in Figure 2.23. This is divided into two parts, the left block being a reference number. To distinguish the storage from the process reference node numbers, a D should precede the data storage node reference number. This will be used to provide a direct reference to the backup documentation, which will include the file layouts, variable specifications, OWDs, and other information to thoroughly define the storage node.

Generally the nodes of a DFD will be linked by data flows to or from

---

*Technically, a storage node could consist of more or less than one file, just as a process node might consist of more or less than one program, especially in the preliminary stages of design.

Figure 2.23 Parts of the Storage Node

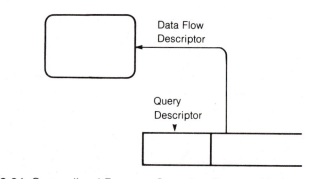

Figure 2.24 Generalized Process Querying Storage Node

each node. Each data flow link will be directed, indicating the flow of the data. Data may frequently flow both ways between any two nodes; in these cases there will be two data flow links, to distinguish between the quality, quantity, or type of data flowing in each direction. There is an exception to this rule in the case of a query. A query, generally addressed to a file, consists of an element of information (called a key) being transmitted to a process which accesses the file. The process then compares the key to a specified field of the various file records in order to find the record or records which qualify. (The process may be generalized to include multiple fields and/or multiple search keys). The symbology presented so far cannot adequately depict this. This type of transaction is so common that some symbolic provision is required to represent this special case. This is given in Figure 2.24.

The query is represented by a combination of the data flow link, in the direction that the major data is flowing, and an arrowhead, pointing toward the data store to indicate the presence and direction of the query. The backup documentation for a query will be included in the documentation for the data store being queried and the program specification of the program making the query. The query descriptor will be used within this documentation to thoroughly define the query-return process.

A final symbolic modification is required for clarity. Quite often DFDs become so involved that a duplication of some nodes is required (links are never duplicated). For example, it might be required to put the same external entity in two places on the DFD in order to reduce

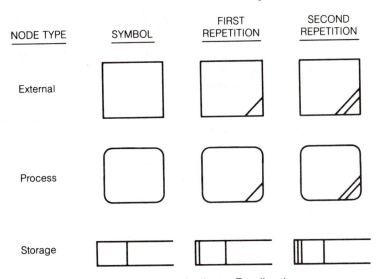

| NODE TYPE | SYMBOL | FIRST REPETITION | SECOND REPETITION |

Figure 2.25 Symbol Modification to Indicate Duplication

the number of links that must cross each other. When this is beneficial it is necessary to indicate within the node that it is a duplicate. Otherwise, a reader might get the impression that a few data flow links interact with the node when in fact it has additional links represented elsewhere. To identify node duplicates, a marking system is given as shown in Figure 2.25. Both the original and the duplicate(s) will be marked as indicated, since there is no real distinction made by the user. Finally, when it is necessary to break up a DFD because it will not fit on one page, generally this is best accomplished by repeating storage nodes while minimizing the repetitions of process nodes.

At this point an example will be introduced to illustrate the symbols and their interactions. Figure 2.26 presents a small portion of an order-handling module of a larger system. Since this example will be used in later subsections to illustrate methods for reading and developing a DFD, the detailed implications of the symbols will be deferred until that time. At this point consider the use of the symbols themselves. There are two external nodes, the customer and the bank. The customer is both a source and a sink of data, while the bank is strictly a sink. There are five process nodes, and the one storage node is called the customer account file. These are interconnected by a variety of links. Unique among these is the query that process P1.5 makes of the customer accounts file using the customer's name as the key.

Note how the labeling rules are observed in Figure 2.26. Each node and each link is labeled. External nodes are adequately defined by one-word descriptors. The descriptors of the storage node will be used consistently throughout the documentation to describe this data storage entity, either by number (D1) or by name (Customer Account File). The link descriptors are all nouns, in that they assign a name to the data that is being transferred. Finally there are three descriptors within each process node: (1) the process number for reference, (2) the verb or action descriptor, and (3) the department or individual responsible for performing the action. It is very important that these last two be kept in order. There is a temptation on the part of students to put departments in the center of the process node. If this is done the entire purpose of the process node is defeated. For example, it would be incorrect to put Customer Service in the center of P1.1.

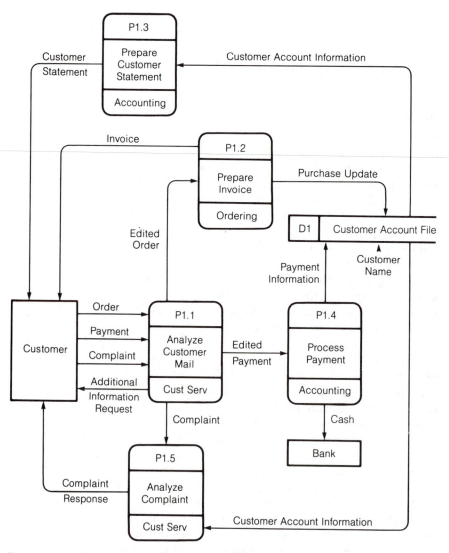

Figure 2.26 Example of a Customer Order DFD

2.2.1.2 DFD Symbol Interaction Syntax

This section will present the syntax rules for data flow construction. The term *syntax* here is used to distinguish from development rules, which will be covered in Section 2.2.1.4. The development of a DFD is much like the computer coding of a procedure: there is no one perfectly correct DFD for a system, just as there are several ways to code a procedure, and each programmer will do it a little differently. However, just as there are syntax rules for coding, the violation of which usually results in ambiguity or unrecognizable code, so there are syntax rules for DFD construction.

We choose to couch the presentation of the rules of DFD construction in the definitions of graph theory. This serves a dual purpose: (1) it provides an unambiguous language by which certain rules can be stated and thus clearly understood, and (2) it provides the language by which the rules for reading and developing the DFD can be stated.

Figure 2.27 presents some example graphs which illustrate the basic terms. Here nodes are represented by points for clarity. A node is a *right neighbor* of a second node if a link between the two is directed toward the first. If the link is pointing from the first to the second, then the first is the *left neighbor* of the second. When a node has itself for a right (and hence a left) neighbor, the link is called a *loop*, and the node is called a *reflexive node*. If links between two nodes point at both nodes, the relationship between the nodes is *symmetric*. A *path* exists between two nodes if links can be traced from one to the other following the direction of the arrowheads. Two points which are connected by both an alternative path and a direct link are called *transitive*. If two points are connected by links irrespective of the direction of the links, then a *walk* exists between the two points. A path which comes back to its originating point is called a *cycle*.

In the uses of these terms which follow, queries are considered to be represented by the primary data flow link. Two types of rules will be given: (1) general rules, which pertain to the general content of the DFD, and (2) relational rules, which pertain to the linkage between nodes.

General Rule #1   (Permissible links rule.)

External nodes and storage nodes may only be linked to process nodes; they may not be linked to themselves or to each other. Process nodes may be connected to each other or to any other node type.

This rule has a number of ramifications. It arises from the practical aspects of DFD construction. It is impossible for an external entity to read from or write to a file without invoking some software subsystem. While it is true that system utilities may be used, they are software and

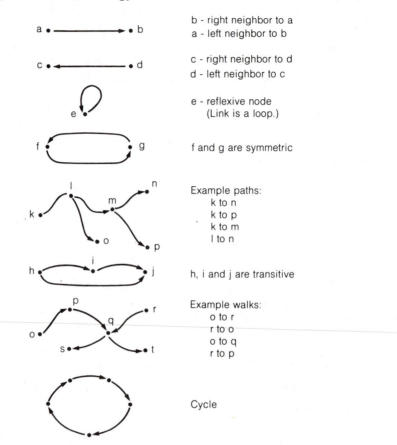

b - right neighbor to a
a - left neighbor to b

c - right neighbor to d
d - left neighbor to c

e - reflexive node
   (Link is a loop.)

f and g are symmetric

Example paths:
   k to n
   k to p
   k to m
   l to n

h, i and j are transitive

Example walks:
   o to r
   r to o
   o to q
   r to p

Cycle

Figure 2.27 Examples of Terms Used in Graph Theory

should not be ignored in the DFD if they are going to be employed by the system. (The process node should reference the utility employed rather than being excluded.) Process nodes are specifically allowed to be linked to the other two node types and also to be linked to each other. This is allowed for clarity, since quite often, for documentation purposes, it is advantageous to divide a long process into several smaller processes.

General Rule #2

A walk must exist between any two nodes of the system, i.e., the graph of the system is a connected graph.

Two totally independent systems might be concurrently represented by a disconnected graph. In systems design work, however, we restrict ourselves to the consideration of one such system at a time. Thus, a DFD is defined as a connected graph with three node types connected by links according to General Rule #1, where the links represent data flows of various quantities, contents, and formats. One final, rather obvious, rule is presented before proceeding to the relational rules.

General Rule #3

DFD graphs are loop-free.

This rules out any node passing data to itself. If there are several components of a node, such as might arise with two subprocesses or multiple subfiles, the analyst has the option to establish additional nodes to more clearly convey the design. A loop tends to obscure rather than clarify and thus it is prohibited. Note that General Rule #3 (GR3) is a corollary of GR1 for external and storage nodes. It also applies to process nodes for the reasons stated.

The rules given above are generally applicable to the entire DFD. Those which deal with particular relationships between nodes are presented next.

Relational Rule #1

Right neighbors are recipients of data from the node under consideration; left neighbors are suppliers of data to the node under consideration.

Thus, the link to a right neighbor is called an *output* link, while a link from a left neighbor is called an *input* link. These definitions are used in Relational Rule #2 below.

Relational Rule #2

If an external node is neither a pure source or sink (i.e., it has both input and output links), a cycle must exist through the system which includes this node.

While this rule is important in the development of the DFD, it is of even greater significance in its reading and interpretation. The fact that Relational Rule #2 is true enables the various paths through the system to be traversed from the source to the sink. In those cases where the same external entity is both, this path becomes a cycle.

Relational Rule #3

If the linkage between two nodes is transitive or symmetric, or if any two paths exist between the two nodes, the data flowing across these two paths must be different in content or format.

This rule is stated to eliminate redundancy. It would be violated only in those cases where redundancy was desirable to increase reliability. However, backup components for this purpose usually employ different hardware and/or software; therefore, separate documentation specifications are required. Most often when two separate paths exist between two nodes, different information is being transmitted, albeit simultaneously.

Relational Rule #4

There must be a path from the external node which represents the primary user of the system to every other user function node in the system.

This rule will become apparent when illustrating DFD development. It should be obvious that in order for users to obtain the output of a process set up for their benefit, they must be able to gain access to it, even though such access might be indirect.

Relational Rule #5

There must be a path from the node which represents the maintenance of the system to every other maintenance process node in the system.

This rule is analogous to Relational Rule #4, but it applies to maintenance functions rather than user functions. Maintenance processes include those necessary functions which are not generally invoked by the user.

Consider Figure 2.26 again, this time in regard to the interactions of the symbols which were described above in terms of their descriptors. The following enumerates the rules in terms of Figure 2.26:

1. **GR1.**   An example of disallowed links would include a link between the customer and the bank or either one of these external entities directly to the Customer Account File.
2. **GR2.**   Figure 2.26 is totally connected; a disconnected subset of nodes would have no meaning since data could not flow between the disconnected subsets.
3. **GR3.**   Figure 2.26 is loop free. A link from any of the nodes to itself would be meaningless.
4. **RR1.**   Data flows only in the direction of the arrowhead. Where information cycles through the system (as in the Customer - P1.1 - P1.5 - Customer sequence), separate links show the flow back to the originating node. Direct interactions in both directions are also permissible, as in the case of "Additional Information Request." This request is made from customer service to the customer for minor omitted details which are necessary before proceeding with processing.

5. **RR2.** The customer external node is neither pure source nor pure sink and therefore at least one cycle through the system must include it. In this case there are five cycles, illustrating that this is the primary external node (hence its placement on the left of the diagram). The bank is a pure sink and therefore it is not part of a cycle. Further development, e.g., the inclusion of monthly bank statements, could introduce such a cycle.

6. **RR3.** The path P1.1 - P1.2 - Customer is made transitive by the link labeled "Additional Information Request," as is the path P1.1 - P1.5 - Customer. While the application of this rule (i.e., that the data flowing along the two separate paths is distinctly different) is trivial in the first example, it is quite meaningful in the second. In essence it necessitates that a clear distinction be made between the data which flows to the customer as a result of a customer complaint, and that which flows as a result of a request by customer service for more information from the customer. The Customer and P1.1 also fit the technical definition of being linked symmetrically. Further, there are a variety of paths which link the same nodes, thus forming paths (examples: P1.1 - P1.2 - D1 and P1.1 - P1.4 - D1; D1 - P1.3 - Customer and D1 - P1.5 - Customer). In all of the cases mentioned the data flows are different either in the content or format, as required.

7. **RR4.** The primary user external node is the customer, and a path can be found from this node to every other node in the diagram.

8. **RR5.** This rule is not applicable since there are no maintenance nodes depicted.

2.2.1.3 Rules for Reading a DFD

Generally we learn to read before we learn to write. It should seem logical that the art of reading a DFD must be cultured before students are in a position to begin developing their own DFD designs. Many analysts, however, failing to understand this, do not properly read their own DFDs for verification. Ideally, the DFD should reproduce exactly what is in the mind of the designer to everyone who reads it. This can be accomplished if some logical rules are applied both in the development and in the reading of the DFD. Since it is desirable to learn to read before writing, the reading rules are outlined in the following steps:

1. Generally read from left to right. Start with the external node to the left which has a link to the lowest-numbered process node.

2. Consider one node at a time. Study all of the outputs of the

node first and get a thorough understanding of each. Then, briefly consider the inputs to the node just to obtain familiarity with them since they will each be studied in more detail later.

3. Select the next node and study it as specified in step 2. Select the process node not yet studied which has the lowest reference number and which has an output link from the node previously studied.

4. Trace through the DFD, repeating step 3 until a cycle is formed back to the original external entity selected for study. This cycle may include storage nodes.

5. Repeat step 4 until all output links and their cycles or paths involving the external entity have been studied.

6. Repeat step 5 for all external entities.

7. Isolate and study the outputs and inputs of each storage node individually.

Because of the multifaceted nature of the DFD, it is impossible for many persons below the genius level to comprehend the entire DFD simultaneously. The procedure given above forces one path at a time to be the focal point of consideration. Within that path only one node or link is being considered and understood at a time. Interacting nodes are given only secondary consideration until their respective paths are considered. While this is time-consuming, it is essential if the DFD is to be understood.

Consider Figure 2.26 once again to illustrate the rules given above. By an enumeration of the rules it should be possible to generate a narrative description of the system. Generally, following the rules given above will result in the following thought process. The external node to the left is the customer. The outputs from the customer are: (1) Order, (2) Payment, and (3) Complaint. These go into the system and are analyzed by P1.1. Inputs to the customer include requests for additional information, responses to complaints, a customer statement, and invoices.

Proceeding to node P1.1 (lowest numbered), there are four outputs, which correspond to the three inputs already studied plus the "Additional Information Request" which is included merely for refinement. Since the node is called "Analyze Customer Mail," this is just an initial review of incoming information with some editing included.

The next node to consider is P1.2. Its input has already been considered, so proceed to the two outputs: (1) Purchase Update and (2) Invoice. The second of these goes back to the customer, thus completing the first nontrivial cycle. A purchase update is made to D1 as a result of the order.

The reader could proceed in many different ways at this point, if it were not for the "lowest number rule." Because of it, the next node to consider is P1.3. This shows Customer Account Information being read from D1 and a customer statement being generated and transmitted to the customer. This completes a second major cycle: Customer–P1.1–P1.2–D1–P1.3–Customer.

Again, the next point of consideration is defined by the lowest numbered node to be P1.4. This is also necessitated by the fact that we have not cleared all of the outputs of node P1.1 from consideration. The outputs of P1.4 include cash to the bank and an update of payment information to D1. The bank is a sink and requires no further consideration. Since D1 is not a process node, it will not be given the same study as the process nodes are given at this time.

The final process node, which completes two cycles simultaneously is P1.5. Its output is the customer complaint response, while the input is the complaint itself and the response to a query from D1. As a final step, D1 will be isolated and studied in terms of its inputs and outputs. While it was considered as part of each cycle, its intensive study is deferred until all process nodes are studied. This is done so that the flow cycles are not interrupted in the original pass. It also gives the reader a chance to consider each storage node as a focal point.

In concluding this example, note how critical the numbering of each process node is. Arbitrary numbering could make the DFD quite confusing to read. This is one critical point that will be considered in the next section.

2.2.1.4 Developing a Data Flow Diagram

Developing a DFD is much more difficult than reading one, since it involves the manifestation of something that exists only fragmentarily in the mind of the designer. Thus, the rules given here will serve more to "break the ice" and "get off dead center" than to actually formalize the thought process employed. It should be recognized that DFD development is an ongoing process throughout design and often into development. As design changes evolve, they will be incorporated into the DFD. Thus, the DFD should be regarded as a constantly improving tool of communication, often from the designer back to himself. It will start as a rough idea, a very gross estimate of what the system will look like, totally devoid of detail. It will evolve with both detail and design improvement into a specification, a blueprint, and a table of contents for the documentation of the final system design.

The rules for DFD development will be divided into preliminary rules (P), operational rules (O), and evolutionary rules (E). The preliminary rules are those which should be followed prior to any drawing up of the diagram itself. These are as follow:

> P1. Identify all external entities. These will not affect or be affected by variations in system design. Any operational changes in the system will be totally transparent to the external entities. In

defining the external entities the scope of the system is tentatively defined.

P2. Identify total system outputs and define the external node(s) to which each output will go.

P3. Identify total system inputs and specify the external entity which is the source of each input.

P4. Tentatively define the files (storage nodes) which will be used for intermediate and/or permanent storage.

P5. Tentatively define the process nodes which will transition data between the external entities, the storage nodes, and each other.

The steps above will yield an initial concept of a design. It is time to begin linking these nodes together in a logical way. The objective is to make this synthesis understandable to all readers of the DFD. This requires the observation of the following operational rules, which reflect the rules for reading the DFD given above. These are as follow:

O1. Place the principal external entity (or entities) on the left of the DFD to facilitate reading from left to right.

O2. Develop the DFD in cycles. Set up a prioritized list of these cycles according to the hierarchy of the system's objectives. Beginning with the top priority cycle, tentatively define the process nodes and data stores to accomplish the objective. Connect these with appropriately labeled data flow links. Number process nodes in the order to be considered by the reader.

O3. Repeat O2 for all cycles, in the order of priority of the objectives which they accomplish.

O4. Add any paths to pure sources or sinks.

O5. Study each node in terms of its outputs and inputs. Read the DFD according to the rules for reading given above. Modify as required.

The first cut of the DFD may be adequate for some simple designs. The tendency, however, is to neglect much necessary detail in the initial DFD. This is not bad, since there are great advantages to considering the generalized overview of the design prior to going into detail. However, as the design process continues these details will need to be addressed. Many of them can be handled in the data dictionary without causing modification of the DFD. When this is not sufficient, the following evolutionary rules aid in further DFD development:

E1. Data Store Modification. Data stores probably undergo more modification than any other component in the DFD. Consider consolidating and eliminating unnecessary files. Consider dividing one data store into two. Redraw the DFD accordingly.

E2. Process Node Modification. Consider exploding one process node into a number of others according to the procedural breakdown within the process. Intermediate data stores, which were previously embedded within the process node, may now be drawn on the DFD. Number the subprocess nodes after the process node being analyzed. For example, subprocesses of process node 5 will be numbered 5.1, 5.2, 5.3, . . .; subprocesses of node 5.2 will be numbered 5.2.1, 5.2.2, etc. In this way the original DFD can serve as a master, and extensive renumbering of all process nodes will not be required. Number the nodes in order according to rule O2.

E3. Reiterate until a level of detail is achieved which is compatible with the backup documentation of the DFD, i.e., the data dictionary. This level should be adequate to enable the programmers to bring the system into existence, but not so unnecessarily detailed as to burden them down with irrelevant considerations.

This last rule requires elaboration. Trivial data flows, such as menu or query responses to invoke programs, should be omitted from the data flow diagram. This follows for three reasons: (1) such menus and queries are much better defined in other areas of the documentation, namely, the User Guide; (2) their inclusion within the DFD leads to such a detailed diagram that it is practically impossible to read; and (3) their inclusion within the DFD obscures the major data flows. As a rule then, user inputs which merely modify the flow of control within the system will not be included as links within the DFD. All significant data flows will be included as links.

Figure 2.26 has been used as an example to illustrate the symbols, interactions, and rules for reading the DFD. Consider now how Figure 2.26 may have been developed from the observation or conceptualization of a real system. The foregoing rules will be used to order the following presentation.

P1. The external entities of this system are the customer and the bank. Both are beyond the direct control of the system. Note that the departments of the company (i.e., Customer Service, Accounting, and Ordering) are not external entities; rather, they will serve in actually implementing the processes.

P2–P3. Total system outputs include: (1) invoices, (2) customer statements, (3) complaint responses, (4) requests for additional information, and (5) cash to the bank. Total system inputs are: (1) orders, (2) payments, and (3) customer complaints. Thus a first gross DFD might have one process node numbered P1, called "Process Customer Orders," and the two external entities with these inputs and outputs linking the three nodes.

P4. The single storage node identified for this example is the customer account file. Many DFDs will have several such files. Here it is assumed that this file can accommodate the data needs of this part of the total system.

P5. The final preliminary step is a tentative definition of the process nodes. Chances are that this tentative definition will not be identical to the final DFD as given in Figure 2.26. It will be known that the incoming customer mail or phone orders will require an initial analysis and editing step. The degree of responsibility to be included within this process might not be clear. For example, P1.1 could range anywhere from a manual sorting operation to an interactive customer information data entry system. While the DFD does not specify this, it does crystallize the design decisions which must be made before system development. An initial definition of process nodes might have P1.1 and P1.5 combined; P1.2 and P1.4 might also be combined. Other single nodes might have originally been defined as multiple nodes.

O1. Note the placement of the customer node to the left so that it is the initial focal point of the reader.

O2–O3. The prioritized list of cycles in Figure 2.26 is: (1) Customer–P1.1–P1.2–Customer; (2) Customer–P1.1–P1.2–D1–P1.3–Customer; and (3) Customer–P1.1–P1.4–D1–P1.5–Customer. (Note that Customer–P1.1–P1.4–D1–P1.3–Customer is also a cycle, but all of these nodes have been traversed by other cycles.) This priority is determined by the numbering scheme assigned. In developing the DFD, the cycles are determined and prioritized first and then numbers are assigned to facilitate reading.

O4. In addition to these cycles there is a path: Customer - P1.1 - P1.4 - Bank. This would be added at this point in the development of the diagram.

O5. The evolution of the tentative DFD begins with a consideration of the data stores. As an example, in Figure 2.26 consideration could be given to setting up a separate customer address or customer information file which is independent of orders (i.e., one record per customer as opposed to one record per order). Decisions regarding the content and structure of D1 will determine if additional data stores must be introduced into the system design. Whenever such a modification is made the DFD should be updated appropriately.

E1. Not applicable.

E2. Chances are good that the introduction of other files would lead to the need for more processes being specified. If the current DFD is valid, such will result from explosion of an existing process rather than the addition of another. For example, in

Figure 2.26, the introduction of a customer address file might lead to the explosion of P1.3 into two nodes: (1) P1.3.1 - compute customer payment due, and (2) P1.3.2 - assign address to customer statement. P1.3.1 would read from D1 while P1.3.2 would use the outputs from P1.3.1 and the new customer address file. A variety of other process node modifications could also be made.

E3. Obviously the process of explosion could continue until there was a process node for every elemental data transition. Figure 2.26 indicates two manual processes (P1.1 and P1.5) and three processes that could be computerized (P1.2, P1.3, and P1.4). The level of detail of the DFD should be compatible with the real system or its backup documentation. Generally each process node will correspond to a different program or a distinct manual operation.

2.2.2 Warnier Flow Diagrams

Data flow diagrams promote a visual understanding of a design at the systems level which cannot be achieved in any other way. The development of the DFD in cycles enables it to simultaneously depict the true interactions between all components of the system. This is particularly useful in the design phase of the system design, development, and operation process. Once the design thought process is completed and the development begins, the DFD capabilities diminish somewhat in importance. At this point two other communication tools become more effective for documenting data flows, namely, the file-program cross-reference chart and the Warnier flow diagram (WFD). The first of these is a very brief table which concisely summarizes the computerized portions of the DFD. Since it is generally considered a part of the data dictionary it will be covered in detail in Section 2.3.5. However, because of the similarity of the WFD to the DFD in content and objective, it will be covered next.

2.2.2.1 Warnier Flow Diagram Syntax

The WFD applies the basic principles of Warnier analysis given above to the analysis of the proposed software system. It has the disadvantage of being in a list format, which cannot depict the interactions between files and programs. A second disadvantage is that the lack of labels on the data flows prevents the reader from getting a feel for the flow of the data. Given these shortcomings it might be wondered why a WFD would ever be used. The reasons are fourfold: (1) it is a more condensed model of the system which shows all program-file interactions in a nutshell and therefore provides a handier reference for maintenance purposes; (2) it is ordered by module and by program number within module, thus providing a more systematic presentation in this regard than the DFD; (3) it is much easier to document and keep updated on word processing equipment since it involves less symbology; and (4) it is consistent with the other Warnier symbology used, thus providing for more effective communications.

The structure of the WFD is presented in Figure 2.28. It is a standard Warnier analysis of the software under consideration. The first level analysis is at the "module" level, which generally has a one-to-one correspondence between its entries and the system supervisory menu options. Each of the modules are analyzed into the process nodes, which in the software system generally correspond to programs. These are ordered according to numbers assigned in the DFD. The third level of analysis is the data interaction of the program either with files, data-entry mats, reports, or other programs.

Figure 2.28 gives some examples of the symbology used. An arrow toward the program indicates that either file(s) and/or data-entry mat(s) provide input to the program. An arrow away from the program indicates that either file(s) and/or reports are being written. Where several inputs or outputs exist for the same program, such as P1.3, a Warnier

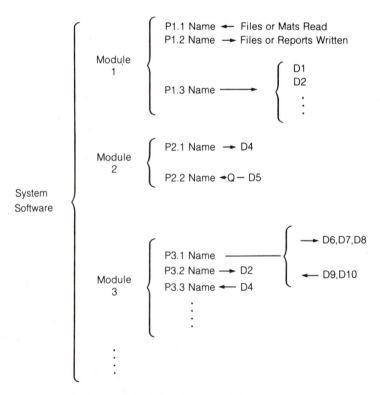

Figure 2.28 Symbology of the Warnier Flow Diagram

bracket is used appropriately. Similar symbology is used when the same program has both inputs and outputs, such as P3.1. Finally, a query is indicated by a Q right behind the arrowhead, as in P2.2.

While program names after the P numbers make the WFD much more readable, labeling of the data flows (as was done in the DFD) is not practical here. In fact, such would make the WFD less useful; if this type of data flow information is required, the DFD is a much better document to use in obtaining it.

The DFD and the WFD provide an interface between the completed system and the system documentation. Each of the files and each of the programs will be documented in the system software maintenance manual. Thus, to perform maintenance on any part of the system, the appropriate program or data file can be identified from the WFD and then referenced in the Maintenance Manual.

2.2.2.2 Comparative Example

It is beneficial to compare the DFD with the WFD for a similar application. Suppose that the example presented in Figure 2.26 were part of a larger integrated system including an inventory control module and a sales module, in addition to the customer order module depicted. The WFD for this system might appear as given in Figure 2.29. Note that only the details corresponding to those given in Figure 2.26 for the customer order module are included.

Note the one-to-one correspondence between the WFD and the DFD. There might be some question as to the need for such redundancy in the documentation, since one of the objectives of structured documentation was the elimination of redundancy. We are not recommending that both be retained in the documentation. The DFD has a tremendous advantage as an aid to design. However, once the design is completed, the WFD might be easier to maintain in the documentation with little loss of informational content at that point. This is left to the designer to determine.

2.3 Data Dictionary Components

A data dictionary is a completely cross-referenced set of documentation which describes the system details. The system flow diagrams serve as its master table of contents, since generally each element of the flow diagrams will have a counterpart backup in the data dictionary. Exceptions to this rule include those cases where the details shown on a DFD are adequate and the component design is obvious. In these cases backup documentation would be self-defeating in that it could obscure other essential information. However, the designer should never assume that the programmer or other system design users are mindread-

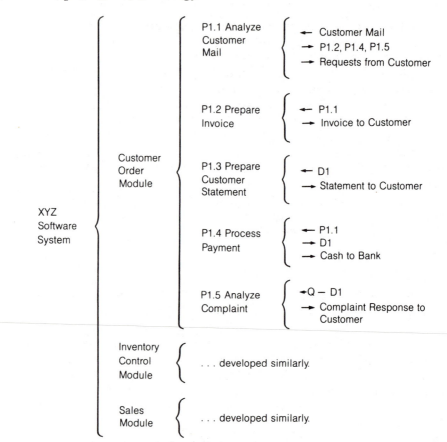

Figure 2.29 Example of a Customer Order WFD

ers; when in doubt, spell it out. The purpose of this section is to introduce some of the basic data dictionary components so that their use in subsequent chapters will not come as a surprise.

The data dictionary will eventually be incorporated into the system documentation. However, prior to discussing the system design methodology it is helpful to consider the data dictionary as consisting of the following five components:

1. Data-entry specifications,
2. File content specification,
3. Report output specifications,
4. Data element narrative references, and
5. File-program cross-reference.

The development of these components is in the following order: 3, 2 and 4, 1, 5. The advantages for this ordering of development will be clearly seen in Chapter 3. However, in introducing these components of the data dictionary, it will be best to order the presentation according to the logical flow of data, as given above. A separate subsection is devoted to each.

2.3.1 Data Entry Specification

This component of the data dictionary will be found in the User Guide portion of the documentation. The details of data entry specification will be presented with the methodology for generating it in Section 6.3. As an introduction to this component of the data dictionary, it is necessary to be aware that the input specifications will consist of illustrated narratives which describe user input requirements in the order in which they occur. Input types are classified according to the following definitions.

Menu—a logically ordered list of choices which appears on the screen, prompting the user to select one; these are usually used for transfer of control as opposed to data entry.

Mat—a data-entry form which appears on the screen, prompting the user to fill in the blanks, analogous to a manual blank form; these are used for large data-entry operations.

Query—a short question (or prompt) which appears on the screen for which there are a limited number of responses.

For data dictionary purposes, the data entry mat is of primary concern in that it is the method most often employed for getting data entered into the system. While menus and queries are at times used for this purpose, this is usually a minor part of their function. In any event, the discussion which follows would apply to any method by which data are entered into the system.

The data-entry specifications will totally define the input information requirements to the user. In the case of the data-entry mat, there will be an item-by-item description of every data element input, so that there is no question on the part of the data gatherer or the data-entry operator as to the source of each data element and its value.

To integrate these specifications into the data dictionary, it is essential that the file destination of each data element be documented with the element specification itself. This is an extremely important concept which is frequently omitted, leading to a loss of system integrity. The method by which this is accomplished is merely to include the file number and the sequence number within the file at the end of the specification. For example, if the third data element on a data-entry mat were the month of the year, its specification might appear as follows:

3. Month - the month of the year that the data is being entered; 01 = January, 02 = February, . . . , 12 = December (D1.3,4).

In this example the data element as well as its coding is totally defined, after which the "(D1.3,4)" indicates that the destination of this data element is file number D1.3, variable sequence number 4. (The exact meaning of this will become clearer after reading Section 2.3.2.)

The addition of this type of reference is of such importance that the reference itself will be called the file, sequence number, or sometimes just (file,seq). Since the file specifications will have been written at this point in the methodology, it will be quite easy to add (file,seq) to the input specifications. This integration of the data element specifications into the data dictionary has two very important functions: (1) it gives the programmer an exact knowledge of the destination of each data element, and (2) it verifies the integrity of the system by assuring that there is a one-to-one correspondence between the file layouts and the data-entry specifications.

As a final qualifier, there are many times when the data entered is not written directly to a file as entered. If the transition made is a minor one which is obvious from the file layout and the input specification, the (file,seq) number will still provide a sufficient linkage. However, if there is a major transition of one or more (or possibly combinations) of the data elements, such that the one-to-one correspondence between data element entered and data element stored is lost, then the program number of the program which performs this transition will be referenced in place of the (file,seq) number. For example, the reference might appear as (P3.5) if program number 3.5 (from the DFD) is transitioning the data to such an extent that a (file,seq) reference has no meaning. In this case the programmer can go to the program specification for P3.5 to determine exactly what the disposition of this variable is to be.

2.3.2 File Content Specification

The specification of file content is an essential component of the data dictionary for two reasons: (1) the data element descriptions detail many, if not all, of the inputs and outputs to/from the process nodes in the data flow diagram, and (2) many of the element descriptions are sufficient within themselves so that no further specifications are required.

The file layout specifications can be viewed as an exception tool. Those variables which are adequately described by the input specifications will require no further description. Those which are not clearly defined within the file layout will be referenced so that immediate access to the more detailed specifications can be obtained (see Section 2.3.4).

FILE LAYOUT FORM - D1.3

FILE TITLE: D1.3 PM Machine File							PAGE 1 OF 1
COMMENTS: One record per machine						RECORD LENGTH: 115 Bytes	
						DSN: ie123db.pm.mach.data	

SEQ	ST POS	END POS	FLD SIZ	FLD*	REF NUM	FIELD DESCRIPTION	SOURCE **
1	1	3	3	N		Machine Reference Number	MDEM-1
2	4	9	6	N		Current Date-mmddyy	MDEM-2
3	10	13	4	X	1	Machine Location Code	MDEM-3
4	14	19	6	X		Charge Code	P2.3
5	20	39	20	X		Brief Machine Description	MDEM-7
6	40	114	75	X		Detailed Machine Description	MDEM-8
7	115	115	1	N	2	Update Type	MDEM-10

*FIELD CHARACTERISTICS: A = Alpha; B = Binary; N = Numeric;
SN = Signed Numeric; X= Alphanumeric

** SOURCE NAMES:
MDEM - Machine data entry mat

Figure 2.30 Example of a Completed File Layout Form

An example of a file layout form is given in Figure 2.30. The file layout form gives a sequentially assigned number (SEQ), the starting position (ST POS), ending position (END POS), field size (FLD SIZ), the field type (FLD) given by the key in the footnote, a reference number (REF NUM) if further specification is required, the field description, and the source of the data element. The methodology for utilizing these areas of the form will be given in Chapter 5. At this point we want to concentrate upon the integration of the file layout form into the data dictionary.

The file layout forms are at the center of the data dictionary in that referencing on both the input destination and the output (e.g., report) source is made to the file layout form. This is done by using the (file,seq) number as described above. For example, the brief machine description would be referenced as (D1.3,5). This reference to the file layout form will be made both for input destination references (Section 2.3.1) or for output report specification, as will be discussed in Section 2.3.3.

While a data element in a file may have several destinations, generally it has but one source. To fully integrate the file layout form into the data dictionary, it is essential that this source be referenced on the form itself. The file layout form enables this under the SOURCE column. In the example of Figure 2.30, the first three variables are defined by the machine data-entry mat (MDEM), variable numbers 1, 2, and 3, respectively. The fourth variable is defined by program number P2.3. It is rare that one file would use several different data entry mats for input source, although this is totally permissible.

As the focal point of the data dictionary, the file layout form enables referencing to it from both its inputs and outputs, and it references the source of each data element. Note the timing of the documentation of each of these. Only the *output* specifications will be available when the file layout forms are developed. Thus, the SOURCE column of the file layout form will have to wait for the input specifications to be developed. The SOURCE can be completed at the same time as the (file,seq) references are posted to the input specifications. Finally, it will be necessary to complete the file layout forms before the output specification references can be made, as discussed in the next section.

2.3.3 Output Report Specification

The specification of output reports is the result of the second analytical step in the design process (see Table 1.3). As such, it precedes the development of the two components of the data dictionary discussed above (input specifications and file layout forms). Procedures for output specification will be detailed in Chapter 3 where the methodology for output specification will be given. The purpose of introducing this component of the data dictionary at this time is to show its integration with the other components, and thus to demonstrate the value of the data dictionary for design.

Similar to data-entry specification, output report specification ultimately gets down to a definition of every variable output in a given report. Generally, a prototype report will be drawn up and the data element types will be numbered within it. For each type of data element output a specification will be given to provide the programmer with a knowledge of the source of the variable and, if necessary, the method for calculating its value. These output specifications will be referenced to the (file,seq) number from which they are obtained. For example, if the fifth type of output on a report is a customer name which is directly moved from a file, the specification might appear as follows:

> 5. Customer Name - the name of the customer for which this line of information is being generated (D1.4,3).

This indicates that the source of this data element is the third variable in data store D1.4.

There are times when variables are combined in such a way that a

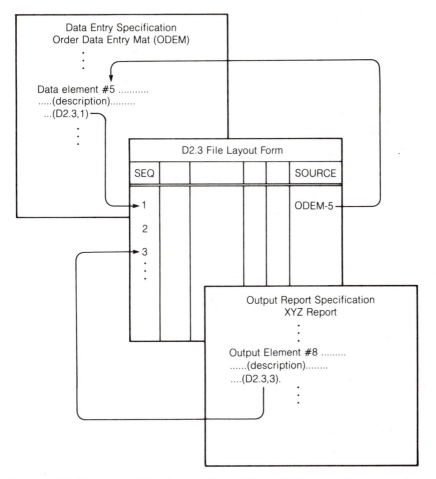

Figure 2.31 Example of the Integration of Data Dictionary Components

one-to-one correspondence between the file variable and the output data element is not possible to specify. In these cases the report generator program will be referenced in lieu of the (file,seq) number. The programmer can then look up the program specification by number to define the output variable. In these cases, the output specification number should be used within the program specification to reference the output.

Since the output report specifications are written before the file layouts, these references will have to wait for the completion of the file layouts. This provides an excellent verification of the design, and it is defined as such in the methodology presented in Chapter 5.

Figure 2.31 reviews the three subsections given above and shows the total integration of these three basic components of the data dictionary. In this example, order data-entry mat (ODEM) data element number 5 is referenced to its destination, file D2.3, sequence number 1. In turn, the file layout form shows ODEM-5 as its source. The output report specification uses the variable in sequence number 3 as the source for its variable number 8. In practice, all of the input, file, and output variables would be cross-referenced in this way. Using these tools, it is possible to thoroughly track data elements through the system. Note that the primary direction of reference is contrary to the data flow, since a data element may have several uses, but usually it only has one source.

2.3.4 Data Element References

The three basic components of the data dictionary were presented above. This subsection elaborates further on the file content specification component. Generally, complete documentation of a file will consist of the file layout forms, each accompanied by narratives and possible other diagrams for defining the data structure techniques (Chapter 4). Up to this point, two ways have been stated for defining a variable within the file layout form: (1) by data-entry specifications, or (2) by program specifications. The reference number for either one of these would appear in the SOURCE column of the data entry form if it played the major role of determining the value of the variable.

Generally, this is all that is necessary to define a variable value. However, we want to leave the door open for additional variable description within the narrative itself. Thus, the file layout form reference number (REF NUM) column provides a means for further variable definition by exception. That is, if this column entry for a variable is blank, as in variables 1, 2, and 4-6 in Figure 2.30, then there is no further backup and the reader is expected to understand the meaning of the variable from the references given. If a number appears, then the narrative for the file layout should be consulted to get more information as to the definition of the variable. This should only rarely occur, however, since the program specification Warnier diagrams which make up the program specifications are quite adequate for most variable definitions. This option is left open so that additional statements with reference to the data elements can be easily located within the file layout narratives.

2.3.5 File-program Cross-reference

Whenever maintenance is being performed on a system, a change will be made either in a program or in a file structure. Such a change can have disastrous effects upon other components of the system if they are expecting the pre-maintenance conditions to exist. In order to prevent such a situation from occurring, a cross-reference list in the format of the example given in Figure 2.32 will be established. This will serve to notify those doing maintenance or debugging of those files and programs which are affected by each change, since each such interaction is presented.

Program	DATA FILES										
	D1	D2	D3	D4	D5	D6	D7	D8	D9	D10	D11
0.1										I	I
0.2					O					I	I
0.3					I					I	I
1.1					I	I				I	I
1.2										I	I
1.3							I			I	I
1.4								I		I	I
1.5				Q						I	I
1.6			Q							I	I
1.7	O	O	O	O		O				I	I
2.1	I	I			I	I				I	I
2.2		I		Q		O				I	I
2.3	O	O	O	O						I	I
3.0	I		I	I	I					I	I
4.0	I		O	I			I			I	I
5.0	B	I	B	B	I		I			I	
6.0	I	I	I	I		I	I			I	I
7.1	I	I			I	B		O		I	I
7.2			B	B							
7.3								I		I	I
8.1	I	I	I	I	I	I			O	I	I
8.2	O	O	O	O		O			I	I	I
M1	C	C	C	C		C					
M2										B	O
M3								B			

Legend: B = Both I & O
C = Creates
I = Input to Program
Mx = Maintenance Program x
O = Output from Program
Q = Query

Figure 2.32 Example of a File-program Cross-reference

The file-program cross-reference is an abbreviated data flow diagram. While lacking all descriptive labels, the file-program cross-reference does show the interaction and the type of interaction between the programs and the files. While this information is redundant with the DFD, it is justified because it is much easier than the DFD to maintain and use for the purpose described above. However, while the DFD is an excellent design tool, it is hard to imagine how the file-program cross-reference could serve as an aid to design. Consequently, the file-program cross-reference is usually established very near the end of the design process.

2.4 Integration of Tools into Methodology

A continuous distinction has been made above between the tools and the methodology for their application. In particular, the ordering of the application of the tools has been emphasized to be inconsistent with their logical presentation. Now that the tools have been presented, a comparison between these tools and Table 1.3 can be synthesized. Table

Table 2.2 Comparison of Methodology Steps and Tools

Step	Tools Applied
1. Analyze the affected management system	Management System Warnier Diagrams Data Flow Diagrams Warnier Flow Diagram
2. Analyze system output requirements	Output Report Specification
3. Analyze input availability	(Review of documentation)
4. Specify file layouts and structures*	File Layout Form Organizational Warnier Diagrams
5. Solidify flow diagrams	Data Flow Diagrams Warnier Flow Diagram
6. Specify input requirements	Data Entry Specification
7. Develop program specifications	Program Specification Warnier Diagrams Decision Warnier Diagrams Decision Tables File-program Cross-reference

*See also Chapter 4 for additional tools.

2.2 illustrates this comparison, showing at which step in the design methodology each tool is applied. There will be no elaboration on this now since this is the purpose of Chapters 3, 5, and 6.

References

David, William S. *Systems Analysis and Design: A Structured Approach.* Reading, Mass.: Addison-Wesley Publishing Co., 1983.

DeMarco, Tom. *Structured Analysis and System Specification.* Englewood Cliffs: Prentice-Hall, Inc., 1979.

Gane, Chris, and Trish Sarson. *Structured Systems Analysis: Tools and Techniques.* Englewood Cliffs: Prentice-Hall, Inc., 1979.

Orr, K. T. *Structured Systems Development.* New York: Yourdon Press, 1977.

Semprevivo, Philip C. *Systems Analysis: Definition, Process, and Design.* Chicago: Science Research Association, 1982.

Van Duyn, J. *Developing a Data Dictionary System.* Englewood Cliffs: Prentice-Hall, Inc., 1982.

Warnier, J. D. *Logical Construction of Programs.* New York: Van Nostrand Reinhold Co., 1976.

Questions and Problems

1. What is the basic rule associated with the use of the Warnier bracket symbol? How does this reflect upon the basic difference in objective between data flow diagrams and Warnier diagrams?

2. Draw the Warnier diagram which satisfies the following Boolean expression: $X = A + B + C$, where $A = DE$, $B = FGH$, and $C = JKL$. E must be performed before D, and G may be performed from zero to 6 times for each occurrence of F and H. Assume inclusive ORs.

3. Define the following: Mission, Decision, Action, Activity, Procedure, Step, Trigger, Condition.

4. Consider the decisions and actions that take place in ordinary grocery shopping. Synthesize this activity with others to form an overall project, and synthesize this project with others to form an overall mission. Obtain parallel missions and projects to establish a Warnier analysis to the activity level, including a variety of example household chores.

5. Analyze the grocery shopping activity of problem 4 by specifying decision, action(s), and procedures.

6. Consider the decision for one machine in a total computerized preventive maintenance (PM) system. Use a Warnier chart, beginning

at the activity level, to analyze the PM process for this machine. The activity is to perform preventive maintenance for an earthmoving tractor. A decision must be made by the PM manager to originate the necessary work orders. He monitors the process on a daily basis in order to make this decision. This procedure consists of obtaining "aging" data daily from the field, i.e., the number of hours that the machine has been in operation. This is entered into the computer system which keeps track of both calendar and machine operation time. A review of these files is made daily to determine if PM is needed. If so, a work order is automatically generated and given to the Maintenance Manager for implementation. For the tractor, the following work orders might be generated:

(a) Oil change. Every 500 hours or every two months.
(b) General lubrication. Every 200 hours or every 3 weeks.
(c) Tune-up. Every 1000 hours or every 6 months.

The procedure for implementation includes the receipt of the work order, the requisition of materials and equipment, and the allocation of labor to the job. Upon completion, the work order is marked with the completion date and returned to the PM manager.

7. Draw the Warnier diagram to specify the logic in the structured English below. Logically document the process logic. Keep it simple. Assume that all unknowns will be obtained by data entry. If a procedure is not detailed, make a bracket for it and leave it blank to the right to indicate that completion is required.

```
PRODUCE INVOICE
   DO COMPUTE-TOTAL.
   DO COMPUTE-DISCOUNT.
   DO COMPUTE-SHIPPING.
   Subtract D from IT to get IN.
   Add H to IN to get TOTAL PAYABLE (TP).
   PRINT INVOICE.
COMPUTE-TOTAL
   Multiply QUANTITY (Q) by UNIT COST (UC)
   to get LINE TOTAL.
   Repeat for all LINE TOTALS.
   Add all LINE TOTALS to get INVOICE
   TOTAL (IT).
COMPUTE-DISCOUNT
   IF         INVOICE TOTAL (IT) is GE $2000
              DISCOUNT is 5% of IT.
   ELSE IF    IT is GE $1000 but LT $2000
              DISCOUNT is 2 1/2% of IT.
   ELSE IF    IT is GE $100 but LT $1000
              DISCOUNT is 1% of IT.
```

```
ELSE        (IT is LT $100)
SO            DISCOUNT is zero.
COMPUTE-SHIPPING (S)
   Shipping Fee (SF) is equal to RATE times
   WEIGHT.
```

8. Illustrate the structure of a decision Warnier diagram. Show how this structure is integrated into the program specification Warnier diagram.

9. Use the Warnier diagram for a decision tree to specify the following shipping rate schedule:

> The air shipping rate is 5 dollars per pound. It is reduced by 2 dollars per pound for every pound over 20 pounds. However, there is a minimum charge of 10 dollars. Truck delivery is 3 dollars per pound for delivery in the city. Outside of the city, truck delivery costs $.20 per hundred-pound mile (i.e., number of hundred pounds times the number of miles shipped), if the parcel weighs over 20 pounds. Delivery of packages up to, and including, 20 pounds is 2 dollars per pound; if it is an express truck delivery, add one dollar per pound to this price (only for parcels under 20 pounds). Any air shipments over 800 miles are charged at double the rate for air shipping given above.

10. Show how the results of problem 9 can be integrated into the solution for problem 7. Perform problem 9 using a decision table.

11. Use an organizational Warnier diagram to specify that records in a file be sorted according to the following specifications (assume fields are known for variables):

a. First by variable XXX, ascending;
b. Within variable XXX:
   1. by variable Y ascending if XXX <5,
   2. by variable X descending if $5 \leqslant XXX \leqslant 10$,
   3. by variable Y ascending if $10 < XXX$;
c. Within variable Y:
   1. by variable W ascending if $Y \leqslant 10$,
   2. by variable A ascending if $Y > 10$;
d. Within variable A by variable Z ascending.

12. Use a Warnier diagram to specify the following menu structure: When the system is booted the following supervisory menu will appear:

a. NEW DATA ENTRY
b. FILE UPDATE
c. REPORT GENERATORS
d. SHUTDOWN/LOGOFF

Option (a) leads to a second menu called a New Data Entry Menu which has the following options: (a1) Work Order, (a2) Inventory Control, (a3)

Purchase Order, (a4) Accounting, (a5) Return to Supervisory Menu. Each one of these options leads to a data entry mat for input. Option (b) also leads to a menu similar to (a), but it is called the File Update Menu (the options are the same). However, when any of these are selected, another menu appears requesting the sections of the form desired. The forms have the following sections:

Form	Sections
Work Order	Header, Materials, Equipment, Labor
Inventory Control	Header, Receipts, Shipments
Purchase Order	Header, Vendor, Specifications
Accounting	Ledger, Receivables, Payables

Option (c) will lead to the same options as option (a), since there are a different set of reports for each system module. The work order, inventory control, purchase order, and accounting modules have 5,3,3, and 8 possible reports, respectively. After any given report is requested there is a query as to the routing and the number of copies of output desired.

13. Draw the organizational Warnier diagram that will specify how to sort randomly placed records in file ZZZ into the order given in Table E2.1. Note: If an ordering occurs, assume that it is to be specified.

14. Compare the symbols used in flowcharting and in data flow diagramming. Use this comparison to distinguish between the objectives of these two techniques.

15. State two ways that a department of a company, such as the shipping department, might be identified within a data flow diagram. How is it determined which of these representations is most correct?

16. Identify the components of a process node.

17. What is a query? How is it distinguished from a typical read operation? How is this distinction identified on a data flow diagram?

18. Define the terms: right neighbor, left neighbor, reflexive, symmetric, path, transitive, walk, cycle. Draw a graph exemplifying each.

19. Which nodes in a DFD may not be linked? Why?

20. Are walks and/or loops permissible in a DFD? Explain.

21. State the five relational rules of data flow diagrams.

22. Why is it essential to master the reading of DFDs before attempting to learn how to develop them? State the rules for reading DFDs.

23. Use a Warnier diagram to analyze the three types of rules that exist for developing a DFD and name the rules within each type.

24. A company receives two types of orders: (a) C.O.D., or (b) credit. C.O.D. orders go directly to the shipping department while credit orders are checked against the accounting file. If the customers' credit is good,

Table E2.1   Data for Problem 13*

Key 1	Key 2	Key 3	CON1	POST	SEQ	V001	V002
---	---	---	1	2	Y	4.4	---
---	---	---	1	2	Y	4.5	---
---	---	---	1	2	Y	4.6	---
---	---	---	1	2	N	3.7	---
---	---	---	1	2	N	3.4	---
---	---	---	1	2	N	3.0	---
1	1	---	2	19	Y	---	---
2	3	---	2	19	N	---	---
3	2	---	2	19	Y	---	---
4	1	---	2	19	N	---	---
2	1	---	2	28	N	---	---
2	2	---	2	28	Y	---	---
3	1	---	2	28	Y	---	---
4	1	---	2	28	N	---	---
1	2	---	2	37	N	---	---
1	3	---	2	37	Y	---	---
4	1	---	2	37	N	---	---
1	1	---	2	45	Y	---	---
1	---	---	3	6	Y	---	3.5
1	---	---	3	3	N	---	3.6
1	---	---	3	7	Y	---	3.7
2	---	---	3	19	N	---	6.1
2	---	---	3	20	Y	---	6.0
2	---	---	3	22	Y	---	5.5
2	---	---	3	21	N	---	4.2
3	---	1	3	---	---	---	---
3	2	2	3	---	---	---	---
4	---	---	3	12	N	---	---
Range 1–4	1–4	1–3	1–3	1–99	Y/N	Cont.	Cont.

*NOTE: Any combination of variable values in their respective ranges may appear in any given record. Assume that missing values are unordered.

their orders are sent to the shipping department. An invoice is prepared and is sent along with the shipments to all customers. C.O.D. customers pay the deliverer while credit customers are billed later. In either event, at each shipment the shipping department notifies the accounting department which updates the accounting file appropriately.

In order to perform a credit check, a direct on-line query is made of the customer's past history from the accounting file using the customer number as a key. If credit is refused, the customer is notified by letter and a C.O.D. order is required.

Each month a billing routine is used to read the accounting file and send bills out to the customers. Payments from both types of customers are received daily which leads to appropriate updating of the accounting file and the deposit of cash to the bank.

Draw a data flow diagram for the subsystem. Set it up so that there is *one* data store, the accounting file.

25. Draw the data flow diagram for the following example.

In a work-order-processing system, let the Works Manager be an external entity, who will be supplying data and receiving data from the system. He will sit down at a terminal and put in new work orders via a work order creation and update program. This program will write the work orders to a work order file. This same program will be used to update work orders, and these updates will also be written to the work order file.

Whenever a series of new work orders or updates is completed, the input/update program will call a separate routine which will read the work order file and produce another file (call it the link list file). This will contain summary information linked in such a way as to provide a prioritized listing of the work order descriptions (along with other key information for scheduling).

When the Works Manager elects to schedule jobs, a program is executed which reads the work order file and the link list file and presents key information to the terminal, one line per work order. The manager can then select any of the work orders for scheduling. The ones selected for scheduling have their reference numbers written to another file (call it the scheduled jobs file). If the manager wants to view the scheduled jobs at any time, this program enables him to do it. Thus, he can check the list and schedule or unschedule jobs until he is satisfied.

When scheduling is complete, the Works Manager can elect to print all scheduled work orders. The system will then print (for the Works Manager) all of the work orders which have their reference numbers in the scheduled jobs file. Once the scheduled work orders are printed this file is cleared and these jobs cannot be rescheduled again.

A final option enables the Works Manager to select any given work order by reference number and have it printed out.

Assume that there is a supervisory menu with the following options:

1. ENTER NEW WORK ORDER
2. MODIFY EXISTING WORK ORDER
3. SCHEDULE JOBS
4. GENERATE WORK ORDERS
5. PRINT SELECTED WORK ORDERS

Number the process blocks and label them as closely as possible to this menu (total conformity may not be possible). Exclude the "responsible" portions from the process nodes since they are not given.

26. Define a data dictionary and state its basic purpose.

27. State the parts of the data dictionary and briefly define the contents of each part.

28. What is the purpose of a Warnier flow diagram? Present its structure by describing its content and function. When would this diagram be drawn and when would it be completed?

29. What is the primary purpose of the file-program cross-reference table? What does its content tell the reader?

30. Present the layout of the data file content form. Diagram how it is cross-referenced to the other components of the data dictionary. How is backup referencing accomplished to obtain more detail on the source/value of a data element?

31. List the time sequence of the methodology from Table 1.3 and assign the tools covered in this chapter to each.

Chapter

The Process of Analysis

Given the analytical tools presented in Chapter 2, the software designer is ready to approach the complexities of specifying the system. The adequacy of these specifications will be measured by the functioning of the final system. Thus, there are no intermediate direct measures of design success prior to system implementation. Only indirect measures can be applied by checking to see if the established steps of sound systems analysis and design have been followed as indicated by their respective outputs. These steps were given as an overview in Chapter 1, Table 1.3. The first three are discussed in detail in this chapter, namely:

1. Analyze the affected management system (AMS),
2. Analyze the output requirements, and
3. Analyze input availability.

These will be given detailed consideration in the sections below.

3.1 Analyze the Affected Management System

In Section 1.3 the concept of system evolution and its ramifications to the systems analyst were discussed. As was shown there, rarely do systems developments "start from scratch." Most of the time there will be some manual or partially mechanized system already in place. This does not excuse the systems analyst from the task of analyzing the management system, but it may de-emphasize the task to one of review. Upon occasion, higher levels of management will totally specify the outputs which are required from a system, making the process discussed in this section seemingly unnecessary. However, regardless of the degree of prespecification, the analyst should always go through this process of management systems analysis, if for no other reason than to verify the output requirements and input availability. While the degree of the management systems analysis may vary, it is essential in all software design projects. Therefore it will be presented in this section as though no current system or prespecifications exist.

The analysis of the management system begins with a statement of higher-level management goals and objectives. The idea of beginning with a clear statement of goals and objectives is often thrown around as a generally good thing and given some lip service, but often it is ignored in the system design. It is necessary to define this step in some detail to prevent this from becoming the case. First of all, the analysis that is being performed at this point is not of the software system that we ultimately intend to design. Rather, it might be of the entire organization, the company, a division of a company, or some function within a subdivision of the company. The term *affected management system* must be defined by the analyst in terms of the proposed software system. Since the term *affected management system* will be used so extensively in describing the steps which follow, the acronym AMS will be used in the discussion.

The distinction between the AMS and the ultimate system design is an important one which should be re-emphasized before continuing. The analysis to be performed at this point must not be allowed to jump into proposals for software. At this early definition stage the analyst may not even know what the tasks of the proposed system will be. It is premature at this point to direct an analysis at the proposed software system, since it does not exist. It is essential that preconceived ideas with regard to the ultimate design not stand in the way of creativity. A sound design can only evolve from that which already exists. The analysis of the AMS involves the study and documentation of two things which do exist: (1) the AMS itself, and (2) a statement of management goals and objectives with regard to the AMS. These things are tangible and they form the foundation upon which this analysis can be developed.

The basis of the analytical technique to be used was introduced in Section 2.1.2. Figure 3.1 presents the basic Warnier model which will be applied. In some cases simplifications will be made to accommodate those activities for which a decision-action analysis is not appropriate. Modifications will be required for mandated output requirements and other situations where this basic model does not fit perfectly. However, it is good to keep the most complete model given in Figure 3.1 before us as a starting point in making each of the detailed applications.

With these basic facts stated, it is now possible to detail the activities involved in an analysis of the AMS, namely:

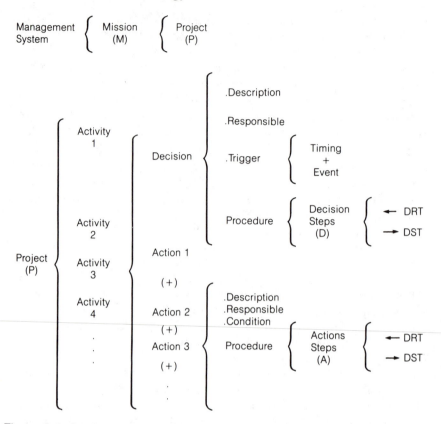

Figure 3.1 Generalized Warnier Application to Management System

1. Define the scope of the AMS,
2. Define the AMS goals and objectives,
3. Define the AMS missions and projects,
4. Define the AMS decisions and actions,
5. Define the AMS procedures, and
6. Define AMS data flows.

These activities proceed from that which exists to that which is proposed. At the conclusion of these activities, the next phase of the analysis will define the software outputs and inputs, which, in turn, will provide a basis for the specification of the software design. These analyses will be discussed in Sections 3.2 and 3.3, after the AMS analysis is detailed.

3.1.1 Define the Scope of the AMS

To summarize the definition of affected management system as used in this context, it is: that part of the company (or organization) which will be impacted by the new software design. It might be argued that we are begging the question since we are defining the AMS in terms of the new (nonexistent) software design, while above we argued against any premature speculation as to that design. This is not the case. We argued against directing this phase of the analysis at the software design itself, not against the use of the project scope to define the subset of the organization which is to be studied in this phase of the analysis. To be realistic, all projects have some specification of scope, albeit poorly defined. The analyst's first task is to determine from upper management what this scope is, and to use this to determine the subset of the existing system which is to come under initial scrutiny (i.e., the AMS). It may include an entire company, or even several companies. But, if only a small portion of the organization is involved, this must be determined early enough to prevent wasted preoccupation with unnecessary detail.

The determination of scope is not as easy as it might seem. Upper-management understanding of data processing almost guarantees that they will overestimate the scope of a project. Hence, project scope must be determined by other surrogate measures, the most common being the budget allocated. A second factor is the number and quality of available personnel for design and development. Management attitude toward quality staff expansion is closely related to this factor. This often becomes of paramount importance when decisions regarding project funding are awaiting preliminary studies. Here the pragmatic experience of the systems design team must be applied using inputs from upper management as they are available.

At this initial stage of the total design project, uncertainty and ambiguity may characterize management thinking. It is essential that the analyst not allow this to stand in the way of performing this first step. In fact, the analyst is the catalyst to clarify management thinking by documenting the project scope. It is fairly safe to err on the side of making the scope of the AMS too broad rather than too narrow. As the analysis continues, those elements which are not relevant can be easily excluded.

Once the project scope is defined, the problem becomes one of identifying the subset of the company which will be impacted. This might be any combination of individuals, departments, or divisions. Usually, groups within certain departments will be affected to the exclusion of others. This determination is not difficult if the scope of the project is adequately defined. If it is not, it might be safer to initially include a doubtful entity into the AMS rather than exclude it.

The procedure for defining the AMS in terms of the organization involves a compilation of a list of all departments of the organization. A brief indication of the relative extent of the involvement of each will be made. Then, those departments which have direct or close involvement will be further analyzed by identifying the individuals involved. Information regarding each individual involved will be documented so that contacts can be made during the analysis of the existing system. This list of involved individuals will define the scope of the analysis of the existing system.

3.1.2 Define the AMS Goals and Objectives

Here again, remember that the goals and objectives of the AMS are of concern, not the goals and objectives of the software. These will be made compatible only if attention is focused upon the AMS now.

Goals are distinguished from objectives in that goals are more general and they are usually only measurable in a binary way. Objectives are more specific, and there is usually some method of measuring their degree of accomplishment so that they can be subjected to more precise evaluation. Sample goals might include making a profit or performing a specific public service. They can be measured only by stating whether the goal was accomplished or not. Goals, by their very nature, often become unmeasurable; for example: to make as much money as possible; or, to provide a maximum level of public awareness.

Objectives, on the other hand, are more specific, and in this context we require that there be a means of measuring the degree of their accomplishment. Thus, "to make a profit of $10 million during the next fiscal year" is an objective. Not only can measurements be made as to its accomplishment, but the degree of accomplishment can also be determined.

The creation of goals and objectives for the AMS is not the prerogative of the systems analyst. However, the documentation of these goals and objectives, and the clarification of their definition which this entails, are clearly the analyst's responsibilities. The analyst must become skilled at interviewing in order to obtain the precision and explicitness required to properly proceed. It is essential that the scope of the management system be defined (see Section 3.1.1) prior to this interview. Focus must be maintained upon the goals and objectives of that AMS. Although problems with the current system (which, hopefully, the new system will solve) are pertinent to this interview, they should not so dominate the discussion as to obscure the ultimate objectives of the organization itself.

As an opener, it might be helpful for the interviewer to state the difference between the AMS and the computer system. Since the interviewer will be dealing primarily with his superiors and those from whom he needs complete cooperation, it may be difficult to redirect the discussion. Thus, the reasons for beginning the analytical process by focusing upon the AMS, as opposed to software requirements, must be made clear from the outset.

The statement of goals and objectives as obtained from higher management will be clearly documented in list form at this point. They should be sent back to upper management for verification. Also, middle managers who are involved should be given ample opportunity to provide input, and, subject to higher management approval, to modify the statement of goals and objectives. The current management system, coupled with the definitive statement of goals and objectives, provides the foundation upon which the remainder of the design process will rest. It is essential that these be understood and accurately perceived in the minds of the analysts as well as all levels of management.

3.1.3 Define AMS Missions and Projects

In the analytical structure given in Figure 3.1, missions and projects form the first two levels of the AMS analysis. Recall from Chapter 2 that a mission was defined to be the name of an operation which embodies a goal. At the same time, projects are components of these operations, each corresponding to an objective. There may be a subtle difference between a mission and its respective goal; similarly, there may be very little distinction between an objective and the name of the project by which it is accomplished. This is the reason that a good statement of goals and objectives was required prior to this step.

While the documented goals and objectives provide the primary input to this step, there is a process of restatement and reorganization required at this point. Restatement may be required since the missions and projects are operations to be executed, as opposed to idealistic statements of end results. Generally this difference is purely semantic and it should not cause difficulty. Table 3.1 gives an example. Note that the original goal is stated in a verb phrase which is not concerned with any operation. When restated as a mission, thought is given as to the

Table 3.1   Example of the Transition of Goals and Objectives Into Mission and Project Descriptors

GOAL	MISSION
To provide customers with all requested information regarding their orders.	Operate a retrieval system that will provide customers with pertinent order information at their request. Descriptor: Customer Inquiry Processing
OBJECTIVES	**PROJECTS**
Retrieve customer order on line within 15 seconds of customer request. Service customers for exceptional cases with customer service specialists within 2 minutes of customer request.	Descriptor: Computerized Customer Order Retrieval Descriptor: Manual Exception Customer Order Handling

operation (i.e., in this case, the retrieval system) for accomplishing this goal. Finally, the mission description is boiled down to an appropriate descriptor which will be integrated into the structure of Figure 3.1. Similarly, the objectives are translated into project descriptors as shown in Table 3.1. The more concise and complete the descriptor, the more information it will convey. Thus, the right choice of descriptors is an essential part of this process, and time is well spent in giving considerable thought to developing effective descriptors.

While the above transition should not be regarded as difficult, some thought as to the organization, or reorganization, of the AMS must be invested at this point. It should be recognized that the mission level is the broadest classification as far as the AMS analysis is concerned. Subsequent modularization in design and implementation will reflect the choices made at this point with regard to the combinations of goals into a given mission. The same thing is true of the combinations of objectives into projects, and this is true to a greater degree since there are fewer goals, and they are usually easier to distinguish.

A serious dilemma often occurs when a common objective appears within two or more goals. For example, in Table 3.1, the Computerized Customer Order Retrieval project, within the Customer Order Requests Processing mission, might be an identical functional component to one used by accounting for checking orders. In this case the dilemma arises as to whether the classification should be made according to the functional classification or by the end user served (i.e., the customer vs. internal requests). In general applications, any number of cross-classifications can cause similar conflicts. These conflicts in organization structure can be quite frustrating to the analyst who wants to exclude all redundancy from the model. Generally there is no one, unique organization that will eliminate all such conflicts and perfectly model the AMS. Thus, the following twofold approach is recommended: (1) reorganize the mission-project structure to eliminate unnecessary conflicts where possible, recognizing that it may be impossible to eliminate all such conflicts; and (2) when no further, obvious simplifications exist, continue the analysis of missions and projects using referencing to eliminate as much redundancy as possible. Identical functions may have to appear in several places. Do not be afraid to scrap the entire model and start over with another one. Recognize that this exercise is being conducted to give everyone, especially the analyst, a better understanding of the current system. The model which best serves this purpose should be sought, not necessarily the one that is theoretically perfect.

The ultimate modularization of the new software design should be the determining factor in transforming goals and objectives into missions and projects. Recall that the analysis of the AMS environment was conducted to facilitate the new design, and especially to obtain a good specification of new system output requirements. Thus, it is beneficial to document the AMS to promote this purpose (i.e., that of new system design modularity), provided that this model does not portray the management system unrealistically.

3.1.4 Define AMS Decisions, Actions, and Mandates

It might have been noticed that the activities given above did not mention the activity level (see Figure 3.1). Rather, once the projects are defined, the next consideration is to define decisions and actions. It should be understood that the key to this definition step begins with an enumeration of the decisions made within each project. Once this list is compiled and organized, each decision can be assigned one or more resultant actions. The collection of each decision with its accompanying set of actions is called an activity. Thus, rather than attempting to define activities first and then assign the decision and actions to each, it is recommended that the opposite be done, and the activity descriptor entry in the documentation will merely summarize the respective decision and its actions.

In the AMS being documented, the question should be asked with regard to each project: What decisions are necessary to make this project accomplish its objective? There is usually a well-defined sequence of decisions required in most projects. Information is brought to bear in making these decisions. The analyst should recognize that little creativity is required at this point. The actual existing system is the basis for this documentation. Although it will be desirable later to modify the documentation to fit new system objectives, at this time the documentation is just a statement of the current AMS structure.

At times the term *decision* might be replaced with the term *mandate*. Certain outputs of an information flow process are not required for decision making. For example, reports may be mandated by government, and they are not used in any way within the company itself. Also, information is often requested from outside of the organizational unit being studied. While the analyst might know that some decisions are based upon this information, access to these decision and action procedures are either denied or are too difficult to readily obtain. In these cases the analysis will be simplified from the structure given in Figure 3.1 to that given in Figure 3.2.

A comparison of Figure 3.2 with Figure 3.1 shows that for those activities which are subdivided into one or more mandates, the mandates replace the decision-action breakdown. Note that the mandate breakdown applies to a given activity, while any other of the activities might be analyzed either into mandates or decision-action procedures (but not both). A mandate will have a descriptor which specifies to the reader the intent, as well as the fact that it is a mandate; for example: Capital Gains Tax Report Mandate. Similar to a decision, the mandate

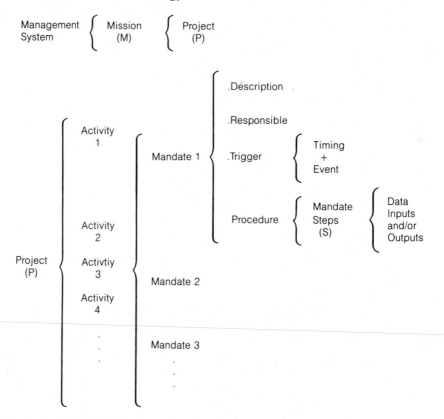

Figure 3.2 MSWD Development of Mandate-Type Activity

is subdivided by procedure and labeled by description, responsible, and trigger. Generally the procedure will both require and generate information. Thus, the data inputs and outputs will be enumerated at each step as noted. Procedures and data flows for decision, action, and mandated procedures are discussed in the next two steps.

While the use of mandates is a convenient substitute for the decision-action analysis, they should only be utilized as a last resort. There is a strong temptation to take the "mandate route" rather than performing the more detailed study required by decision-action analysis. However, the analyst should recognize that when mandates are used unnecessarily, the entire structure of the analysis is severely weakened. One of the primary objectives of this step is to clarify the true information requirements. When information is required for internal decision-making, it is essential that those decisions and actions be defined in terms of their procedures, data requirements, and information generation. Only when there is no such decision-action sequence, and information is still required by an outside source, should the mandate breakdown be employed.

3.1.5 Define AMS Procedures

Up to this point, three types of procedures have been established as referenced in Figures 3.1 and 3.2: decision, action, and mandate procedures. In all three cases the method for determining the procedures currently being employed are the same. Direct contact must be made with the individuals who are actually performing the procedures. A stepwise sequence of events will be developed. Each step in the procedure should be chosen such that its size is small enough to conveniently define the data inputs and outputs to that step. It should not involve so many actions or combinations of data sources that the step descriptor is more than two or three lines in length. If so, consideration should be given to further breakdowns. On the other hand, the steps must not be so small that the detail makes the documentation of the procedure unnecessarily long. The documentation at this point is a stepping stone to data flow and data file content specifications, and the optimal step size is one which best meets this objective.

Decision procedure steps are those steps necessary for making the decisions defined in the analysis of the AMS. Generally information will be required from one or several sources which will be processed and formatted to present the information to the decision maker. The stepwise specification of this procedure will lead naturally to the major output of this analysis—a rough statement of data needs. The output of the decision process is a specification of action. This, along with other secondary outputs of the decision procedure, will ultimately be documented as data outputs.

Action procedure steps are those steps required to accomplish a given action which is a direct result of the implementation of a decision. Usually actions generate data, but occasionally additional data are required for their accomplishment. The retrieval of data, the production and storage of information, and the particular action which is performed will be documented at this point.

Mandated procedures do not involve decisions and actions. Rather, the particular steps required for generating the mandated report or information will be documented. These will generally involve a combination of input and output data, both of which should be assigned descriptors at this point.

3.1.6 Define AMS Data Flows

The analysis of the AMS discussed above documents all the procedures of the organization affected by the new software system. It presents the system in a sequential order with activities grouped within project, and projects grouped within mission. While this organization of the documented procedures is ideal from the viewpoint of understanding the system in terms of its goals and objectives, it does not readily lend itself to an understanding of data flow. This is because data may emanate from any of the decisions or actions, and it may be required by any other decision-action or mandate.

The first step in defining the data flows is to add Data Storage Transactions (DSTs) or Data Retrieval Transactions (DRTs) to each procedure step in the analysis which is either generating or retrieving data, respectively. Exceptions will generally be made for those steps which merely receive data from a previous step and/or pass it to the next step. DRTs and DSTs need only be assigned a descriptor at this point, since the primary emphasis here is data flow, not data flow content (this will be detailed when output requirements are analyzed). Given this stage of the system analysis, it is now possible to improve upon the clarity and conciseness with which the data interactions can be displayed by the use of a data flow diagram (DFD). The rules for DFD development given in Section 2.2 should be followed. Existing external entities and files will provide the starting point. For an initial broad overview, the activities or projects in the Warnier analysis may be converted to process nodes in the DFD. In the explosion process, the procedure steps ultimately might be put within the process nodes, or even more detail might be developed using the DFD. However, it should be recognized that this data flow diagram is being developed for overall understanding of the current system, not for design specification. As such, it is not recommended that time be wasted on unnecessary detail.

3.1.7 Traffic Safety System Example

It is essential to present an example of the analytical techniques described above despite the dangers involved. In reviewing this example, recognize that a large application would, of necessity, consume as many pages as this book, and a small application does not require all of the tools that need to be exemplified. Thus, the example must be a part of a large system—similar to a cross-section cutaway. A second warning that bears consideration by the reader is to keep from getting "locked in" on the example. In other words, recognize that the example involves a specific application of the theory which may not be totally pertinent to another application. The important point is to learn the theory (not the example). The example is merely presented to aid in this regard.

A second set of problems associated with examples is that a word statement is required to set the stage for them. A *good* detailed word statement of a situation almost solves the problem, again making the application of the tools described above seem convenient for presentation but not essential for analysis. The true value of these tools is in making sense out of a confused and confounding situation, not in modeling an already concise and accurate word statement. On the other hand, if we intentionally make the word statement confusing we are in danger of misleading the reader. In a real-world application, the analyst might go back to the operating personnel several times to resolve ambiguities, and it is the application of the steps described above which aids in this effort. Unfortunately, such cannot be simulated within the confines of a book.

With these shortcomings in mind let us proceed, recognizing that the value of the example is merely to add some substance to the theoretical presentation. Suppose that an allocation of $100,000 has been made to your group to design, develop, and install a mechanized system for determining an optimal allocation of federal government funds dedicated to improving the safety of the roadways in a given state by means of small roadway construction projects. This system will be used to allocate approximately $5 million per year to 150–200 projects. These range in price from a simple sign modification costing a few hundred dollars to a major realignment worth several hundred thousand dollars. There is a basic breakdown in funding according to the following categories mandated by legislation: (1) railroad-grade crossing improvements, (2) high-hazard locations, and (3) roadside improvements.

Obvious resources are available to aid in the accomplishment of the allocations. The Highway Department initiated a data collection and processing system some years ago which has functioned well. The new system software will operate within this environment and take advantage of as many of the established procedures as will prove beneficial over the long term. These include the following:

1. Data collection and central entry of all accident reports, which include location and all codable information of interest;
2. An inventory of roadway characteristics by location which includes most pertinent factors relating to safety, such as sight distance, roadway roughness, and types of traffic control;
3. An organized team of accident location investigators within each of the eleven divisions of the state who are trained to identify problems and propose countermeasures; and
4. Sufficient on-line hardware available for any additional software or processing of the magnitude anticipated.

While the above statement is quite sketchy, it sets the stage for the process of analysis which will be described in detail below. In the AMS analysis procedure below, each activity is covered in correspondingly numbered subsections (for example, activity x is described in subsection 3.1.7.x).

3.1.7.1 Example of an AMS Definition of Scope

The input to this activity mentioned above was the $100,000 available for the total design and development effort, as well as the other available resources. This is a rather modest sum for the total project; indeed, the Federal Government has been known to spend several times this figure just for general design specifications. Of necessity, the scope must be limited to the particular problem at hand. At the same time, it is recognized that several "departments" (called *bureaus*) within the Highway Department will be affected, and these must be given at least peripheral consideration in the analysis of the affected management system.

Remember that the objective of this step is to define the elements of the organization that are going to be subjected to analysis. Since this is a project of selection, it might begin with a consideration of the entire organization. Table 3.2 presents the various bureaus within the Highway Department. To each is assigned a measure of the effect that the bureau will have upon the system (or vice versa). Those which are listed as directly affected will be the primary subjects for detailed analysis, while those which are listed as indirectly affected will have strong representation in this analysis. Those with peripheral involvement may have minimal representation, and the negligible rating indicates that a bureau can be excluded. Obviously, the assignments made in Table 3.2 for this example require familiarity with the organization. This must be obtained either through the direct experience of the analyst or by interviewing those who have such experience.

Table 3.2 is a rough cut by broad organizational categories. A further definition of scope in terms of the bureau subcomponents is necessary. This need not be elaborate, but it should be documented if for

Table 3.2 Highway Department Organization Breakdown

Bureau	Extent of Effect on/by New System*
Legal	Negligible
Accounting & Finance	Peripheral
Bridge	Direct
Computer Services	Direct
Construction	Indirect
Maintenance	Direct
Materials & Tests	Peripheral
Personnel	Negligible
Planning & Programming	Direct
Research & Development	Indirect
Surveys & Plans	Direct
Operating Divisions	Direct

*Relative effect, from greatest to least:
1. Direct
2. Indirect
3. Peripheral
4. Negligible

PROJECT: Optimal Allocation of Safety Spot Improvement Budget

DEPARTMENT OR BUREAU: Maintenance

EXTENT OF INVOLVEMENT: Direct

INDIVIDUAL INVOLVED (PHONE)	FUNCTION
Asst. Maintenance Engineer (836-4443)	Has ultimate responsibility for developing plans for safety spot roadway improvement projects.
Division Maintenance Engineers (11) (827-1234, 838-3121)	Have responsibility for submitting proposals for safety spot roadway improvement projects; are ultimately responsible for constructing and implementing approved projects.

Figure 3.3 Example of a Format for Specification of Organizational Involvement

no other purpose than to plan the information-gathering requirements. To further exemplify this process, let us concentrate upon one bureau—maintenance. In actual practice the same process would be performed for all of the bureaus which are directly involved as well as with some or all of those indirectly involved.

Figure 3.3 gives a sample specification of scope for the maintenance bureau. At this level of detail the scope is defined by the persons who are involved. Generally these are the ones who will be interviewed to obtain the details of the current operation. Upper level management, or their staff assistants, will be consulted to identify the key individuals involved. By documenting these individuals within each subcomponent of the organization, the scope of the AMS is thoroughly identified. As the study continues and these individuals are contacted for information, they may make reference to other individuals more directly involved with the problem at hand. When this occurs, additions and modifications to the specifications will be made.

Both Table 3.2 and Figure 3.3 represent working documents that are assembled merely to aid in the further development of the analysis. Figure 3.3 contains the department or bureau and the extent of involvement from Table 3.2. It goes on to specify the individual by position. Name and phone number will aid in the ultimate contact which will be made. Note that in the case of the 11 Division Maintenance Engineers (one per division), only one or two might need to be contacted in order to determine the roles of all the engineers. In a decentralized organization where procedures between divisions are inconsistent additional contacts might be required. Finally, a brief description of function is included to explain how the individual relates to the project.

3.1.7.2 Example of an AMS Definition of Goals and Objectives

At this point a subset of the organization has been carved out which is significantly involved with the proposed new system development. We have called this the affected management system. In this second activity a formulation of the goals and objectives of the AMS is in order. It is essential not to overly influence this formulation by predefined ideas concerning the new system developments. That is, the goals and objectives of the software must be made subject to the goals and objectives of the AMS, and not vice versa. However, the very specification of a need on the part of upper management for the budget allocation system in this example establishes two things: (1) it must already be a requirement of the AMS to perform a function with the same objective as the new system function, and (2) a deficiency in method, practicality, or efficiency must have been detected in order to warrant further mechanization. Nevertheless, it is essential that the analyst not jump to conclusions, so that the AMS can be subjected to proper analytical study.

The study is initiated by presenting the findings of the AMS scope definition for higher management review. Higher management will usually be more than ready to articulate their perceptions of the goals and objectives that they desire to be accomplished. For this example assume that the goals and objectives have been formulated as stated in Table 3.3. In reality the list of objectives might be ten times as long.

Table 3.3 Example of AMS Goals and Objectives

Goals

1. To reduce to a minimum the number of accidents due to engineering deficiencies.

2. To reduce the severity of those accidents which cannot be prevented by engineering modifications.

Objectives

1. To maintain the necessary data by which the following can be clearly determined: (a) the extent of the accident problem at each location throughout the state, and (b) the extent to which this problem is reduced through countermeasure implementation.

2. To process the data collected to determine the most critical locations in the state in need of improvements, and to select those locations and improvements for implementation which return the maximum benefit to the state highway users.

3. To perform research and evaluation to develop countermeasures and to determine the extent to which these countermeasures reduce accidents.

Table 3.4 Example of Transition of Goals to Missions

Goals	Mission Descriptor
Allocate new construction funds to provide the maximum utility to state roadway users.	Build New Highways
Reduce the number of accidents due to engineering deficiencies to a minimum. Reduce the severity of those accidents which cannot be prevented by engineering modifications.	Improve Existing Highway Safety
To maintain roadways at or above those standards established by the State Standard for Roadway Maintenance	Maintain Existing Highways

Goals and objectives other than those related to safety and budget allocation should be stated, so that the analyst can obtain the proper perspective of the environment in which the design must fit.

Executives, concerned with broader company policies, generally will not present a concise set of AMS objectives. While it is not the analysts' job to originate the goals and objectives, they do serve to define and clarify them by documenting them. Several iterations involving upper-management approval (or disapproval) will generally be required before agreement is reached on an appropriate set of goals and objectives for the affected management system.

3.1.7.3 Example of an AMS Definition of Missions and Projects

The sample output of activity 2 will provide the input to this activity. Continue to bear in mind that we are only revealing a small subset of the total process. Table 3.4 illustrates how those potential goals of the Highway Department might be turned into mission descriptors. The second of these is of primary concern to our example here. However, the other two also have relevance to the safety function and thus to our defined AMS. They may either provide data to or obtain data from the parts of the organization under direct consideration. To the extent that they were identified as part of the affected management system in step 1 (the AMS scope definition), they will be analyzed further in subsequent steps. They are presented here merely to broaden the example. Note that there need not be a one-to-one transition of goals into missions. In the process of modularization several reiterations may be required, involving the combination and, possibly, the subdividing of goals.

Table 3.5   Example of Transition of Objectives to Projects

MISSION: Improve Existing Highway Safety	
Objective	Project Descriptor
To maintain the necessary data by which the following can clearly be determined: (a) the extent of the accident problem at each location throughout the state, and (b) the extent by which this problem is reduced through countermeasure implementation.	Maintain Inventories of Roadway Characteristics
To process the data collected to determine the most critical locations in the state in need of improvement, and to select those locations and improvements which return the maximum benefit to the state highway users.	Implement Optimal Roadway Improvements
To perform research and evaluation to develop countermeasures, and to determine the extent to which these countermeasures reduce accidents.	Perform Research to Advance Safety Technology

Table 3.5 shows how the objectives of Table 3.3 are translated into project descriptors. Very much the same process is employed, again giving organizational thought to the ultimate modularization of the existing system. Generally, organizational lines provide an excellent guide to aid in the definition of modules for the current system. However, the analyst should not feel restricted to these boundaries. To prevent this, organizational boundaries generally do not appear in the Warnier analysis or the mission and project descriptors.

3.1.7.4 Example of Definition of Decision, Actions, and Mandates

In order to illustrate this step in terms of the example above, we will concentrate our attention on one of the projects: "Implement Optimal Roadway Improvements," since this is the primary activity for analysis in the example. In practice, however, all projects will be further analyzed into decision-action sequences and/or mandates. The extent of the analysis will depend upon the definition of the scope of the AMS.

Figure 3.4 shows the Warnier diagram with the mission and project descriptors included (as they were defined above). At this point study will be performed to determine what decisions are made. The example shown in Figure 3.4 illustrates that within the "Implement Optimal Roadway Improvements" project there are three decisions, one for each funding type. Also, there is a mandate, which in this case is a report required by the Federal Government which is not used for local decision making.

3.1.7.5 Example of an AMS Definition of Procedures

Again, to make our example meaningful, it is necessary to concentrate upon just one of the decision-action sequences developed above. In reality each of the decision-action sequences would be developed, and procedures would also be developed to state how the mandated information was generated, as given in Figure 3.2.

Figure 3.5 presents the procedures for the "Allocation of High-Hazard-Elimination Funds" activity, following the pattern established in Figure 3.1. Note that this is an example of a continuous decision process (as opposed to a discrete process). That is, since there are virtually an infinite number of ways to allocate the budget, no attempt is made to list each individual one. Since the same general procedures apply (only the specific locations and countermeasures change), this documentation of the procedures is adequate.

MISSION	PROJECT	ACTIVITY	DECISION-ACTION
Build New Roads	Maintain Inventories of Roadway Characteristics	Allocation of Railroad-Grade Crossing Funds	Determine Allocations of Railroad Funds Allocate Funds
		Allocation of High-Hazard-Elimination Funds	Determine Allocations of High-Hazard Funds Allocate Funds
Improve Existing Roadway Safety	Implement Optimal Roadway Improvements	Allocation of Roadside Improvement Funds	Determine Allocations of High-Hazard Funds Allocate Funds
Maintain Existing Roads	Perform Research to Advance Safety Technology	Generate Federal Cost/Benefit Report Mandate	Mandate

Figure 3.4 Example of Definition of Decisions, Actions, and Mandates

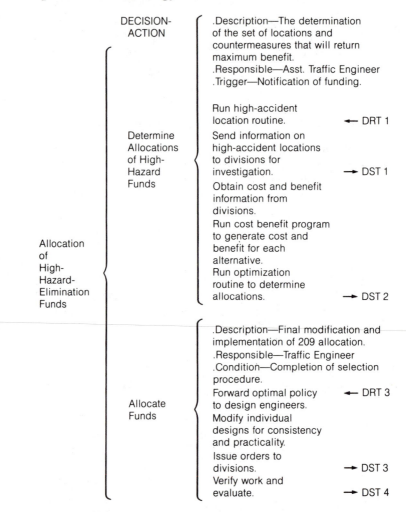

DECISION-
ACTION
.Description—The determination of the set of locations and countermeasures that will return maximum benefit.
.Responsible—Asst. Traffic Engineer
.Trigger—Notification of funding.

Determine Allocations of High-Hazard Funds

Run high-accident location routine.   ← DRT 1

Send information on high-accident locations to divisions for investigation.   → DST 1

Obtain cost and benefit information from divisions.

Run cost benefit program to generate cost and benefit for each alternative.

Run optimization routine to determine allocations.   → DST 2

Allocation of High-Hazard-Elimination Funds

Allocate Funds

.Description—Final modification and implementation of 209 allocation.
.Responsible—Traffic Engineer
.Condition—Completion of selection procedure.

Forward optimal policy to design engineers.   ← DRT 3

Modify individual designs for consistency and practicality.

Issue orders to divisions.   → DST 3

Verify work and evaluate.   → DST 4

Figure 3.5 Example of Definition of Procedures

3.1.7.6 Example of Definition of Data Flows for the AMS

The primary reason for performing the analytical activities above is to accurately define the data flows. Thus, this activity is crucial, but it cannot be performed if the previous activities were not thorough and accurate. As a first approximation, the DRTs and the DSTs are added to the Warnier analysis. This has already been done within Figure 3.5.

Table 3.6 presents a definition of the DSTs and DRTs in terms of their descriptors. No attempt is made to elaborate on the DST and DRT content, or to worry about file structure or content at this point. This should be put off until the general data flows of the AMS are understood and documented. The goal at this point is to determine the boundary of the forest, not to ascertain the size and types of the trees. Note that if room were available the descriptor could have been included within Figure 3.5; however, the DRT and DST reference numbers must be included since they provide the key to the reference material that will be developed in the next phase of the design process.

In addition to the descriptor, which gives the reader an intuitive feel for the data which is flowing, Table 3.6 presents the source of the DRTs and the destination of the DSTs. In our example most of the transactions are to and from entities external to the primary bureau of concern, namely the Planning & Programming Bureau (P & P). However, there is also information passed within P & P which is still documented using a DST and DRT (namely, DST 2 and DRT 3); this is done since there is a definite physical file employed, as will be shown in the data flow diagram.

We must emphasize again at this point that only a very small piece of the total system is being shown here. In the full analysis of the total AMS the number of DSTs and DRTs will range into the hundreds. (The wisdom of neglecting trivial data pass-alongs becomes quite obvious.) As the total AMS is analyzed, a large portion (possibly all) of the DRTs

Table 3.6 Example of Definition of DRTs and DSTs

Data Retrieval Transaction	Descriptor	Source
DRT 1	Accident Records	Department of Public Safety, Master Accident Records Tape
DRT 2	Investigation Data	Highway Department, Division Maintenance Engineers
DRT 3	Project Specification	Project File

Data Storage Transaction	Descriptor	Destination
DST 1	High Accident Location Reports	High Accident Location File
DST 2	Project Specification	Project File
DST 3	Work Orders	Highway Department, Division Construction Engineers
DST 4	Work Verification	Evaluation File and Federal Highway Administration

will match with DSTs. Table 3.6 is an example of a working document to keep track of the data flows; it should be updated as the data paths through the AMS become clearer.

The transition from Figure 3.5 and Table 3.6 to the data flow diagram for the AMS should be quite simple at this point. Figure 3.6 presents the part of the AMS described in detail above. Note that there are many data stores that could be integrated into the data flow diagram. Technically, a data store could be put between each of the processes, indicating that the data is batched or collected, forwarded, and then used in the next process. However, such would obscure rather than clarify. We choose to reserve the use of this symbol for either: (1) an instance when more than one process draws from the data, or (2) where a formal computerized or manual file exists. Thus, D1 in Figure 3.6 is employed because a physical computerized tape file exists within the current system. D2 is a pool of information used by the various divisions. D3 is a continuously updated file which reflects the current status of all projects. Finally, D4 is utilized since evaluation information requires storage for a variety of subsequent uses in the future.

In closing this example, note that we have included the DST and DRT numbers within the data flow diagram (DFD). This ties the Warnier and DFD models together. They both model the same process, yet in different ways. While the Warnier diagram shows the sequential procedures employed and the actual physical modular organization, the DFD shows the interactions between the processes, data stores, and external entities. (Note the appropriate use of queries where needed.) The DST and DRT numbers link these two models together and demonstrate how the structural information for the DFD can be derived from the Warnier analysis. The DFD will be developed using the rules given in Section 2.2.1.

3.1.8 AMS Documentation

Let us assume that the procedure given above has been followed for the entire affected management system. Thus, a total set of cross-referenced Warnier diagrams and DFDs will exist at this point for the current AMS. They should be assembled and logically ordered in a loose-leaf binder to provide a quick reference to any part of the AMS of future concern. These figures must be numbered and indexed to provide this reference capability, for it is not unusual for their number to range into the hundreds.

In Chapter 1 it was recommended that, if this documentation be distributed at all, it be included as an appendix to the design manual. It should be recalled that the ultimate target documents are an Overview, User Guide, and Design Manual, with the Design Manual ultimately becoming the Maintenance Manual after conversion. The doc-

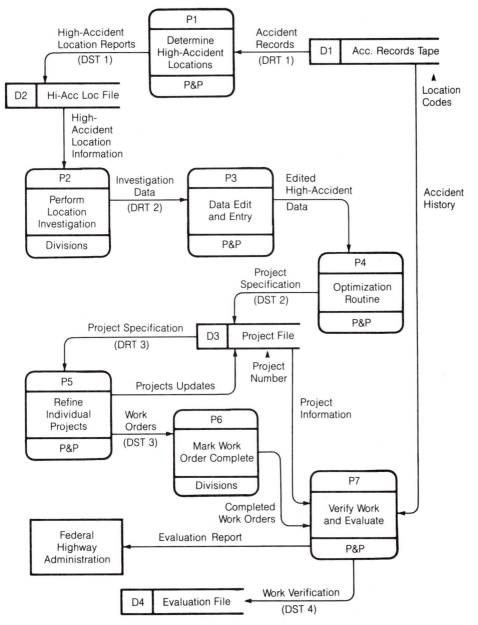

Figure 3.6 Example of an AMS Data Flow Diagram

umentation discussed above is a part of the foundation upon which these three documents can be developed. It should be viewed as a tool and not an end in itself. In fact, the imposition of the new system will make the documentation obsolete and in need of update. However, the value of this documentation as a tool in the further development of the User Guide and Design Manual will be obvious. Thus, although the AMS documentation may never be used outside of the design effort, its importance cannot be overestimated.

Finally, at this point in the documentation process an Overview, albeit tentative, should be developed. Chances are good that a brief proposal was already written for project development approval. This is the time for updating that document. The objective here is to emphasize the timing of the Overview development. Sufficient information is available at the conclusion of the first analytical step to assemble the Overview, which by its nature must exclude most system details and concentrate upon general features. Higher-level review and approval of the Overview at this point is a key element in the success of the design effort.

3.2 Analyze System Output Requirements

Section 3.1 above was quite long, and there might be a tendency for the reader to get bogged down in details and lose sight of the overall picture. Section 3.1 merely detailed the first of nine steps given in Table 1.3, namely: Analyze the Affected Management System (AMS). This is the first of three analysis steps which are described in this chapter. The other two are: Analyze the Output Requirements, and Analyze Input Availability. While not requiring nearly as much detailed explanation, these are quite important steps, and they will be given consideration here and in Section 3.3.

It should be quite evident that a proper analysis of system output requirements will depend heavily upon the analysis of the AMS discussed above. Some, possibly all, of the information requirements of the AMS will be subjected to analysis. In addition, those features of the system which are not in the AMS will also be subjected to analysis. The analysis of the output requirements can be subdivided into the following activities:

1. Transform the AMS Warnier analysis into an idealized analysis of the new system,
2. Transform the AMS data flow diagram into an idealized data flow diagram for the new system,
3. Identify all possible outputs of the new system and assign a name to each,
4. Document each of the output reports identified, and
5. Document any generalized capabilities of the new system.

Each of these activities will be discussed in separate sections below.

3.2.1 Transform Warnier Analysis: AMS to New System

Obviously this cannot be done without the following: (1) a good Warnier analysis of the AMS, and (2) a good specification of the scope of the new system. Section 3.1 was presented with the objective of furnishing the procedure for satisfying the first requirement. The astute student will recognize that it also provides the procedure for clearly delineating the new system scope. In fact, a reiteration of the steps given in Section 3.1 *in terms of the new system* will lead to the following:

1. Define the scope of the new system,
2. Define the new system goals and objectives,
3. Define the new system missions and projects,
4. Define the new system decisions and actions,
5. Define the new system procedures, and
6. Define the new system data flows.

Now, it is not necessary to describe these activities in detail since they have already been discussed in terms of the AMS in Section 3.1, and the majority of the details which apply to the application to the AMS readily apply to the new system. Those which do not are evident. However, the remainder of this subsection will be dedicated to showing the major distinctions between the application of these activities to the new system as opposed to the AMS.

First, recognize that the scope of the new system has not been, and probably never will be, clearly defined for the analyst. It is always in a state of flux, and the best we can do is put a tentative fence around it with all six of the activities given above. So, although activity 1 is a definition of scope, consider it highly tentative and subject to further specification as this process continues.

Second, the reiteration of the six activities in terms of the new system should be considerably less difficult and time-consuming than the original analysis of the AMS performed in step 1. The process is now one of refinement rather than original documentation, and it goes much faster. The new system generally will not include the entire AMS, so those components not directly involved can be eliminated. Some additional components may need to be added to accomplish those goals and objectives of the new system which were not being met by the AMS. Generally this will not be an extensive task, however, since a familiarity with the existing system has been attained during step 1.

While the reiteration process should not be complex, given a good output from step 1, neither should it be taken lightly. This is the first activity in which consideration is isolated to the new system. Decisions must be made as to which of the AMS goals, objectives, decisions, actions, and procedures are to be included in, and which are to be excluded from, the new system. Further, if the AMS is lacking in any essentials (which obviously it is or the project would never have been undertaken), these must be added to the new system documentation. While the documentation procedures are no different, the analyst now has entered into the realm of design—specifying something that does not yet exist.

Finally, note that here, as in Section 3.2.2 below, the transformation of the AMS documentation does not imply that the AMS documentation is to be destroyed. A copy must be kept to provide valuable reference information in the future.

3.2.2 Transform Data Flow Diagram: AMS to New System

Technically, and for completeness, this was covered in Section 3.2.1, and the comments there as far as the use of the procedures given in Section 3.1 are applicable. However, the data flow diagram is so critical to this step that for emphasis it is considered as a separate activity. Both the new system Warnier diagram and the new system data flow diagrams are tools. Their proper development will assure that they function to produce the desired output; conversely, any development methods which effectively lead to the desired output should be considered as proper. Thus, this activity will focus upon the desired use of the tools, with the hope that if the analyst keeps these in mind while applying the techniques described above, the probability of success will be increased.

This transformation activity is within the step called "Analyze System Output Requirements." Looking ahead, this will be accomplished by systematically traversing the Warnier diagrams and data flow diagrams for the new system in order to determine each output requirement. It is essential that both the Warnier analysis and the data flow diagrams be transformed with this objective in mind so that they are complete and accurate for the new system. This will assure that no output reports or capabilities are omitted.

3.2.3 Identify New System Outputs

From the outset of this activity it is essential that the distinction be maintained between two types of outputs. The first will be called an output *report*. These will have standard formats which present the values of certain numerical or string variables. They may be output to a printer, a CRT, or any other device or storage media. The major distinguishing characteristic, however, is that they can be described in a sequence of variable definitions, where a variable definition would give the source of the variable value and its location within the report output.

OUTPUT IDENTIFICATION FORM				
TYPE R or C	NAME (Acronym)	DESCRIPTION	REFERENCE	
			WD	DFD

Legend
  R = Report
  C = Capability
MSWD = Management System Warnier Diagram
  DFD = Data Flow Diagram Node Reference

Figure 3.7 Output Identification Form

A second type of output will be called an output *capability.* This output type may also be sent to any peripheral device. However, it is not preformatted and prespecified. It is not possible to state for this type of output just what its printout will look like or what the source of the variable values are, except in a very general way. Thus, data base manipulation systems and certain flexible report generators must be described in terms of their capabilities as opposed to specific reports which they generate.

In the identification of new system outputs, it is essential that the new system be traversed twice: once using the management system Warnier diagrams (MSWD) and once using the data flow diagrams. During these two traversals, an enumeration will be made of all the output reports and output capabilities that are required at each point. Each will be given a brief name and a short description. The format for this identification procedure is given in Figure 3.7. Outputs will be listed on this form in the order in which they are identified during the systematic review of the new system documentation. It is recommended that the MSWD be reviewed first. This will generally keep the report listing according to the modular design of the system, although modifications for this purpose can be made later. The most important consideration is that of completeness. Several reviews by all involved will help to assure that no essential outputs are excluded.

Figure 3.7 requires the listing of the output type: R for report, and C for capability, as described above. Next, a brief descriptive name for the report or capability will be given, and its acronym will be listed in parentheses for future reference. The output should then be given a brief description, which will communicate its objective and target user. Finally, a reference will be given to the Warnier diagram (WD) and the data flow diagram (DFD) as to the source and the destination of the output.

In addition to providing a list of outputs, which will be used in the next activity, Figure 3.7 also validates the Warnier and data flow diagrams. If these are inconsistent with each other, there will be no reference of the output in one or the other. This deficiency should be resolved before continuing to the next step.

One final word of caution. The designer should not try to force a one-to-one correspondence between outputs and DRTs or data flow links. At times there will be a one-to-one correspondence, and when it occurs it will be self-evident. However, sometimes one data flow link might refer to several outputs, and at other times it may refer to none at all. The same thing is true of the Warnier diagram DSTs and DRTs. These are all very effective tools for bringing the analysis to a point where the output requirements can be conceptualized. After this exercise it will almost always be required to return to the previously established documentation and update it to reflect the further developments that have evolved out of the output identification.

3.2.4 Document Output Reports

During this activity each of the output reports identified in activity 3 will be documented according to the format given in Figure 3.8. Note that Figure 3.8 specifies content but it is not a form; it is recommended that this be documented using a text editor in a free-flowing format, with each output report assigned to a separate subsection of the Design Manual. Figure 3.8, of necessity, repeats a portion of the Output Identification List. The name and acronym are repeated, as will be the brief description. However, considerable additional information about the output report will be included under the purpose and availability sections. The purpose should reference the Warnier diagram decision-action sequence or mandate in which it is used. This will clarify the target of the output report. This will be spelled out further under Availability, which will define both the media and the individuals to whom this report will be addressed.

Under the SORT/ORDER heading, the order of the report will be specified. Most reports are in the form of lists, and these must be output in some order, the default usually being the order in which they appear in the file. Additional specification of this will be discussed in Chapter 4.

The next heading references a prototype. It is essential that sample output reports be developed. If done manually, 132-column layout forms are recommended. The prototype will have each item numerically referenced (usually a manually circled number indicates that this "item number" is not part of the computer output). This prototype may not fit within the documentation or on the word processor being used to generate the documentation. For this reason, the Prototype Reference is used to designate another volume used for prototypes.

The final item given in the format of Figure 3.8 is the list of variables referenced by the number assigned on the prototype. Generally, this number will not appear on the output report itself since it is strictly for internal documentation reference. The variable name will be the same as given on the labeled output. A fuller explanation of this label meaning will follow, if necessary, including placement, format, and source. In parentheses at the end of this description appears the destination (file,seq) number, which was introduced in Section 2.3.3. Since the file layouts have not been developed yet, the (file,seq) number is mentioned only in prospect. This will be completed once the files are designed, and it will serve as a validation of the comprehensiveness of the file design.

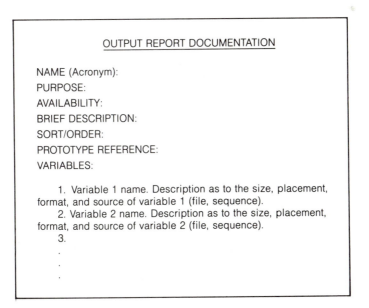

Figure 3.8 Format and Content of Output Report Documentation

3.2.5 Document Capabilities

An output capability (see definition in Section 3.2.3) cannot be defined and documented as easily as an output report. However it is essential that all required output capabilities be documented if adequate software is to be developed. To aid in this endeavor, consider the contents of the output report documentation given in Figure 3.8. Most of the information in these specifications is applicable to the definition of output capabilities. Certainly the name, purpose, availability, and brief description are in order. It is not until the specification of the actual output report format and content that problems arise, usually because they are specified by the user (or by system-determined parameters in the case of exception reports). This can be indicated as such. For example, if the report list ordering is determined by the user, this might be indicated as follows:

> SORT/ORDER: Determined by the user in response to query immediately following request for output.

The prototype reference cannot refer to one unique output type; rather, it should refer to one or more examples of potential outputs resulting from the exercise of this capability.

The greatest problem in adapting the documentation format of Figure 3.8 to the specification of output capabilities is associated with the variable specifications. In most generalized report-generation systems the user specifies the variables and the formats for their output. Thus, rather than enumerating a list of variables with reference to their location on the prototype, it is necessary to specify the method by which the user will structure the output produced. This will take the form of a stepwise procedure specification, which, if necessary, will be repeated for each variable output.

3.2.6 Management Information System Example

Before continuing, please review the introductory remarks in Section 3.1.7 with regard to the problems inherent in presenting examples, since they are still applicable here. To provide variety and aid in the appreciation of a different application, a different type of example will be represented here—that of a management information system (MIS). The purpose of an MIS is to facilitate decision making within an organization. While the advantages of introducing a new application are obvious, there is a significant disadvantage in having to "start from scratch" in describing the application. This we will attempt to do by introducing slices of the total system design, as was done in Section 3.1.7.

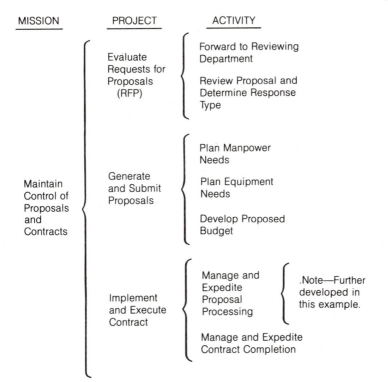

MISSION   PROJECT   ACTIVITY

Evaluate Requests for Proposals (RFP)
- Forward to Reviewing Department
- Review Proposal and Determine Response Type

Generate and Submit Proposals
- Plan Manpower Needs
- Plan Equipment Needs
- Develop Proposed Budget

Maintain Control of Proposals and Contracts

Implement and Execute Contract
- Manage and Expedite Proposal Processing   .Note—Further developed in this example.
- Manage and Expedite Contract Completion

Figure 3.9 Analysis of the "Maintain Control of Proposals and Contracts" Mission to the Activity Level

To set the stage for this example, consider a research and consulting organization with a budget of $20 million annually which handles approximately 400 research and technical assistance contracts annually. This organization has a good computerized accounting system for tax and cost control purposes. However, there are certain management information requirements that are not now available from this accounting system which management wants. Since a long-term contract has been negotiated for the maintenance of the accounting package, a modification of this system is not considered expedient (although its replacement has not been ruled out).

The affected management system must first be analyzed according to the activities presented in Section 3.1. For this example a cutaway view of some of the results of that analysis will be presented to give the reader a foundation for defining some of the system output requirements for the new system. Figure 3.9 presents an analysis of the activity level in a portion of the overall AMS which involves the tracking of contracts and proposals through the funding process. Each of these activities will be analyzed into decision and action procedures as in-

ACTIVITY	DECISION- ACTION	

Determine
Status
of
Proposal

.Responsible—Contract Administrator
.Trigger—Need for proposal information.
Obtain updated staus list of proposals.   ◄— DRT 1
Determine currency of most recent update.
Find proposal of interest.
Determine proposal status, allowing for
updates not yet posted to proposal status list.

Expedite
Completion

(+)

.Condition—Proposal not completed at due
                date.
Determine principal investigator and contact.
Determine additional resource requirements.
Secure authorization for increased effort.

Manage
and
Expedite
Proposal
Processing

Expedite
Internal
Approval

(+)

.Condition—Proposal completed but not
                signed.
Determine responsible vice-president and
contact.
Determine cause for delay, if any.
Resolve cause & obtain signature.

Send to
Sponsor

(+)

.Condition—Proposal signed but not mailed.
Determine due date.
Mail, UPS, or hand carry as required.

Verify
Status
with
Sponsor

(+)

.Condition—Proposal overdue for sponsor
                decision.
Determine sponsor and contact.
Determine status of approval process; obtain
decision or expected date of decision.

Activate
Contract

(+)

.Condition—Proposal approved by sponsor.
Convert proposal record to contract record.   —► DST 1
Set up contract accounting records.   —► DST 2
Inform principal investigator to initiate work.

Deactivate
Proposal

.Condition—Proposal rejected by sponsor.
Remove proposal from proposal status list.
File proposal in "Rejected Proposal" file.

Figure 3.10 Example of MIS Decision-Action Procedures

dicated in Figure 3.1. To further set the stage for this example, attention is concentrated on one of the activities: Manage and Expedite Proposal Processing. The AMS of this activity, as required by the analytical steps specified in Section 3.1, might yield the decision-action procedures given in Figure 3.10. Assume that this is currently being done manually and that management has stated the following deficiencies within this phase of the operation:

1. Slowness in updating the status list of proposals due to the manual procedures employed and the inaccessibility of the master list to a wide variety of potential users;
2. Inability to quickly retrieve the names and access (i.e., addresses and/or phone numbers) of necessary contacts for control. These include the principal investigator, sponsor, and responsible vice-president; and
3. Inability to easily convert "proposal" information to "contract" information, thus necessitating additional manual work.

As a further aid to establishing the basis for this example, assume that the portion of the data flow diagram corresponding to Figure 3.10 is given in Figure 3.11. Note once again that this is only a very small slice of the total system under consideration, and, as such, the integration with other similar components of the system has been eliminated for this example (this should not be true of a fully developed application). We have indicated the following three external entities: (1) Sponsor, (2) Principal Investigator, and (3) Accounting. Of these, the second

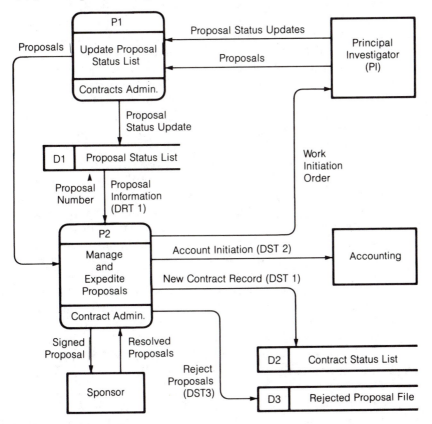

Figure 3.11 Data Flow Diagram Corresponding to Figure 3.10

two might disappear when this component of the AMS is integrated into the full AMS data flow diagram. However, with these basic specifications of the example subsystem, it is possible to illustrate the activities described within Section 3.2 above. Note that Subsection 3.2.6.x below corresponds to the activity discussed in Section 3.2.x.

3.2.6.1 Example of the Transformation of a Warnier Diagram

Within this activity the AMS Warnier diagram is transformed into that required for the new system. Thus, Figure 3.10 was presented for the MIS not only to define the example, but also to provide the input to this activity. In many cases modifications to the existing AMS will be minimal and changes within the Warnier diagrams will be unnecessary. They should, however, be reproduced and integrated into the set of revised Warnier diagrams for the new system. On the other hand, there will be times when entire activities are combined, created, merged, and modified in specifying the new system.

A considerable amount of judgement must be exercised on the part of the designer here, and no two analysts would generate exactly the same result. But our emphasis is not on dictating the exact detailed result as much as it is on providing the methodology whereby the results obtained will be useful, regardless of their potential variations. The most important aspect of this activity is that the resulting Warnier diagram adequately models the proposed new system. Consider one possibility, that given in Figure 3.12. Within the new computerized system the Contract Administrator will still have the task of managing and expediting proposal processing, which will still require the same decision-action structure. However, the means employed for such will change considerably, requiring a modification at the procedure level. (Note that this is not universally true; some applications may require modifications as high as the mission level.)

The first modification in Figure 3.10 is in the decision procedure. The new system will provide for immediate updating of the proposal file as soon as it is known. There will be no "hard copy" maintenance or retrieval required, as with the old system. Figure 3.12 shows the modified and shortened procedure for determining a proposal's status. Note that DRT 1 in Figure 3.12 is not the same as DRT 1 in Figure 3.10. These two diagrams are of two different systems, and the specifications of the old and new systems should be maintained as *two separate sets of documentation*. Thus, references within one set of documentation will not necessarily correspond to references within another.

The other modifications are itemized as follows:

1. The addition of DRT 1 to several of the action procedures to indicate that this information requirement is satisfied by the same query as required by the decision procedure,

ACTIVITY	DECISION-ACTION	
	Determine Status of Proposals	.Responsible—Contract Administrator .Trigger—Need for proposal information Enter Proposal Status option on menu(s). Enter proposal number when queried. Obtain status of proposal.
	Expedite Completion	.Condition—Proposal not completed at due date. Determine principal investigator and contact. ← DRT 1 Determine additional resource requirements. Secure authorization for increased effort.
	(+)	
	Expedite Internal Approval	.Condition—Proposal completed but not signed. Determine responsible vice-president and contact. ← DRT 1 Determine cause for delay, if any. Resolve cause and obtain signature.
	(+)	
Manage and Expedite Proposal Processing	Send to Sponsor	.Condition—Proposal signed, not mailed. Determine due date. ← DRT 1 Mail, UPS, or hand carry as required. Microfiche and file proposal. → DST 4
	(+)	
	Verify Status With Sponsor	.Condition—Proposal overdue for sponsor decision. Determine sponsor and contact. ← DRT 1 Determine status of approval process; obtain decision or time of decision.
	(+)	
	Activate Contract	.Condition—Proposal approved by sponsor. Enter proposal acceptance option on menu(s). Enter proposal Number when queried. System: set up contract record. → DST 1 System: set up accounting records. → DST 2 System: Issue work order to principal investigator.
	(+)	
	Deactivate Proposal	.Condition—Proposal rejected by sponsor. Enter proposal rejection option on menu(s). Enter proposal number when queried. System: write proposal record to "Rejected Proposal List". → DST 3

Figure 3.12 New System Decision-Action Procedures for "Determine Status of Proposal" Activity

2. A modification of the "Send to Sponsor" and "Deactivate Proposal" actions resulting in the filing of all proposals within the former action due to a decision to microfiche all proposals at that point,

3. Within the "Activate Contract" action, the computerization of the procedure steps: (1) set up contract record, (2) set up accounting records, and (3) issue work order to principal investigator, the first two of which produce data storage transactions into computerized files,

4. The establishment of a file which lists the rejected proposals instead of maintaining a separate file of them, and

5. A variety of other detailed procedure changes to accommodate the anticipated methods under the computerized system.

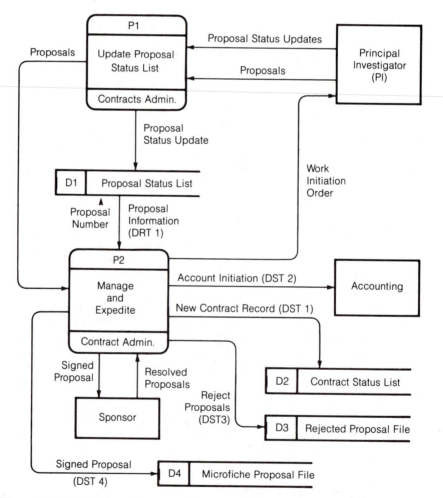

Figure 3.13 Data Flow Diagram Corresponding to Figure 3.12

Note that it is within this activity of AMS Warnier diagram transition to the new system that the new design is formulated. Decisions are required to answer questions regarding the scope and capabilities of the new system. Thus, Figure 3.12 cannot be regarded as unique or "correct." Depending upon organizational needs and computer resource availability, decisions might be made to mechanize considerably more, or possibly none, of the functions involved within this activity. This example presents one possibility in order to illustrate the process involved in making the transition.

3.2.6.2 Example Data Flow Diagram Transformation

Given that the AMS Warnier diagram has been transformed, the basis now exists for the transformation of the data flow diagram. Figure 3.13 reflects the changes made within Figure 3.12. Note the following changes for this example:

1. A new file of completed proposals, D4, is established, and
2. Rejected proposals are no longer stored separately but are maintained within D4, while a list of those rejected is maintained in D3.

The important thing demonstrated by this example is not the changes themselves but the absence of such. Indeed, Figure 3.13 is not significantly different from Figure 3.11, and this is characteristically the case. The reason for this is that the data flow diagram depicts manual operations in exactly the same way that it represents computerized operations. Only structural changes, such as the addition of a new file or process, will change the data flow diagrams. For this reason the data flow diagram must be accompanied by its backup documentation in order to thoroughly define the new system.

3.2.6.3 Example of Identification of New System Outputs

The designer is now in a position to itemize the system output requirements for this module. This will be done in conjunction with all other modules, and the two transformation activities performed above will be completed prior to any specification of output. To keep our example manageable we will perform this activity just for that module specified in Figures 3.12 and 3.13.

Figure 3.7 provides a form in which the outputs can be listed. The completed form for this example is given in Figure 3.14. In both cases the output is a report (not a capability). The names given correspond closely to the descriptors given in Figure 3.13. The brief description gives the basic purpose of the report without detailed elaboration. Finally, the reference to the Warnier diagram (WD) or the data flow dia-

		OUTPUT IDENTIFICATION FORM		

TYPE R or C	NAME (Acronym)	DESCRIPTION	REFERENCE	
			WD	DFD
R	Proposal Status Report (PSR)	All necessary details for the status of any given proposal.	DRT1	D1,P2
R	Work Initiation Order (WIO)	All details for principal investigator to initiate work on contract.	—	P2,PI

Legend
R = Report
C = Capability
WD = Management System Warnier Diagram
DFD = Data Flow Diagram Node Reference

Figure 3.14 Example of an Output Identification Form

gram (DFD) is given. In the case of the WD, the DRT or DST numbers (or both) may be given, allowing for the fact that not all outputs will be so referenced in the WD (see Figure 3.12 for reference). In the DFD reference, two items are presented: (1) the originating node, and (2) the receiving node of the link which references the report. In the case of the proposal status report (PSR), it originates at D1 and is used within process 2. Similarly, the work initiation order (WIO) originates within process node 2 and is sent to the principal investigator (PI) external entity.

Now questions may arise regarding the other links in Figure 3.13. Why is there not a system output required for each link? The answer would become obvious if the entire system DFD were completed prior to performing this identification, which was the reason for the insistence upon this early in this subsection. The following considers the other data flows within Figure 3.13:

1. Proposals. These are manually generated by the Principal Investigators and thus are not outputs of the computer system.

2. Proposal status updates. These are lists of information on project status which are transmitted to data entry (via forms or checklists); they are used to update the proposal status within file D1.

3. New contract record (DST 1). This is strictly written to the file, D2; the output reports generated from the data with D2 will be documented as an output from the system.

4. Account initiation (DST 2). If this were considered to be a report sent to accounting it should be documented as an output; however, Figure 3.12 indicates that DST 2 will actually set up the accounting records within the computer. It is obvious that a considerable amount of documentation will be required here, but not in the form of output specification. Note that the development of this is not beyond the scope of the system under consideration, it is merely beyond the scope of the example being presented.

5. Reject proposal data (DST 3). As with DST 1, this is data being written to a file.

6. Signed proposal (DST 4). This is a filing process of the hard copy microfiched proposal. It may also include the magnetic medium for the word processor which contains the text of the proposal. This will be determined when the file specification is written. No output specification is required.

3.2.6.4 Example of Documentation of Output Reports

Assuming that the activity of identifying all output reports has been completed for all modules, the next activity involves their full specification and definition. We will continue the example above by presenting the detailed documentation of the proposal status report. The data flow diagram and the new system Warnier diagram will be reviewed several times in the process of designing and specifying the output report content to assure that it satisfies all user requirements. The following is in the format given in Figure 3.8 for specifying the proposal status report:

NAME: Proposal status report (PSR)
PURPOSE: To provide the Contract Administrator with a worksheet to determine the status of any and/or all proposals.
AVAILABILITY: On CRT, or hard copy to the Contract Administrator upon request from his terminal.
BRIEF DESCRIPTION: This report contains all necessary details for determining the status of any given proposal. Access to a particular proposal is by proposal number.
SORT/ORDER: Chronologically by date of entry if the entire list is requested.

```
┌─────────────────────────────────────────────────────────────────┐
│            SCIENTIFIC COMPUTER APPLICATIONS, INC. (1)             │
│                                                                   │
│  08/02/84 (2)                    PROPOSAL STATUS REPORT (PSR) (3) │
│  15:57:15 (5)                             REPORT PAGE 1 (4)       │
│  PGM = PCR (7)                              USER ID 48.1 (6)      │
│                                                                   │
│          SUMMARY OF PROPOSAL STATUS AS OF  07/31/84 (8)           │
```

(9) PROPOSAL NUMBER	(10) PRINCIPAL INVESTIGATOR	(12) TITLE
ER82-C-60	Day ARO (11)	ANALYSIS AND SIMULATION OF THE MARINE AIR TRAFFIC CONTROL
ER82-C-61	Radney ERG	EVALUATION OF THE EFFECT OF HEAT STRESS ON WORKER PRODUC- TIVITY. FOR SELECTED SOUTHERN FOREST HARVESTING
ER82-C-62	Chandler MCH	ELASTIC STABILITY RESEARCH OF NONSTABLE MEMBERS
ER82-C-63	Gaines ARO	DEVELOPMENT AND VALIDATION OF A SUBMISSLE AERODYNAMICS PREDICTION
ER82-C-64	Maddox MCH	IMPROVING THE EFFICIENCY, SAFETY, AND UTILITY OF WOOD- BURNING UNITS
ER82-C-65	Collins ARO	NEAR-GROUND TORNADO WIND FIELDS
ER82-C-66	Salmon PET	OIL RECYCLING PROPOSAL DEVELOPMENT PROJECT

Figure 3.15a Prototype Proposal Status Report—Screen 1

PROTOTYPE REFERENCE: See Figure 3.15.
VARIABLES:

1. Major heading (hard copy only).   This will be constant for all printed reports.
2. Computer date (hard copy only).   The date stored as the current date within the computer.
3. Report title.
4. Report page (hard copy only).
5. Time of day (hard copy only).   The time stored as the current time within the computer.
6. User identification number (hard copy only).   The user identification number of the user requesting the hard copy report.
7. Program code.   The reference code of the program which produces this report.

SCIENTIFIC COMPUTER APPLICATIONS, INC. (1)

08/02/84 (2)   PROPOSAL STATUS REPORT (PSR) (3)
15:57:15 (5)   REPORT PAGE 1 (4)
PGM = PCR (7)   USER ID 48.1 (6)

SUMMARY OF PROPOSAL STATUS AS OF 07/31/84 (8)

(13) AGENCY	(14) AMOUNT	(15) DURATION		VP FOR MAILED TO	RESEARCH REC'D FROM	PROPOSAL STATUS (18)
		FROM	TO			
DEPT AERO	48,497	12/10/82	12/31/84	5/05/84 (16)	5/05/84 (17)	APPROVED
USDA FS	24,531	1 YEAR		5/04/84	5/04/84	APPROVED
ABC Corp	1,800	05/01/84	11/30/84	5/07/84		COMP,N/SIG
US ARMY	649,016	07/01/84	09/30/84	5/07/84	5/07/84	COMP, SIGNED
DOE	532,792	1 YEAR		5/18/84	5/18/84	AT SPON, PD
BMD ASSN.	66,000	6 MOS.		6/08/84	6/08/84	REJECTED
RSCH CORP	16,000	1 YEAR		6/10/84	6/11/84	IN WRITING

Figure 3.15b Prototype Proposal Status Report—Screen 2

8. Current date (preceded by constant descriptor). The date at which the last update was made to the account file.
9. Proposal number. Internal number assigned by SCA, Inc.; 12 alphanumeric characters (D2.1,3).*
10. Principal investigator. The individual primarily responsible for performing the work defined by the proposal (D2.1,4).
11. Department. Three-character departmental abbreviation (D2.1,5).

---

*The reference given in parentheses is a file number followed by the variable sequence number within the file of the variable to be output. This defines the source of the information to be output. It is recognized that this reference will not be available until the files are specified (Chapter 4). They are included here for completeness and future reference.

12. Title.   Project title (D2.1,6).
13. Agency.   Sponsoring agency (D2.1,7).
14. Total budget.   Total amount of budget independent of source (D2.1,10).
15. Duration.   The start and finish dates in the form mm/dd/yy, or alternatively the number of months and years of anticipated contract duration (D2.1,11).
16. Date mailed to approving VP.   mm/dd/yy (D2.1,12).
17. Date received from approving VP.   mm/dd/yy (D2.1,13).
18. Proposal status.   One of the following status codes will appear under this column heading, according to the most recent status update (D2.1,14):

IN WRITING	Proposal being written and not past due.
WRITING PD	Being written and past due date.
COMP, N/SIG	Written but not signed by VP.
COMP, SIGNED	Completed and signed.
AT SPONSOR	Sent to sponsor, not past expected response date.
AT SPON PD	At sponsor and past due date.
APPROVED	Approved by sponsor.
REJECTED	Rejected by sponsor.

The example given above, in practice, might require additional modification as file structures are developed. Additional variables may become available which could enhance the report. This basic specification, however, should be considered as minimal in its further use in the development of file designs.

3.3 Analysis of Input Availability

After the considerable amount of attention given to the first two analytical steps, this third step of input analysis might seem to be a neglected stepchild. If neglected, however, considerable problems usually arise. Thus, it is not the intent to de-emphasize the importance of this step, either by putting it third in this lengthy chapter, or by making it brief. This is an essential step that must be performed prior to the system synthesis steps, even though the activities associated with this step are not complex. They are:

1. Review the new system output requirements and generally determine the input required for each,
2. Determine the source for each input,

3. Determine if the necessary data collection effort is cost/beneficial, and
4. Reiterate the analysis of new system output requirements for those outputs for which the necessary input is not available.

These activities will be described in separate subsections below.

3.3.1 Define General Input Requirements

Emphasis must be placed upon the word "general" since there is a danger of overspecification at this point. Input specifications will be given considerable attention when the system design is much more mature; at this point it could be counterproductive. However, it is essential now to determine if the input required for the system is available in order to assure the integrity of the design effort. Thus some preliminary definition of input requirements is essential.

Usually this activity is performed concurrently with the specification of output requirements, although this is not recommended. Rather, a process of reiteration which separates the two thought processes is preferable. Since the output requirements have been specified, the designer can now systematically proceed through these outputs and determine what input is required for each. This most certainly should be documented by exception. Only those outputs which are doubtful as to origin need be noted for further review. Many of the output items will be quite obvious as to their origins. However, all those which are not readily available will be listed for further consideration.

3.3.2 Determine Input Sources

During this activity all data inputs identified as questionable by the activity above will be subjected to additional study to identify and/or verify their availability. Only those that are obviously infeasible to obtain should be eliminated from consideration at this time. This points out the validity of not prejudging the availability of certain inputs prior to a detailed study. With recent advances in technology, data are available which were previously infeasible to even consider. Bar code readers, remote terminals in the field, telemetry, and even robotic sensors can be used to provide data. The analyst must recognize that if certain data are necessary for total system control, chances are some means can be devised to capture it. The determination of whether this means is justified is the subject of the next section.

3.3.3 Perform Cost/Benefit Analysis

That the required data *can* be obtained at any cost is not the question. Rather, it is: are the efforts and resources required to obtain the data justified by the information which the system will produce? Can this data collection and maintenance effort be justified by the new system? While the techniques of engineering economy might be applied to solve this problem, in most cases there is not sufficient information available to perform such a sophisticated analysis. The information requirements are twofold: (1) the cost over time to establish and maintain the particular data elements of concern in a manner amenable to processing, and (2) the benefit to the organization that will accrue over time with the processing and utilization of these data elements. Of these, the second is by far the most difficult to estimate. However, this early in the design process even the first might be difficult to assess.

If hard cost information is available to estimate the costs and benefits, then by all means use it to determine if the additional data collection effort is justified. Generally, if the present worth of the benefit exceeds the present worth of the cost, then the effort is justified. However, if the information is not available the analysis will have to be based upon softer estimates. Generally these will take the form of management attitudes toward the data elements. These attitudes may be articulated in expressions ranging from "what do you need that for?" to "I could really use that information." Unfortunately, the data elements which might be thought most useful often prove disappointing. This is the reason that we began this analytical journey with a statement of relevant missions and continued to analyze from that point, as opposed to just sitting down and "thinking up" the data requirements. We dare not sacrifice the fruits of our logical approach at this point.

To regain the proper perspective, it is fair to say that these hard decisions regarding data availability will be required only in rare circumstances. Most of the data elements required will either be readily available or totally infeasible to obtain. Where this is not the case, formulate the problem as quantitatively as possible in terms of cost and benefit, and present it to upper management to obtain their opinions. Either obtain their decision or obtain the delegation of their authority to make the decision with respect to its inclusion. As long as all of the ramifications of this decision are known to all, the decision should not be a difficult one.

One additional qualifier. Sometimes it might be known that a given data element is of relatively little value compared to its cost. However, if there is strong middle and upper management support for its retention, this "psychological" value should be factored into the cost/benefit equation. In fact, there are times when catering to the data entry operator or other users can pay handsome rewards as far as their acceptance of the new system. Since the acceptance of the users and their respective management will have far more to do with determining the success of the new system than any other factor, these considerations are far from trivial.

3.3.4 Reiterate the Design Steps

At this point a decision has been made to either accept or reject some of the marginal data elements. Obviously, if some of the inputs are not going to be available, it will be necessary to redesign the system to accommodate this. In some rare cases a system redesign might enable the same information (output reports) to be generated by utilizing other data that are available. However, in most cases the output reports and/ or capabilities themselves will require modification. It is essential that all prior documentation be updated before continuing, in order to eliminate the inefficiency that could result from inconsistencies.

Questions and Problems

1. What are the first three steps in the design process? What do they have in common? When would these be used for verification and review rather than for design?

2. Present the structure of the generalized Warnier diagram applied to the analysis of a management system.

3. What are the six recommended activities for performing an analysis of the affected management system (AMS)? What is the output of these activities (generally), and how will this output be used?

4. Show how these initial activities of the design process are not "leaps of faith," but rather, they are based upon certain existing source materials. Define these source materials.

5. How is the scope of the AMS defined? What difficulties and possible contradictions might need resolution?

6. What danger is there in confusing the subject system when defining goals? Which goals are of concern here?

7. What is the difference between goals and objectives?

8. Does the analyst define the goals and objectives? What is his role in this process?

9. How do goals and objectives relate to missions and projects? State the differences as well as the similarities.

10. In response to problem 2 above, the structure included an "activity" level. Why is this level seemingly skipped in the six activities presented (response to question 3)?

11. What is the difference between a discrete set of actions and a continuous action? Action here refers to the resultant selection of the decision process.

12. Distinguish between a decision and a mandate.

13. How are the procedure steps of decisions and actions determined? How should their magnitude be determined?

14. Define what is meant by a Data Retrieval Transaction (DRT) and a Data Storage Transaction (DST).

15. In defining AMS data flows, why is it important not to get involved with unnecessary detail?

16. Enumerate the problems associated with textbook examples and state how these may be overcome.

17. At the completion of the six activities referenced in question 3, what documentation should result and how should it be organized?

18. Analyze a computerized security system using the management system Warnier diagram analysis. The system has a burglar alarm and a fire alarm component. The fire alarm component continually checks for smoke and excess heat. If either is present it calls the fire department and rings a fire alarm. The burglar alarm component checks all locks, monitors the windows, and checks movement with ultrasonic sound. Any doors that are not locked are reported to the computer console. If any windows are broken or if there is movement in any room, a burglar alarm goes off and a call is automatically placed to the police department.

19. Draw a management system Warnier diagram to analyze an inventory control system. Start at the decision level. The decision is to determine the amount of each part to order of each of the 1152 items in the inventory. The procedure for determining the amount to order is:

A. Obtain inventory listing from data processing.
B. Check exception report generated for the following:
    1. Less than one week supply.
    2. Greater than one week but less than order level (OL).
    3. Greater than order level (OL).
C. If (1) take physical inventory and place emergency order.
D. If (2) normal order the difference between what is on hand and the restock level (RL).
E. If (3) do not order.

The assistant controller is responsible for inventory control and ordering, and a review is made by his staff on a monthly basis. Emergency purchase orders are hand carried through the purchasing department and called in to the supplier by phone, and the purchase order is sent out for verification later. Normal purchase orders are sent to the purchasing department and mailed to the suppliers.

Make the Warnier diagram as complete as possible with the information given.

20. Set up a management system Warnier diagram to the DRT and DST level to analyze an inventory control system. When completed, determine if the system is self-sufficient by matching DRTs and DSTs. Let there be five activities as follow:

1. Determination of reorder quantities for perishable goods.
2. Determination of reorder quantities for durable goods.
3. Determination of reorder levels.
4. Determination of EOQ (economic order quantity).
5. Determination of physical inventories.

For this exercise ignore the "responsible" and the "description" entries and concentrate on the process logic. Be sure to include, however, the trigger and condition entries and all frequency and logical qualifiers.

There are two sets of suppliers, one set for perishable goods and another set for durable goods. The procedure for perishables involves first a weekly physical inventory of all 86 perishable items. The quantity on hand (fit to use over the next week) is compared against the reorder level.

Reorder levels for all items are calculated and verified monthly using the following formula:

$$RL = 1.25((\text{Monthly utilization})/4.3)$$

If the number in stock for the item is less than its RL value, then a purchase order is issued in an amount determined as follows:

$$\text{Order amount} = \text{EOQ} - \text{amount in inventory}$$

The EOQ for each item is calculated utilizing a standard formula every six months.

For the 110 durable items in stock, a perpetual inventory is maintained, i.e., no physical inventories are taken except for verification. Whenever an item gets below its reorder level it is automatically reordered in its EOQ. These inventory items are scanned and checked on a daily basis.

The formula for the EOQ is:

$$EOQ = SQRT ((2*Z*CR)/(C*CC))$$

where

Z = the demand
C = the cost per item
CC = the carrying cost (percent interest rate)
CR = cost per order.

Note: SQRT is the square root function.

21. Draw a data flow diagram for a preventive maintenance system for a fleet of automobiles. The system's objective is to assure that preventive maintenance (PM) is performed for each vehicle. Its primary function is to generate work orders for PM by performing a daily review of the "age" of each vehicle. Age includes both mileage and calendar age considerations. Whenever the age exceeds certain predetermined values (parameters) a work order is automatically generated.

The following is a summary of the operational menu:

1. Update machine age
2. Update PM completion
3. Print PM work orders
4. Print jobs scheduled but not completed
5. Print work order duplicate

There is one program to perform each of these functions.

The primary user of the system is the PM Works Manager, who can be considered as an external entity. The system has two files: (1) a vehicle file, one record per vehicle, and (2) a procedure file, one record per PM procedure. Several vehicles may share the same procedure (e.g., oil change, tune-up, etc.), and several procedures may apply to the same vehicle. The vehicle record for a given automobile contains information as to which procedure it requires and how frequently those PM procedures are to be performed.

The user of the system selects one of the options given above and performs the following:

1. Update machine age. Machine utilization information is obtained from the drivers and this information is used to update the Vehicle File, such that periodically the age of each vehicle is updated.
2. Update PM completion. The PM Works Manager obtains maintenance completion information, in the form of a "marked completed" work order from the maintenance department. He uses this information to update the Vehicle File such that the last performance of each type of PM is recorded.
3. Print PM work orders. When the user selects this option a total scan is made of the Vehicle File. Any vehicles requiring maintenance are identified (if age exceeds predefined parameters). The procedure required will then be identified from the Procedure File. (Assume that *updates* to the Procedure File are outside the scope of this DFD.) The combined machine record and procedure record will be used to generate each work order, which is sent to maintenance for performance.
4. Print jobs scheduled but not completed. When a work order is printed the machine record will be updated to indicate this. This "switch" will be kept "on" until function #2 above (Up-

date PM completion) turns it off. This function merely goes through the Vehicle File and lists out all jobs which are in progress.

5. Print work order duplicate. The user will specify both a vehicle number and a procedure reference number and the system will produce the work order for that job.

Be sure that all *descriptors* are present and correct. Put down the responsible persons (if performed by computer, put *DP*). Do not break any of the data flow diagram rules.

22. Review the second analytical step (question 1) and list the five activities associated with this step.

23. What is the procedure for transforming the AMS Warnier analysis to that for the proposed new system? Why is this not a difficult process?

24. Is the scope of the new system totally defined at this point? Explain. What are some advantages and disadvantages of having a perfectly rigid definition of scope?

25. What is the primary question to be answered in transforming the AMS WD and DFD to the new System WD and DFD?

26. What two types of outputs need identification? What is the distinction between them?

27. What is the content of the form for enumerating output requirements?

28. Is there a one-to-one correspondence between DRTs, DSTs, data flow links, and outputs? Explain.

29. What is the necessary content of detailed output report documentation?

30. How are capabilities documented, as opposed to output reports?

Chapter

File Design Concepts

The next step in the systems design procedure presented in Chapter 1 is to specify file layouts and structures. Prior to performing this step it is essential that the designer understand some basic file design concepts. The objective of this chapter is to present some fundamental information storage and retrieval concepts followed by a strategy for determining which file accessing capabilities are most appropriate for the system being designed. This will enable the designer to complete the file specifications necessary for system software development, which are detailed in Chapter 5. A large body of literature has been devoted to this topic; therefore, it is appropriate to define the scope of this chapter before proceeding.

Since the basic objective of this book concerns the applications software design effort, attention will be directed toward single-process access of files residing on direct-access mass storage media, particularly disk. It is true that magnetic tape has many applications in contemporary data processing, such as providing backup, storing very large data sets, and providing an interface between machines. However, since tape capabilities are a subset of direct-access device capabilities, the discussion here will not preclude these applications.

The other restriction mentioned was that of single-process access. This means that only one application will utilize a given set of files for the duration of that application's execution. While it is true that this precludes the consideration of many capabilities which are currently available, it conveniently limits the scope of the discussion to one particular application, rather than global system requirements. This is necessary especially when files are being updated on a frequent basis. Designers should bear in mind, however, that the concepts and techniques in this chapter are certainly available in more complex environments, but that these will require considerations beyond the scope of the application itself.

There are two basic activities involved in documenting the file specifications: (1) file content specifications, and (2) file structure specifications. The first of these will proceed directly from the output of the analysis discussed in Chapter 3. File structures are more complex, and they will include consideration of storage organization, record layout, and accessing techniques. This chapter continues by presenting the concepts necessary for understanding the alternatives available to the designer. Once these basic concepts are understood, Chapter 5 provides the activities for taking the outputs of Chapter 3 and completing the file layout and structure specifications. The basic concepts will be presented at a level which the designer can understand, while circumventing the rigorous treatment that is often required of the technician.

4.1 File Content Considerations

It is hard to imagine a designer who, after faithfully following the steps given in Chapter 3, would have enough self-control to keep from producing a tentative file layout at this point. We have warned against jumping ahead before thoroughly completing each step, since such effort could be wasteful of time, and it might also serve to lock the analyst's mind onto an inferior design. However, as long as the initial layouts are truly regarded as preliminary and tentative, no harm will be incurred, and the activities to be discussed in the early part of Chapter 5 would then be dedicated to modification and verification.

The decisions as to the number of files and the content of each are not finalized until the structure and organization of the files are resolved. For example, a decision might be made based upon structural considerations to move a data element to a different file, or to divide or combine files. These decisions will be based upon processing considerations, which will be addressed in the remainder of this chapter. During the file content specification, emphasis must be placed on assuring that all data elements are present so that they will not be lost in the shuffle later. Also note that such changes may require updates in the data flow diagrams.

Concerning redundancy, it is strongly recommended that if at all possible the same data elements do not appear in two or more different files. Sometimes this is not possible, and at times such redundancy can lead to great gains in processing efficiency. However, redundancy not only consumes valuable storage space, it also requires duplication in record creation and maintenance. Therefore, it should only be present when processing advantages mandate its use.

Without getting too bogged down with technical terminology, the designer should be aware of the notion of record blocking. A *logical record* is defined to be a collection of data that represents one case or entry in the file. Generally when the term *record* is used without qualifiers, it refers to a logical record. There is usually an advantage in grouping these logical records together (i.e., concatenating them) and storing the result as a physical entity known as a *physical record* or *block*. This is quite useful when the block size is chosen properly, since an entire block might be read or written in as little time as it takes to read a single logical record. The designer may want to investigate the specification of access-method software that transparently deblocks physical records into logical records to support application programs. Such software is usually hardware or operating-system dependent.

The considerations above for file content are given to set the stage for the data structure concepts given below. Obviously the contents must be defined before too much consideration can be given to the other structural specifications. The activities employed in specifying file content are given in Chapter 5. However, before attempting to perform these activities, it is essential that the concepts presented in the remainder of this chapter be understood.

4.2 Concept of Indexing

Given that the file contents are specified, some designers might believe that their job is finished. They delegate to the programming staff the job of determining file details based upon this content and the output report requirements. They might even reason that a sequential reading of these files will be adequate to produce the data required for processing. Unfortunately, though, this is rarely true. In most situations the order of output reports in no way reflects the ordering of the data within a file. Also, the need to maintain records, i.e., add, change, and delete, presents a problem if purely sequential storage and processing are assumed. For these reasons, the other aspects of file specification will begin with a strong consideration of the concept of indexing. Given that the file content specifications have been completed, this will prepare the designer for determining the best means to store and access the data.

4.2.1 Keys

In previous sections, the term *file* has been used to refer to a collection of related records. In this section, the definition will be enlarged to refer to both the aggregate grouping of these data records and any auxiliary information necessary to access them, such as directories and indexes.

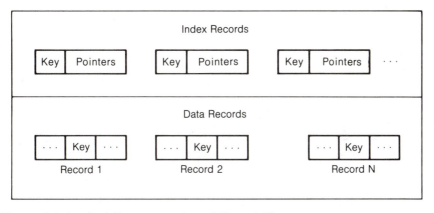

Figure 4.1 Logical Representation of Keyed File

The *data record* itself is made up of several fields, one or more of which may be *key* fields that are used to identify the record either uniquely or generically. Similarly, an *index record* will generally contain a value of the record key, and a reference to the respective location of the record or records which contain that value in the key field. Figure 4.1 shows the overall structure of a file containing both index and data records. Of course, all files will contain an area for data records, while the index area is an optional portion of the file used to facilitate access of the data records. The adjacent *pointers** in each index record identify the data records related by the respective key. Multiple pointers are allowed if multiple data records have the same key value.

An example of a keyed file is shown in Figure 4.2. In this example there are five nonempty index records having respective key values of 101, 105, 107, 289, and 305. The fields in each index record are pointers to the relative positions of the records. Thus, the index record with the key field value of 101 points to this key value being in the third data record. The data records are structured to have their first field as the key field, which is followed by whatever other data is to be stored. In this example, each record has a unique key value with the exception of records 6 and 10, which have a key value of 305. Note the presence of empty records for expansion in both the index and data records. Methods to keep track of the empty data records include the establishment of one index record containing pointers to all empty data records (other methods will be introduced below).

---

*The terms *pointer* and *link* will be used interchangeably to reference a field whose value is a relative storage location. Some authors distinguish between the two by applying the term *link* to references within or between data files while applying the term *pointer* to the reference of a program to a record. No such distinction will be made here.

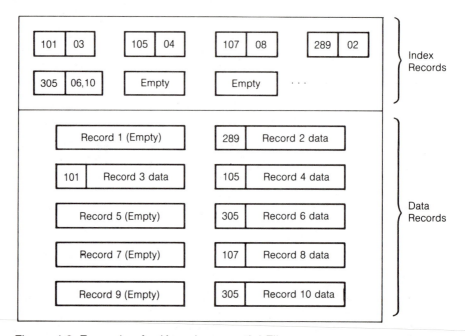

Figure 4.2 Example of a Keyed-sequential File

4.2.2 Keyed-sequential File Organization

Any file which has the components of both index and data records can be called an *indexed* file or a *keyed* file, since both indexes and keys are used to access the data records.* To make Figure 4.2 more meaningful, the typical arrangement of the data and index records are assumed, as follows: (1) the index records are arranged in order of the *key* value which they index, and (2) the data records are displayed *sequentially*, the arrangement being determined by the order of entry and the location of empty, available records. We define the term *keyed-sequential* to apply to the organization of files which have this general arrangement. Keyed-sequential files greatly facilitate record access as can be seen in Figure 4.2. For example, to get to the record with a key value of 107, it is possible to quickly read the index records and determine that the key value of 107 is in record 8. At that point record 8 can be directly accessed to utilize the data in that record. A nonindexed arrangement

---

*It is imperative to recognize that computer terminology has a life of its own, and anyone who wants to conduct a semantic fight over this terminology is blessed with a huge battlefield. The intent here is to avoid terminology which has become local to a given hardware or software vendor, and to utilize that nomenclature which is most inherently descriptive or generally accepted in conveying the concepts which are being presented.

would generally require every record prior to the desired record to be read and each key value to be tested.

To summarize, the main considerations for the keyed-sequential arrangement are as follow:

1. Index records are ordered in the sequence of ascending unique key value sequence,
2. Empty index and data records exist for expansion and contraction of the dynamic file,
3. Data records are essentially unordered in the data area, and
4. Some provision must exist for identifying empty data records.

The reader should readily see that this arrangement provides a great deal of flexibility subject to only a few restrictions, discussed in Section 4.2.3 below. Moreover, the actual retrieval process need not require multiple file-read operations to obtain a data record if the index is loaded into an array in main computer memory when the file is initially opened for access.

Figure 4.3 depicts the record retrieval arrangement using the terms introduced above. The application program specifies a key value goal that is sought in the records to be retrieved from the keyed-sequential file. This is used by the accessing software to search the index records of the file. Given that at least one index record exists with this key value, the remaining fields of the index record provide the pointers to the data records, which are read by the accessing software, and then used by the application program. Figure 4.4 shows how the designer would specify this process within the data flow diagram. A comparison

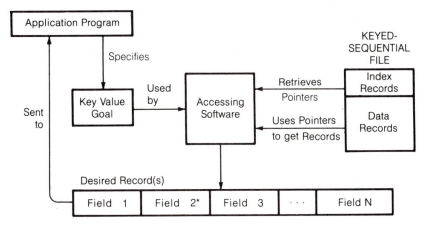

*Key Field

Figure 4.3 Accessing a Keyed Record

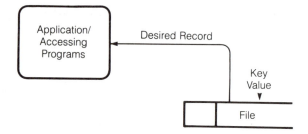

Figure 4.4 Specification for Keyed Record Access

of Figures 4.3 and 4.4 illustrates the need for additional file structure specification, since the data flow diagram specification is independent of file structure. Section 4.8 will present the mechanism for specifying keys and indexes, after some other basic concepts are introduced.

4.2.3 Searching Through the Index

Although searching strategies have not yet been discussed, readers should begin to recognize that since the index is kept in key value sequence, it need not be exhaustively searched one entry after another. A much more powerful technique, known as the binary search, should certainly be considered for this task, as will be described in Section 4.6.2. However, even if a serial search of the index were required, it would be more efficient than a serial search of the data records, since it involves a much smaller subset of data. As mentioned above, the index portion of the file is usually small enough to be loaded (completely or in part) directly into main memory. This has tremendous advantages when the index pointers are loaded into an array, the subscript of which is the key value—a situation not uncommon in practice. In any event, the search is radically accelerated since individual records do not have to be read for each comparison.

4.2.4 Types and Levels of Indexes

Large keyed-sequential files sometimes warrant considerations beyond a single index that contains an exhaustive key/pointer accounting for every data record. As stated above, maintaining the index in computer memory during file accessing greatly reduces the number of read operations. An important objective, then, is to keep the index small enough to use it directly in main memory.

One popular, alternative indexing method which may be considered involves the use of *block anchors*, where index entries are kept for only one data record in every physical block (usually the first or last). The index is searched to locate the block which must contain the record corresponding to the search goal. The block is then fetched and serially scanned for the desired record (or point of insertion for updates). The main drawback of this technique is that data records must be kept in their proper block relative to key values, which makes updating more complex. Updating either requires much shifting or a separate overflow area which must be serially scanned. (Overflow areas are discussed further in Section 4.2.5.) While this alternative increases maintenance complexity, there are some applications which lend themselves quite favorably to this approach. Indexed files having these characteristics are sometimes called *indexed-sequential.*

Inverted files are data files for which many indexes have been created. These indexes may contain exhaustive entries for every data record and every variable *(completely inverted),* or only selective subsets of records might be indexed, perhaps in accordance with specific case criteria (e.g., those customer records with unpaid balances, or those highway accident records involving pedestrians). Each index provides a unique attribute to access records in the file. The flexibility and ease of this method are obvious when storage resources exceed search-time availability. However, maintenance problems can be introduced unless careful consideration is given to the file-updating strategy. This is discussed further in Section 4.5.3.2.

4.2.5 Overflow Areas

Additions to some indexed files are made in places called overflow areas. Many strategies exist for specifying these overflow areas; among them are the following, which may be implemented independently or in combination:

1. An entirely separate physical file that is connected to the original file by a pointer,
2. A separate partition of the file below the data partition, and
3. Reserved records at the end of each physical data block.

The necessary considerations are mainly time trade-offs; the first two strategies keep the main part of the file contiguous and dense, so that accessing the original records will consume a minimum of time. However, under these strategies the access of overflow areas generally requires considerably more time. On the other hand, the third strategy provides for uniform growth at only modest access-time expense. The problem with this strategy is that most files grow in a skewed manner. For instance, any simple means to reserve space within the primary data area (when the file is built) will probably not reserve it in the locations and positions that will match future growth patterns. This is often true, even where nonuniform growth patterns are predictable, like files keyed by people's surnames or company names. In these instances clustering will occur around certain letters which are most often used as first characters in surnames or company names, respectively. Thus, it is not easy to preallocate growth areas within a file. A good way to deal with irregular file growth of this nature is to periodically copy and reorganize the entire file.

4.3 Record Types

Having discussed file content considerations and introduced some basic concepts of indexing, the various record types can now be put in perspective. The reader should realize that all of these factors enter into the design process simultaneously. While this chapter is ordered to provide for the presentation of these concepts, Chapter 5 is ordered according to their application in specifying the design. Three record types will be described in the following subsections, followed by a summary comparison of the record format types.

4.3.1 Fixed-length Record Files

Many applications utilize data records that are fixed in length, which are the simplest to understand. These records have a fixed number of data fields which occur a definite number of times, and in a statically designated location in each record. The simple fixed-length record accommodates the repeating of a particular field only by allowing for the maximum number of occurrences in advance. For example, a highway traffic accident record may need to contain a variable number of involved vehicles or victims. Thus, a dedicated area of the record may be set aside to hold, for example, up to three vehicles and five victims. The presence of this dedicated area, however, generally results in a degree of guaranteed wasted file storage area, since each record must be large enough to hold the maximum number of occurrences for each field in the record. In many applications the maximum does not occur very frequently, resulting in wasted space. The alternative is to reduce this dedicated space at the expense of not storing all of the information for certain records.

While the limitation of fixed-length records has been accentuated above, it should hastily be added that there are a large number of applications in which field repetition is not required, and for these applications any record structure other than the fixed-length, fixed-format structure would be inappropriate. This is, by far, the easiest type of record to maintain and process, and, possibly more importantly, it is quite easy for programmers and users to understand.

4.3.2 Variable-length Record Files

Some of the deficiencies cited above for fixed-length record files can be overcome by allowing the length and the specific format to vary. The

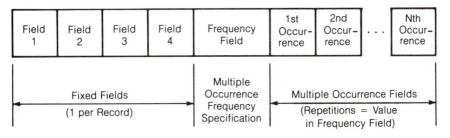

Figure 4.5 Example of a Format for a Variable-length Record

general format of the record will remain constant in that it can be specified in a generic way. Consider, for example, the simple specification given in Figure 4.5. This example record has four fixed fields which have the same basic qualifications as the fields in a fixed-length record. However, this record also contains a frequency field, an integer specification of the number of occurrences of the fields to follow, namely, N. Following this, any number of fixed fields may appear; however, the set of fixed fields will be repeated in the record N times. The presence of the frequency field is essential to tell the accessing program how many occurrences of the repetitive fields exist. Obviously this example could be generalized further to present a record with several frequency fields and several corresponding groups of fields which are to be repeated.

In order to interpret the data in a record as described above, it is necessary for the application program to use the frequency field as the upper boundary on a loop index. Access can be accomplished in many cases within a single record read. However, when the use of variable record length becomes extensive, nonrepetitive fields, along with the frequency fields, can be organized into a case header record which precedes the Multiple Occurrence Fields, each set of which is organized by a separate record layout, and each occurrence of which is contained in a separate record. This is illustrated in Figure 4.6. This collection of records is generally called a *hierarchical file* and the records are called *hierarchical records*. In Figure 4.6 it is assumed that frequency field 1 contains the value of N1, frequency field 2 the value of N2, and frequency field 3 the value of N3. It might be noted here that to make a "case" synonymous with a record and call this a variable-length record, or to consider the various classifications of fields as records within themselves (as is done in Figure 4.7), might be purely a matter of user semantics. That is, there is little, if any, difference as far as the physical storage of the information is concerned.

The further generalization of Figure 4.6 to allow any of the multiple occurrence records to assume "case record" status for its own set of repetitive records should be obvious at this point. As an example of such, consider a traffic accident file which is structured to allow for

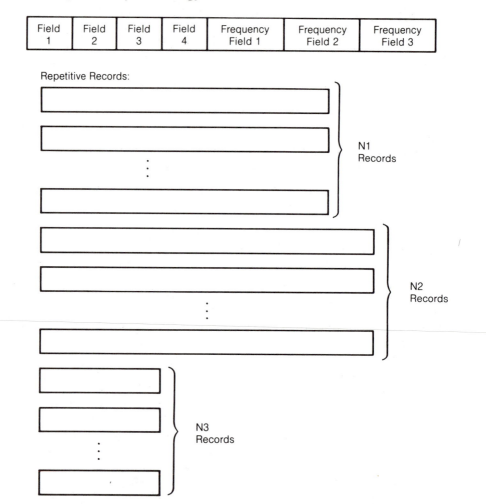

Field 1	Field 2	Field 3	Field 4	Frequency Field 1	Frequency Field 2	Frequency Field 3

Repetitive Records:

N1 Records

N2 Records

N3 Records

Figure 4.6 General Hierarchical File Layout

multiple vehicles in an accident and multiple passengers in a vehicle. An example for a two-vehicle accident (case) is given in Figure 4.7. Note that the first vehicle has three passengers; hence, the records for these follow. The second vehicle record has one passenger record associated with it.

The presentation of Figures 4.5 through 4.7, and their accompanying narratives, show how a hierarchy can be established such that a given case record may have several different types of repetitive records associated with it, each of which may have several repetitive records associated with them, and so forth. Thus a hierarchy can be established between the different record types, as depicted in Figure 4.8.

The fields in Figure 4.8 are denoted F1,F2,...F5, where the appearance of a frequency modifier (such as N3 or N5) indicates that there will be a repetition of either the field (as in the case of F3) or a set of subfields (as in the case of F5). This concept can be further applied to any of these subfields by making them frequency fields, as is exemplified by the first and fifth subfields of F5. This representation of a hierarchical file is often likened to a family tree where each field may have both ancestors and descendants.

Hopefully, at this point, it is evident that no fixed-length record with any static field arrangement could represent groups of household members (including ancestors and descendants) without incurring large amounts of wasted file space in all of the nonmaximal records. The degree to which levels of subfields should be used is, of course, at the designer's discretion, and they are presented here to communicate

Case Record:

Accident Number	Time of Day	Date	· · ·	· · ·	Number of Vehicles (2)	· · ·	Other Fixed Fields

First Vehicle Record:

License Number	Driver Name	· · ·	· · ·	Number of Passengers (3)	· · ·	Other Fixed Fields

Passenger Records - Vehicle 1:

Psgr. name	Injury Type	Age	Sex	Seating Location	· · ·	Other Fixed Fields

Psgr. name	Injury Type	Age	Sex	Seating Location	· · ·	Other Fixed Fields

Psgr. name	Injury Type	Age	Sex	Seating Location	· · ·	Other Fixed Fields

Second Vehicle Record:

License Number	Driver Name	· · ·	· · ·	Number of Passengers (1)	· · ·	Other Fixed Fields

Psgr. name	Injury Type	Age	Sex	Seating Location	· · ·	Other Fixed Fields

Figure 4.7 Example of a Hierarchical Accident Record

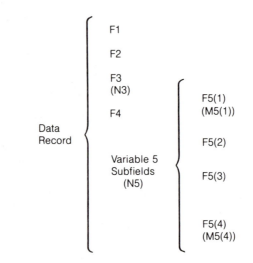

Figure 4.8 Variable Occurrence Field Hierarchy.

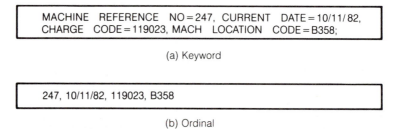

(a) Keyword

(b) Ordinal

Figure 4.9 Examples of Pile Records

the basic concept so that the designer can specify their use. Certainly, consideration must be given to features of the programming language being used as well as the application. Therefore, it is highly recommended that the decisions related to file structure be made in conjunction with the technical software development specialists.

4.3.3 Undefined-length Record Files

Files containing records having no defined length are used in a small number of applications. The data being transferred in this case are generally in a keyword or ordinal-pile format. In the former case, data fields are encoded beside their respective variable names, as shown in Figure 4.9a. The term *ordinal* implies a predetermined order separated

SUMMARY OF RECORD FORMAT
MERITS AND DRAWBACKS

CHARACTERISTIC	FORMAT		
	FIXED	VARIABLE	UNDEFINED
Ease of use	Easiest to use	Easy to use unless several levels are implemented	Special routines generally required
Speed	Fastest	Slightly slower than fixed format	Tolerable for some applications
Storage utilization	Excellent, except when multiple field occurrences are required	Good for multiple occurrences of fields or subfields	Generally quite efficient

Figure 4.10 Comparison of Record Formats

by delimiters, as shown in Figure 4.9b. The pile record format features ultimate flexibility, but requires extensive preprocessing before the data can become usable. The pre- and post- processing can also become comparatively time consuming to the CPU. It can also be a serious problem to a programmer not using a language equipped to handle data in this form. Neither COBOL nor many versions of FORTRAN are well suited for pile-file input or output, but the stream input-output routines provided by PL/I and some versions of BASIC handle this data quite readily.

Figure 4.10 presents a summary of the three subsections above by showing the advantages and disadvantages of each of the record types discussed.

4.4 File Organization and Access

A typical early use of computers for commercial or business data processing included the organizing and merging of various data files, followed by retrieval and formatting for some end use. An experienced programmer will be amused when reminded of the number of sort/ merge steps that made up an average jobstream in the 1960 to 1970 timeframe. Such requirements existed for several reasons, among them:

1. Prevalent use of tape storage for economic reasons,
2. Predominant batch mode for information processing, and
3. Restricted availability of file accessing software.

In short, a great deal of application master file updating was performed by collating the update file(s) according to some key field and then merging the result with the existing master in an add/change/delete fashion, producing a distinctly new master file.

More recent technology has shifted emphasis from tape storage to the more cost-effective disk, and has simultaneously provided interactive computer access in lieu of batch processing. These two factors necessitated file accessing capabilities far beyond simple-sequential record retrieval; thus, files now contain records stored in correspondingly more flexible manners.

Before studying nonsequential techniques further, consider the difference in meaning between the terms *file organization* and *file access*. Earlier in this chapter, Figure 4.1 depicted a logical file that consisted of two parts: an index area and a data area. The file organization depicted there was indexed, or keyed-sequential (as defined above). Correspondingly, this file *organization* provides for random or sequential record *access*. Simple-sequential files do not contain an index, just as they do not permit traditionally effective random retrieval.

The meaning of *file access*, then, is the method by which the software stores and retrieves the logical records in a file. Correspondingly, *file organization* refers to the arrangement of the logical records in a file. This arrangement may be a physical arrangement within the storage media, or it might be a logical arrangement induced by indexes or pointers. Although distinctly different, the terms *file organization* and *file access* are frequently confused primarily because identical words are often used in their specification. For example, a keyed-sequential file is said to be accessed either randomly or sequentially, while a sequential file is generally described as being accessed sequentially. With this potential confusion, it is essential to define some terms in order to distinguish among these rather simple concepts.

For an initial set of definitions, consider the following file organization techniques:

Simple-sequential—records are physically ordered, one following the next; logically, there is no linkage or pointers between records.
Sorted-sequential—records are physically ordered according to some specified field or fields; logically there are no linkage or pointers between records.
Keyed-sequential—data records are physically unordered, but they are referenced by sets of pointers which are ordered by some specified key (see Figure 4.2).
Direct-linked—records are physically unordered, but links within the records themselves establish an ordering. (The linkage or ordering is directly within the record itself; no index records are present.)

Although these file organizations will be considered independently, it is understood that hybrids may also exist. An example of a useful hybrid will be discussed at the end of Section 4.6.2.

Next, consider the following definitions of access methods:

Sequential—in order to read a given record, all of the "previous" records in the file must be read first. Here "previous" is defined as the record which exists physically in the position prior to the position of the given record.

Random—the system is able to go directly to a given record without reading all of the records from the physical beginning of the file. This is sometimes called "direct" access, although this term is usually applied to the type of hardware that enables random-access software and techniques to be employed. Thus, a disk drive is a direct-access device since its hardware can keep track of the location of any record in terms of rotation and head movement positioning.

Note that while keyed-sequential and direct-linked-type files cannot be stored on tape (or similar, strictly sequential-read devices), there is nothing to prevent simple-sequential or sorted-sequential files from being stored on direct-access devices. Thus, it is important for the designer to maintain a clear distinction between the access method, the file organization, and the device type. The access method and file organization are determined by software, subject to the available hardware constraints, while the device type is a description of the hardware itself.

Figure 4.11 presents the various possible combinations of the above-defined file organization types and the two access methods. Note that the only disallowed combinations are that the keyed-sequential and direct-linked organizations are not amenable to sequential access. This is because there is no capability within sequential access to get to another record other than that of incrementing to the very next record.* Generally, sequentially accessed files may be updated by adding records to the end of the file. In most sequentially accessed files, a new record may not be written between two previously existing records. Further, it should be clear that finding a record with a unique characteristic (key field value) within a simple-sequential file requires reading, on the average, half of the records in the file. To determine that a record of this characteristic does not exist will require that all of the records be read. The same is true of the sorted-sequential file, with the exception that, on the average, only half of the records will need to be read to verify the nonexistence of the record sought.

*Some sequential-access methods allow skipping of records, called *skip sequential.*

FILE ORGANIZATION TYPES	FILE ACCESS METHODS	
	SEQUENTIAL	RANDOM
Simple-sequential	YES	YES
Sorted-sequential	YES	YES
Keyed-sequential	NO	YES
Direct-linked	NO	YES

Figure 4.11 Means to Access Files of Various Organizations

The clearest example of both simple-sequential file organization and sequential access is that generally associated with magnetic tape. Tape can be crudely manipulated in a *stack* fashion, whereby the top of the stack is the last data block; hence, physical records may be sequentially added to, or collectively removed from, the end of the tape file if no other file physically follows it. However, any other operations on a tape file (i.e., insertions, deletions, or modifications) generally require that the entire file be completely rewritten on a separate tape *volume* (i.e., reel). Furthermore, tape files can be concatenated on a single (or set of) tape volume(s) to maximize storage utilization. This arrangement, though, like the records within a file, exhibits the qualities of a stack, thereby eliminating any modification of all but the last tape file on the set of volumes. For this reason, magnetic tape is generally used for archival information that may be retrieved at some distant, later time, or for storing large files that would be uneconomical or impossible to store on disk. In any event, tape files can be stored in several schemes on tape volumes, as summarized in Figure 4.12.

The final type of file organization to be considered here is direct-linked, which has many interesting characteristics that are useful, but not always obvious, to the designer. This file organization is effectively a stored "linked list" which takes advantage of the random retrieval capability of disk to jump from record to (nonadjacent) record. The understanding of the concept of linking, and the power and efficiency obtained thereby, is so important that a separate section on data-structure concepts will be devoted to it immediately below.

4.5 Data-structure Concepts

This section will present some basic concepts of data structures. The term *data structure* applies to the establishment of relationships among various components of data and the manipulation of these to relationships that achieve specific processing objectives. While this section can-

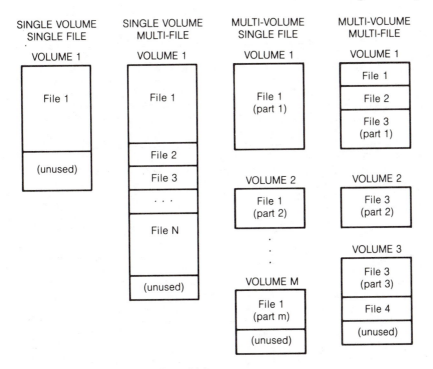

Figure 4.12 Tape Files on Tape Volumes

not hope to cover the entire data-structures field; an understanding of some of the basic concepts is essential to the designer. Once designers have an understanding of these basic concepts, they will be in a position to interact with specialists in working out the details if more complex structures are required. This section also provides a foundation and appreciation of data structures to motivate students to study further to enrich their background in this area.

4.5.1 Concept of Linking

Consider a fixed-record layout as given at the top of Figure 4.13. One of the fields of this record (called Next Link) is not typical data, but instead contains the *address* of another record. This may be the absolute address of the starting point of the record, or it may be a relative value which can be converted to this absolute address. Generally, the designer is not immediately concerned with the mechanics of actually obtaining the value residing in this field. Therefore, a relative position may be assigned, as in Figure 4.13. The diagram used to depict a linked structure for a file will be called a data-structure diagram or just a *structure diagram*. Generally, structure diagrams will show a link between any link field in a given record to the record which is given by the location

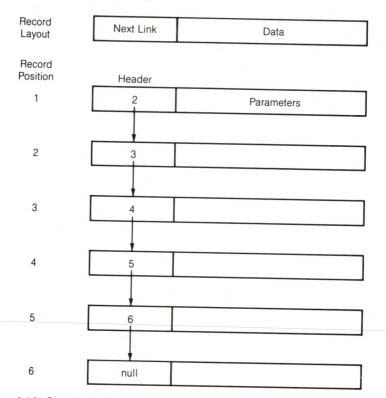

Figure 4.13 Simply Linked Records

contained within that field. At the same time, since the actual physical location is irrelevant to the structure being depicted*, it is generally omitted.

The values of the Next Link field within each record will be demonstrated for an initial example given in Figures 4.13 and 4.14. These two figures show how an additional record obtained from any location can be added logically between records 4 and 5 of the original file given in Figure 4.13. Suppose that this new record is at location 19. The new structure diagram will appear as in Figure 4.14. This demonstrates how the new record can be logically inserted between records 4 and 5 without being physically inserted there. In fact, the numbers 1 through 6 in Figures 4.13 and 4.14 could be replaced consistently (i.e., in all occurrences) with any other numbers without affecting the structure. Similarly, in Figure 4.14, record 19 could be depicted between records 4 and 5 without changing the structure, since the position of a record in the structure diagram bears

---

*They are also generally unknown to the applications programmer and constantly changing, as will be shown.

no relation to its physical location. Thus, Figure 4.14 depicts the same structure as Figure 4.13, only one more record is used to portray the structure.

Note that, although its basic objective is to communicate the necessary record linkages to the programmer, the structure diagram also enables the mechanics of logical record insertion to be visualized. Conceptually it is quite simple, as can be exemplified by Figures 4.13 and 4.14: (1) determine the logical position of the new record; (2) change the Next Link of the record, which is logically before the new record, to the location of the new record; and (3) change the Next Link of the new record to the location of the record logically following the new record. For the example given above these steps become: (1) logical position between 4 and 5; (2) change the Next Link of record 4 to a value of the new record, 19; and (3) change the Next Link of record 19 to the location of the record logically following, location 5. Finally, an

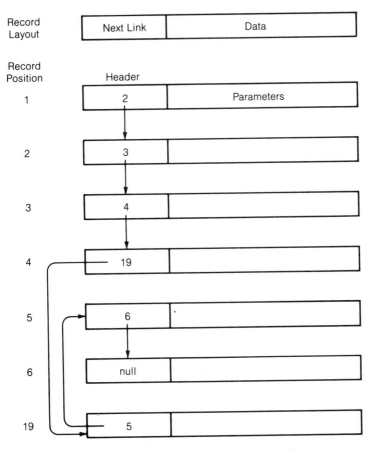

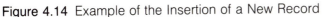

Figure 4.14 Example of the Insertion of a New Record

insertion procedure can be depicted by a series of structure diagrams to aid in communicating the structure of the direct-linked file organization to be maintained.

Three other basic concepts of structure diagrams need explanation before leaving this elementary example. The first is the existence of the record layout. Frequently records will require more than one link. For example, there might be a forward link to the next record in a sequence, a backward link for finding the previous record in that sequence, and another link field, possibly to a record in another file. Also, there might be several different sequences for which output (or other processing) is required, each of which will require link fields. Thus, a brief record layout diagram should appear at the top of each structure diagram, as given in Figures 4.13 and 4.14. This will tell the reader the names and relative positions of the various links, such that the structure diagram is readable without being cluttered with labels. The details of the data and nonkey fields need not be included in this brief layout.

A second concept requiring emphasis is that of the header record. Generally, this is the first record in a file which is read whenever the file is opened (although the presence of multiple header records is certainly acceptable). It then becomes the immediate linkage from the application program to the direct-linked file, since the values of the header-link field will tell the location of the first data record which the application program will consider. The header is unlike the other records in the file since it is an exception to the record layout in that it contains no application data. However, the header record will contain all of the link fields, and it should be the same length as all other records. Thus, to the computer the header is just another record in the file. However, in place of data, file parameters will be stored here. These may contain such information as: (1) the date and time of last update; (2) the location of the last record inserted, deleted, or modified; (3) general format information; (4) entry pointers to special records; and (5) any other parametric information related to the entire file. A final, very important use of the header record is in maintaining the integrity of the file when there are no data records in the file.

A third and final concept regarding direct-linked files is the use of the null character in a link field. This is shown in Figures 4.13 and 4.14. Accessing programs will recognize the null character as a signal that the sequence of records as specified by the links in a particular field is completed. Any other entry might be interpreted as a storage location. The null can generally be any "sacred" character which is uniquely recognizable as not being a storage location, i.e., it cannot be mistaken for a link. Generally, the Greek character *lambda* (Λ) is used to indicate the null character in data structures literature.

Hopefully, now that the basic concepts of linked lists have been established, an appreciation has been gained as to their value. These will be further discussed in the subsections which follow.

4.5.2 Multiply Linked Structures

Consider the data structure depicted in the structure diagram given in
Figure 4.15. This figure shows a record structure of two link fields, the
first of which designates the beginning of a list of data records. The
second link designates the location of a larger data record, usually in
another file. Each record in the linked list also contains key/order data,
which may be several independent variables that are used to determine
the linkage of the records at any given time. The purpose of such an
arrangement is to enable fast modifications, insertions, or deletions
with regard to the key/order data without having to read the other data,
as would be required if this were structured with the data as part of
the record as in Figure 4.14. The list is singly linked, in that access may
proceed in a forward direction only.

The Next Link field of the header record contains the location of
the first record in the chain of records. The data shown in each linked
record consists of key and other critical fields for linked-list ordering.
Any of these fields may be used as search arguments, or for maintaining

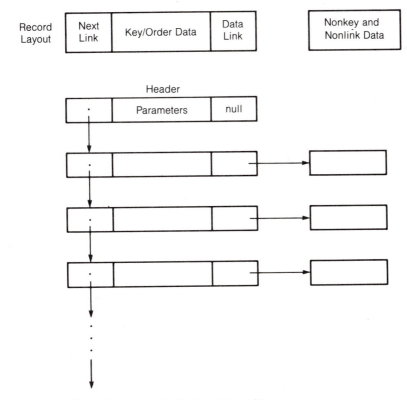

Figure 4.15 Two-dimensional Singly Linked List

Table 4.1   Comparison of Linked Data Record with Concatenated Data Record

	Linked Data Record	Concatenated Data Record
Access Time	—	Reduced
Maintenance Time	Superior	—
Security Potential	Improved	—
Ease of Modification	Improved	—
Implementation and Use	—	Easier to program
Utilization of External Storage	—	Better
Utilization of Internal Memory	Better	—

the linked list. For example, if the list were to be ordered by the date and time of a transaction, these would be included for purposes of maintaining the linked list. On the other hand, if the list were to be searched to obtain the records for a given customer, then customer number would be included within the key/order data. Several examples will be given below in Section 4.5.3.

Now consider the link in each record of the linked list, which is labeled Data Link in Figure 4.15. This is a record pointer which indicates the location of the data record that contains more information about this case. The contents of this data record, however, are considered augmentative in nature as far as the data structure is concerned. It is generally a relatively large amount of data, such as a customer history. The general idea is to keep track of this information without involving it in each pass of a key search retrieval process until it is specifically required. It can be shown that requiring a second access for the horizontally linked data record has cost/benefit trade-offs, with the alternative of concatenating that data with the linked list record. This would form a larger data record, eliminating the linked access to it. The factors to be considered are summarized in Table 4.1. Generally, the horizontally linked structure will be implemented when all or part of the linked list may be brought into internal memory and manipulated, modified, and relinked directly before being stored externally for future reference, given that this same operation would be impossible under the concatenated structure.

The structure discussed above, which is diagrammed in Figure 4.15, will form the basis for the access techniques to be introduced in the next subsection. However, there are obvious generalizations that can be made at this point. At times there are requirements for records to be linked so that a traversal can be made through several different paths. As simple examples, Figure 4.16 presents three example possibilities. Figure 4.16a shows a simple backward reference scheme using a previous (Prev) link. This is generalized to an additional third alternative path presented in Figure 4.16b, called an alternative (Alt) link. Such might be required, for example, if reports were required both in order by date and in order of customer number. Note also the multiple entry points which can be obtained directly from the header record. Finally, 4.16c shows that the data records of Figure 4.15 might also be linked to other records, some of which might share common left neighbors. Such might be motivated by a parts-explosion representation for an assembly operation where several different subcomponents may utilize the same parts.

In summary, these data structures are introduced to give the designer an appreciation for the various possible alternatives when specifying file design. Since file design is one of the most critical areas within the systems design process, serious consideration should be given to assuring that the proper structure is being employed before proceeding. Designers who do not have a solid background in this area would be well advised to consult with someone who has experience in the use of various structures.

4.5.3 Linked File Access Operations

An understanding of file access operations is required at this point so that direct-linked files can be more intelligently specified. The terms

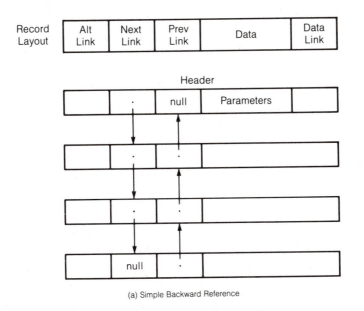

(a) Simple Backward Reference

Figure 4.16 Examples of Other Multidimensional Structures

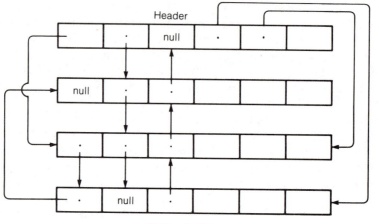

(b) Alternative Linking

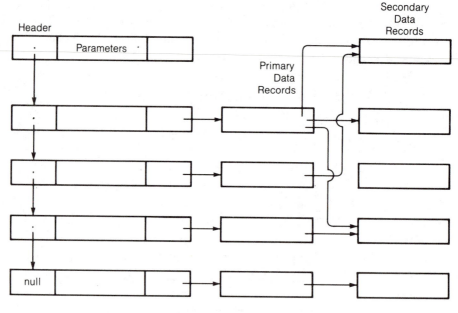

(c) Linkage to Other Records

Figure 4.16 Examples of Other Multidimensional Structures (continued)

used to name the operations are descriptive of the operations themselves, namely: (1) fetch, (2) insert, (3) update, and (4) delete. Before exploring the nature of these operations specifically with regard to direct-linked files, consider the access methods necessary to perform these access operations for the various file organizations described above. Table 4.2 presents this in summary fashion. To fetch a record from a simple- or sorted-sequential file requires sequential access since there

Table 4.2 File Access Requirements for Each Access Operation

	File Organization	
Access Operation	Simple-sequential Sorted-sequential	Indexed-sequential Direct-linked
Fetch	Sequential Access	Random Access
Insert	Sequential* Access	Random Access
Update	Sequential** Access	Random Access
Delete	Not Possible	Random Access

*Insertions are permitted only if the records are added to the end of the simple-sequential or sorted-sequential file.

**Updates-in-place are generally supported for direct-access devices such as disk; this is called a record "rewrite."

is no index or linkage established. Note that this does not preclude such files from being stored on direct-access devices; Table 4.2 is emphasizing the logical organization and the access method, not the hardware used. The fetch operation for keyed-sequential or direct-linked files generally requires random access, as do the other access operations. This is not to say that these file organizations cannot be accessed sequentially; indeed, this is preferable for some applications (such as file dumps). However, it is difficult to imagine an application in which the fetch operation would be performed using sequential access exclusively, since this would defeat the purpose of indexing or linking.

While random access is generally required for all keyed-sequential and direct-linked files, Table 4.2 indicates that simple- and sorted-sequential files are restricted to sequential access for the insert and update operations. This is further qualified, however, in that insertions are only allowed at the end of these files (technically, these are additions or augmentations). Further, updates can only be made if the hardware device employed is direct access, even though the accessing method is sequential. Finally, the delete operation is generally not possible for simple- or sorted-sequential.

With this review of old terms, and the introduction of the file access operation terms, it becomes clear that simple- and sorted-sequential files are quite limiting in their ability to be maintained easily without frequent copying of the entire file. This brings us to a point where a close examination of the file access operations for direct-linked files can be appreciated. Each of the four operations will be considered in the next four subsections.

4.5.3.1 Fetch

A fetch operation is employed when records are retrieved for informational display (perhaps as the result of a user query). A fetch is also

Relative Record Number	Status*	Key Data	Other Data	Other Link	Next Link
0	0	—			0003
1	1	—			0006
2	0	289			(null)
3	0	101			0004
4	0	105			0008
5	1	—			0007
6	1	—			0005
7	1	—			(null)
8	0	107			0002

*Note that a status of 1 indicates that the record is not being used for data and is thus available.

Figure 4.17 Example of a Direct-linked File

employed for field updates such as might be required to decrease a quantity-in-stock count in a parts record of an inventory file after a piece has been sold. This retrieval process has been described previously by Figure 4.3.

The following steps will accomplish a *serial search** through a file of linked records, based on a user specified search goal (key value) for a given key field:

1. Get the header record.
2. Chain through the linked list, comparing each key data field value with the search goal.
3. Terminate when the search goal is matched with a record key (successful completion), or when it is clear that the desired record does not exist (e.g., at end of file).

This simplified overview serves to convey a fetch strategy at the very highest level. Consider how the three steps above would be applied to Figure 4.17 to find the record with the value of 105 in its key data field. Note first that this file is linked in ascending order of the key data field. In the first step the file would be opened and the header record, in the relative position zero, would be read. The retrieval software will not check the header record but will immediately chain to the first data record, in this case, in relative position 3, as given in the Next Link

---

*Note the difference between a serial search, in which all of the records logically preceding the target record must be read, and a sequential access, in which all of the records physically preceding the target record must be read.

field. The key data value of record 0003 (101) is checked against the target value (105) and found to be not equal. Thus, the chaining procedure continues to relative position 0004, as given by the Next Link field of record 3. The key value of this next record is a value of 105, which is identical to the target value; thus, this record will be retrieved and processed further. If none of the key data values had matched, the null in record 2 would have been detected signaling the end of the file. Alternatively, since it is known that the linkage is in ascending order by key data field, the search could have been abandoned when a key data value exceeded the value of the search goal. In Figure 4.17 the values in the other data and other link fields have been omitted for clarity; these values would be present in a real application.

In the example above, the serial searching is performed on a direct-linked file wherein records are actually being *randomly* accessed. This is important to observe, since serial searching is sometimes called "sequential searching," a term which unfortunately confuses the concepts presented thus far.

4.5.3.2 Insert

The next file operation presented in Figure 4.15 is insert. A record insert occurs when a new record is to be stored between two existing records or added at the beginning or the end of the file. Inserting is often done when some order is being maintained among the records in a file. The specific nature of this order is momentarily unimportant, other than the fact that the records are linked by a key field(s) in some order which the insertion process must recognize and accommodate. The insertion steps are summarized as follows:

1. Fetch the record which is immediately *prior to* the desired point of insertion. (Note that this record points to the record which is immediately *behind* the desired point of insertion.)
2. Prepare the new insert record with the desired data and key values.
3. Move the value of the Next Link field from the "prior to" record fetched in step #1 to the new insert record.
4. Change the Next Link field of the "prior to" record to the new insert record and rewrite it.

This process was presented above in Figures 4.13 and 4.14 in terms of the original and resultant structure diagrams. For an additional example, consider how the actual values of the link field would be modified by inserting a record into the file given in Figure 4.17. Suppose that the new record value of the key data field is a value of 106. To maintain ordering according to this key, this will be inserted "between" those records with key values of 105 and 107. The file would be opened and the header record read according to the serial-search fetch procedure given above. This fetch will proceed by reading the value of the Next Link (0003), and then proceeding to record 3. Its key value is 101, which is less than 106. Reference to this record will be maintained (in main memory) while the next record, given by the Next Link value (0004) is checked. Record 4 has a key value of 105, which is also less than 106. Thus, the reference to record 3 can be replaced with that to record 4. Continuing, the next record to be read, as determined by the Next Link, is record 8. This record has a key data value of 107, which exceeds the search argument of 106. Thus, it is determined that the new record will be *logically* "placed" between records 4 and 8. Physically, the storage location of the new record does not matter. For the purposes of this example, let us arbitrarily place it in the first available storage location, which is in record 1. We have fetched the record (record 4) immediately prior to the desired point of insertion. The second step is to prepare the new record. Its status value will become 0, its key data field will become 106, and all other data fields will be set to their appropriate values. For step 3, the new record Next Link value will be set to 0008 (the value in the Next Link field of record 4). Finally, while record 4 is still available, its Next Link field value will be changed to the relative record number of the new record, 0001. The file after the insertion is given in Figure 4.18.

The insertion process just outlined only considers linkage dependent upon one key field. It should be clear that insertion with regard to several key fields follows the identical stepwise procedure. Also, if several link fields exist, there should be no problem in inserting the same record simultaneously in several logical "places" according to their respective defining key fields. This is left as an exercise for the student.

Relative Record Number	Status*	Key Data	Other Data	Other Link	Next Link
0	0	—			0003
1	0	106			0008
2	0	289			(null)
3	0	101			0004
4	0	105			0001
5	1	—			0007
6	1	—			0005
7	1	—			(null)
8	0	107			0002

*Note that a status of 1 indicates that the record is not being used for data and is thus available.

Figure 4.18 Example of a Direct-linked File After Insertion

The number of links in any particular record will depend on the number of keys being supported for a given direct-linked file as well as the necessity for backward referencing. The simple case of one key with forward direction access explains the presence of one link field. Should bidirectional file manipulation (i.e., forward and backward linkage) be desired, another pointer would have to be maintained. A small modification in Steps 4 and 5, required to support bidirectional linkage, should be easy to understand. Step 4 would be extended to point the newly inserted record back to the "prior to" record as well as the forward reference. Further, Step 5 would include a similar operation on the record behind, after which it too would be rewritten. This is also proposed as an exercise in the problems at the end of this Chapter.

4.5.3.3 Update

An update consists of fetching an existing record, changing the information in it, and then writing this record back to the file. If the size of the record does not change, and the file is on a direct-access device, generally an update-in-place will be allowed. This type of update consists of rewriting the modified record in its old location. Updating is quite useful, and this operation can be far less complex than the insert operation, as the following steps indicate:

1. Fetch the existing record upon which an update is to occur.
2. Modify the existing fields in this record.
3. Rewrite the modified record in place.

As long as none of the key fields for linkage determination are affected, this simple mechanism does not necessitate remeshing any pointers, either in the updated record or in other referenced records. However, it should be recognized that if a key field that is used to determine linkage is modified then a relinkage would have to take place, essentially by repeating the insert operation given above. For example, if a linkage were being maintained in order to list output according to priority, then whenever a user changed a key field affecting priority there would have to follow a modification of the linkage in order to maintain the priority listing.

Relinkage is also required in those situations where the size of the record changes, or where an update-in-place is not automatically supported on the system being used. In these cases, the following more general procedure must be followed in order to accomplish an update:

1. Fetch the existing record upon which updates are to be made.
2. Rewrite all fields, including link fields of this record to a new record, with the exception of those fields to be modified.

3. Write the new values of those fields to be modified to the new record, including link fields.
4. Set the link fields of any records that referenced the original record to the location of the new record.

Step 4 above can become exceedingly complicated unless there is a backward reference from each record to every record which references it. If not, the process will require a complete search of all link fields to determine those which reference the modified records. For certain applications this can result in an extremely large amount of thrashing. This shows the advantages of both (1) a strategy which leads to rewriting in place, and (2) the separation of the fixed-length (possibly the link fields and their corresponding keys) from the variable-length portions of the record (see Figure 4.15).

4.5.3.4 Delete

File access operations may be concluded with a description of the delete process. Here, an existing record is to be logically removed from the file as the following steps describe:

1. Fetch the record to be deleted.
2. Mark the record status as "empty" and rewrite it.
3. Update any records that point to the deleted records by pointing the "prior to" record's link field to the record that logically follows the deleted record.

The old record will now be available for reuse, and the file will essentially ignore it until it is used again.

4.5.3.5 Closure on File Access

A summary of these file operations will point out some additional considerations. First, it is best to consider these file operations as complete logical units. Although they may reference one another (e.g., UPDATE references FETCH), they should do so only within the bounds of one iteration (or nesting) of the original operation. This means that during a file operation, the designer and programmers should be aware that this operation should be considered an essentially uninterruptable logic path. It should be clear that the attempt to perform several access operations simultaneously could have unpredictable results.

It is appropriate, now that linked file access operations have been discussed, to consider the trade-offs involved in introducing an additional link into a file. Some key fields are used purely for retrieval as opposed to ordering. However, every key field is a candidate for the establishment of a link field. Thus, designers who are overcommitted to the idea of using links can create some very complicated structures if the number of key fields is large (greater than three or four).

The alternatives to linkage include indexing and sorting. Indexing was discussed in Section 4.2.2, and it has its own linkage and maintenance ramifications. Sorting emulates a next link with respect to the key fields used to specify the sort. This "linkage" costs nothing in storage space, since it can be exploited by sequential file access. A decision on whether to sort before output or to maintain a linkage is a function of the following: (1) availability of fast-sort utilities, (2) availability of central storage, (3) speed of accessing required, and (4) availability of temporary storage space. The cost of sorting is definable in terms of the CPU time consumed, and the main and peripheral storage required.

The cost of linkage is also definable, and an analysis of the costs is useful in resolving this design decision. To calculate the cost in terms of physical space, multiply the size of the link fields times the number of link fields times the number of records. Processing time cost is more difficult to calculate. The addition of one extra link is only of serious concern if updates to key fields are very frequent and they must be performed quickly. Usually some experience is required on the hardware-operating-system configuration to determine the tolerable limits. Here it is recommended that actual trials be made on dummy sets of data to determine access times. This can be performed at little cost since real data are not required; rather, the key and link fields can be emulated, the remainder of the record being zero-filled. A few trials of this type will quickly show the feasibility of a given structure, and thus save a large amount of wasted effort in mistaken file specification and false starts.

While storage space and linkage maintenance time are significant factors, programming effort might have the largest impact upon cost, especially in efforts where the applications programmer does not have sort or access utilities. The expansion of the amount of applications code tends to produce a disproportionately larger number of errors. Further, locally written sort and access routines usually lack the efficiency of utilities designed and written by data structure specialists. Thus, while multiple links might seem, on the surface, to solve all access problems, they also require tolls in terms of storage, processing, and development. Judicious use of links is therefore recommended.

Finally, the procedures given above for the various file operations were presented to provide the designer with the concepts of maintaining direct-linked files. The considerations commonly called *boundary conditions* have been omitted. These include many special cases, such as: (1) no room left in the file, (2) search argument equal to the key value, (3) insert to the end or the beginning of the file, (4) no records in the file, and (5) a host of other seemingly trivial yet inevitable exceptions. Unfortunately, the detailed consideration of all of these boundary conditions is beyond the scope of the current discussion, and the reader

FILE ORGANIZATION	SEARCH METHOD	
	SERIAL	BINARY
Simple-Sequential	Yes	No
Sorted-Sequential	Yes	Yes
Keyed-Sequential	Yes	Yes
Direct-Linked	Yes	No
ACCESS METHODS		
Sequential	Yes	No*
Random	Yes	Yes
*An exception exists for the sequential representation of the binary tree.		

Figure 4.19 Search Methods by File Organization

is referred to the many excellent texts which exist, some of which have been listed at the end of this chapter. However, the many possible complications in programming caused by boundary conditions, some of which are not readily anticipated, are given emphasis here so that they will not be lightly discarded. Indeed, quite often the majority of the programming effort must be dedicated to error checking and boundary condition resolution. For this reason it is strongly advised to use standard time-proven packages to maintain data structures wherever possible. Where it is not possible, do not underestimate the effort required to handle all contingencies in even the simplest of structures.

4.6 File Search Techniques

Once information has been stored in a file it obviously must be retrieved (fetched) for straightforward query or complex update processes. Two techniques of searching through a file will be presented next: serial searching and binary searching. Figure 4.19 relates these search techniques to the file organizations to which these may be applied. Finally, an introduction will be provided to the more sophisticated concept of tree searching.

4.6.1 Serial Search

The serial search boils down to little more than systematic reading of all the records in a file until the desired record is found. Since simple-sequential and sorted-sequential files have no assisting index or linkage, serial searching is often the only means to access the records. However, exhaustive serial searches are required only for simple-sequential files.

Generally, sorted-sequential files can take advantage of the record ordering to eliminate about half of the searching. To understand this search notion, recall that sorted-sequential files are ordered by some key field; therefore (assuming key value uniqueness), every record has a specific location within the file relative to every other record. This fact can be used to shorten the duration of a serial search through a sorted-sequential file, since (in almost all cases) every record need not be fetched and compared with the goal of the search. The only case where this is not true is when the desired record is the last one in the file or has a key value greater than the last one in the file. This guideline pertains to the average number of searches required for a file containing key values which are evenly distributed or otherwise correlated with search goals.

Two serial-search examples are given in conjunction with Figure 4.20 which show the merit of searching through a sorted-sequential file as compared with searching through a simple-sequential file. Suppose

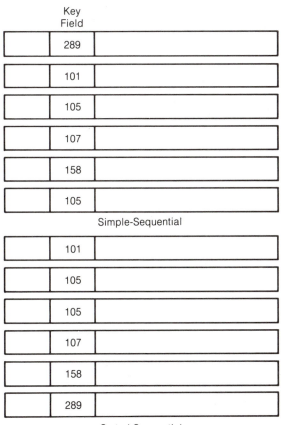

Figure 4.20 Example of Simple-sequential and Sorted-sequential Files

that the record which was specified had a target key value of 104. To verify that it is not in the simple-sequential file, all of the records would have to be read and their key values compared to the target value. However, with the sorted-sequential file, searching could stop once a record key field value exceeded the target key value. Assuming a random distribution of both key values and search arguments, on the average, half of the file would require search in both cases given that the target value exists in one record. However, if the key value does not exist in any record, all of the simple-sequential file would need to be read while, on the average, only half of the sorted-sequential file would require reading.

It will be shown below that the sorted-sequential file also has advantages in performing a binary search. For certain applications, sorting can also lead to tremendous benefits in facilitating output, since rarely are random or entry-sequenced output listings required. Finally, the sorted-sequential arrangement might facilitate other types of processing, such as the identification of record subsets which have identical key values. Figure 4.20 illustrates how easy it would be to do this type of search using the sorted-sequential as opposed to the simple-sequential organization. An example application is the determination of all accident records by location, which is performed quite easily given a sort by location (see Figure 2.20).

Serial searching can be used to locate particular records in any file organization, as indicated by Figure 4.19. However, records in sorted-sequential files can exist only in one order, while direct-linked file rec-

RRN	Status	Order Number	Due Date	Link 1	Link 2
0000	0	– – –	0002	0001	0003
0001	0	155	04/01	0003	0004
0002	1	–	–	0007	– –
0003	0	201	04/01	0016	0001
0004	0	203	04/15	0006	0006
0005	0	404	05/11	0012	0013
0006	0	208	04/15	0009	0009
0007	1	–	–	0008	– –
0008	1	–	–	0010	– –
0009	0	216	05/01	0013	0005
0010	1	–	–	0015	– –
0011	0	320	06/01	0005	0012
0012	0	503	06/01	(null)	0016
0013	0	310	05/15	0014	0011
0014	0	312	06/17	0011	(null)
0015	1	–	–	(null)	– –
0016	0	202	06/10	0004	0014

Figure 4.21 Double Link Search Example

ords may be linked together in several ways according to several keys. Therefore, in a direct-linked file, searching can take place along any path of linkage defined for any key. To illustrate this with an example, consider Figure 4.21. Here there are two key fields; namely, Order Number and Due Date. They are used to link the records together in ascending order using Link 1 field and Link 2 field, respectively. To search for the record with order number 203, the sequence of the search would be: Header, 0001, 0003, 0016, *0004*. To search for the record with due date 05/01, the sequence of the search would be: Header, 0003, 0001, 0004, 0006, *0009*. In the case of due date, several records may qualify, so provision would have to be made within the software to continue searching until all target records are found.

In Figure 4.21, serial searching can take place on the direct-linked file shown, either by Order Number or Due Date fields. Random access provides successive logical records in either chain of linkage, and serial searching enables the search to quickly terminate when the goal does not exist in the file. As a final illustration from Figure 4.21, note that the header has a value of 0002 in the Due Date field. As stated above, this value is not data in the header record; rather, the space is used to store parameters of the file. In this case the parameter is the first free (empty) record. Note further that all of the remaining empty records are linked together by the Link 1 field. This gives the software access to all empty records, and thus makes the status field unnecessary.

4.6.2 Binary Search

Serial searching, even on a limited basis, has definite performance restrictions. When files are larger than roughly 50–100* records, binary search techniques generally begin to show significant improvements over serial techniques. The binary search is only useful on random-access files that contain ordered information, particularly in keyed-sequential files. These files provide for random retrieval by key goal, a feature the other file organizations cannot accommodate without serial searching. The reader should recall the components of this file, which are the index and data partitions (see Figure 4.2). The index is actually a small sequential file which is constantly kept in order according to the key field. Thus, while data records are put in any empty slot in the data area, the corresponding index entry is put in the physical order of the key field. The index, then, is constantly being rearranged and perhaps rewritten (entirely or partially) as update activities occur in the file.

Unlike link-field maintenance in a direct-linked file, which affects only a couple of records per update, keyed-sequential file index main-

---

*The reader should note that this record-count estimate should be tempered by factors including size of each record, intensity of file use, and specific performance objectives.

tenance has more global implications. However, this organization provides a large search advantage, since the index is always a completely accurate directory of information contained in the file. The combination of this accurate map and random-access capability make the binary search the fastest way to access the keyed-sequential file.

The understanding of the binary search process begins with the recognition that it is used intuitively by most people in searching through ordered information. For example, in a phone book or a dictionary we turn somewhere to the inside of the book, check to see what the respective information (record key value) is, and then proceed either forward or backward (binary) to another page, depending upon whether the information on the page was greater than or less than our target. This process is repeated until the target information is found. In the process of one search, at least half of the remaining candidate elements can be eliminated. The advantage of the binary search over a sequential search is obvious, and this is the reason few, if any, would use a sequential search to look up a word in the dictionary.

Of course, this process must be structured more formally for computerization. Note once again the prerequisites for the file: direct-access capability and search-key ordering, either by physical record arrangement or by links. Given this structure the procedure for performing a binary search is as follows:

1. Open the file and determine the logical "middle" of the file with respect to the search key. This will be approximately "half way through" the file, logically speaking.*
2. Compare the key field of the center record against the target value sought. If they match, the record has been found and the search can be terminated; if not, continue to Step 3.
3. If the key field value is not equal to the target value, determine the next record to use for comparison. If the key field value is greater than the search target, discard the greater half of the undiscarded records and fetch the middle record of the undiscarded subset of records. If the key field value is less than the search target, discard the lesser half of the undiscarded records and fetch the middle record of the undiscarded subset.
4. Proceed to Step 2, where the center record is now that of the undiscarded subset of records.

The power of a binary search can be expressed in terms of the maximum number of comparisons required to search a file for a given goal.

---

*In the sorted-sequential arrangement the logical and physical distinctions vanish. These distinctions will not be made in further steps. Note that if the exact center is not produced by a direct division, consistent rounding will be employed to determine the location of the approximate center record.

This is equal to the base two logarithm of the number of index records (with fractions rounded up to the next higher integer). For example, the maximum number of comparisons within a file containing 128 records is seven, while a file containing a million records would require, at most, twenty comparisons. This can be compared with the previous maximum for the serial search (i.e., the number of index records). Clearly, the serial method is the least effective means to search for information by key comparison, while it can be shown that binary searching is the best.*

To conclude this section and demonstrate the potential value of careful data structuring, consider what has been called a sequential representation of a binary tree. The file presented in Figure 4.22 will be used to first perform an example binary search and then to show how this same thing can be performed without actually computing the midpoints of each "undiscarded" subset of records. To illustrate the search as outlined above, suppose the record which has a key field value of S is to be retrieved. This would take 11 searches sequentially. The binary search would begin by examining record 008 which has a key value of M. Since S is greater than M, the upper half of the file is discarded and the midpoint of the undiscarded subset is record 012, which has a key value of T. Comparing this with the target value of S leads to a discarding of the lower half of the undiscarded subset of records. The midpoint of the undiscarded remaining subset of records is record 010, which has a key field value of R. This leads to the discarding of the upper portion and the final retrieval of the sought record. Four searches were required (the maximum for a file of 15 records).

Now suppose that the data within Figure 4.22a were stable and needed to be subjected to a large number of searches, as would be the case in a customer file which is used constantly to satisfy customer and internal inquiries. Is there some way that this data could be organized such that these searches need not have to use division to find and access the "middle record" on each iteration? For the answer, consider the linking arrangement given in Figure 4.22a. Here the low link would be used in those cases where the target is less than the key field, while the high link would be employed when the target is greater than the key field. The graph for this structure is called a binary tree, shown in Figure 4.22b.

Now consider one final innovation which might further facilitate the process. Consider the rearrangement of the key field values of Figure 4.22a into that given in Figure 4.23. This arrangement is obtained by listing the middle record (008) first, followed by the two links which it generates (004 and 012), and then recursively listing the links for each

---

*Knuth, Donald E. *The Art of Computer Programming, Volume 3: Sorting and Searching.* Reading, Mass.: Addison-Wesley Publishing Co., 1973, page 411.

Relative Record Number			Key Field	Low Link	High Link
000	Header Record			008	008
001			A	null	null
002			B	001	003
003			D	null	null
004	Data		F	002	006
005		Fields	I	null	null
006			J	005	007
007			L	null	null
008			M	004	012
009			Q	null	null
010			R	009	011
011			S	null	null
012			T	010	014
013			U	null	null
014			V	013	015
015			W	null	null

(a) Example of a Sorted File for Binary Search

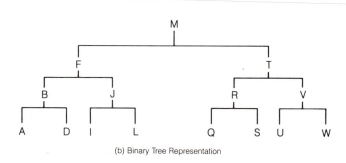

(b) Binary Tree Representation

Figure 4.22 File and Tree Structures for Binary Search

of the elements generated in order until all records are represented as in Figure 4.22b. Go down Figure 4.23 systematically and confirm that the ordering is unique by referring to the data link column, which is the original relative record number. Note that the center is record 008, which produced records 004 and 012, 004 generates 002 and 006, and 012 generates 010 and 014. Now go back up the list to 002, which is the next to be used for generation and which generates 001 and 003. This process of generation continues as 006 generates 005 and 007, 010 generates 009 and 011, and 014 generates 013 and 015. Note that this ordering, which is unique, is essential. Finally, the easiest way to manually produce this arrangement is to perform a left-to-right traversal of the binary tree given in Figure 4.22b.

What is so special about this new generated ordering? With it, a binary search can be performed sequentially, i.e., without ever having

Relative Record Number	Key Field	Data Link*
001	M	008
002	F	004
003	T	012
004	B	002
005	J	006
006	R	010
007	V	014
008	A	001
009	D	003
010	I	005
011	L	007
012	Q	009
013	S	011
014	U	013
015	W	015

*This is the value of the relative record number in Figure 4.22.

Figure 4.23 Example of a Sequential Binary Search Argument

to "back up" to a record. Further, the accessing can be performed without links and without determining the mid-record of the undiscarded subsets. To see this, note that the first record in Figure 4.23 is the first to be evaluated in the binary search (i.e., it is the middle record of the original file shown at the top of the tree). Let the relative record position of the sequential binary search arrangement be given by i, (at this point i = 1), and let N be the number of records in the file. The binary search will continue as follows:

1. If i > N, end the search with no record found. Otherwise, compare the key field value of relative record i with the search goal. If they match, stop and retrieve record i and end, having found the desired record.
2. If the key field value is less than the target value, reset the value of i to be equal to 2 times the current value of i and proceed to Step 1; otherwise, continue to the next step.
3. Since the key field value is greater than the target value, reset the value of i to be equal to 2 times the original value of i plus 1 (i.e., 2i + 1) and proceed to step 1.

This procedure has all of the advantages of accessing by computation (as opposed to looking up the location in a table). Further, since each subsequent search is performed on a record with greater relative record number than the previous search, a sequentially accessed file could be employed. Even if direct-access is available, the advantages of such an arrangement should still be obvious. However, no approach has all the advantages. It is important to recognize the disadvantages inherent in the structure above. Namely, it requires considerable setup. This is no big problem if the file is stable, i.e., if it is not being frequently updated with additions, deletions, and changes to the *key field* value. However, for every such change, the entire reordering process of Figure 4.23 would have to be performed.

Before leaving Figure 4.23, note that the original relative record number, which was used to show how this relates back to Figure 4.22, was also called a Data Link to imply its further use. While considerable rearrangement may be necessary to produce a file that can be subjected to binary search, it is not necessary to rearrange the nonkey fields. Rather, they can be broken out into a separate file as was done in the example transition from Figure 4.22 to Figure 4.23. Now the data link in Figure 4.23 provides the linkage for retrieving the data. Thus, the binary search can be performed on the file given in Figure 4.23 and the data retrieved from another file, as given in Figure 4.22. For example, if the record with the key value of Q were to be obtained the sequence of the binary search would be 001→003→006→012 in Figure 4.23. Then, using the data link value for record 012 (which is 009), the data for this record could be retrieved from the file in Figure 4.22.

In concluding this section the advantages of sequential access should be emphasized. Designers and programmers often get infatuated with the use of links to such an extent that all other alternative structures are ignored. Often the reordering of output can be more easily accomplished by setting up a temporary file and performing a utility sort on it prior to output. Sequential access is often much faster and easier than nonsequential methods since no links need be read, compared, or maintained. Thus, the designer should structure the files to take advantage of this implied ordinal link whenever possible. Search techniques given in conjunction with the example of Figure 4.22 would favor the first for sequential files sorted on key field. However, if the file could not be sorted by a unique key field, e.g., in the case of multiple search keys, the linked arrangement would become essential.

4.6.3 Closure on File Search Techniques

The previous discussion on file search techniques was given for two reasons. First, it provided an introduction to those who have not studied file search techniques. Rather than detailing all of the methods and complications of these techniques, the discussion remained on a general level to convey an overall appreciation for the larger field of study. The reader to whom this material is an introduction should obtain a broader background in this area through additional study. A second objective, directed at those who have a broad-based background in this area, was to provide a brief review of some of the basic concepts and to describe terminology for continuity with the remainder of the text. Thus, some practical applications to the principles that are usually studied on a more theoretical basis have been shown. In both cases, it is hoped that this presentation has further enhanced the capability of the designer to communicate with the data-structures specialist in order to develop the best possible file design.

4.7 File Type Selection

After discussing the details of the various types of files and their accessing techniques, it is essential to address the mechanism for putting this information into practice. In other words, how can this body of information and experience be translated into design specifications that can be used by the programmers to bring the intended system into existence? This section will attempt to answer this question based upon the information presented above, and thus it will serve as a review. The next chapter provides the means for documenting these results. Note that the term *file type* includes the general considerations of format, organization, and access.

Like any other decision in the design specification process, file-type selection should be formulated in terms of alternatives which are available. Usually, by eliminating all of the unfeasible or obviously bad alternatives for a given application, the total number of alternatives is sufficiently reduced so that detailed study can be performed upon those that remain. Table 4.3 presents the alternatives in terms of the various file-type considerations of: (1) hardware, (2) sort/organization, (3) search technique, (4) access method, and (5) record format. These have been presented in detail above, along with the alternatives presented in the second column. Note that these alternatives are not exhaustive; rather, they reflect the scope of this chapter. However, they will cover the field adequately enough to enable the methodology given below to be exemplified. That is, if additional alternatives are introduced, then the methodology for selecting among them will not be significantly altered. Note also that the alternatives given are not necessarily mutually exclusive either within or between the various considerations.

Enumerating and understanding the meaning of each alternative is the first step in determining the file type to specify. Once these are determined, the solution is defined by selecting the alternative within each consideration which will best accomplish the objectives of the system. This selection is based upon a consideration of the constraints and requirements of the system. The following constraints and requirements will be discussed in order below:

1. Hardware,
2. System output requirements,

3. Data input and maintenance, and
4. Nature of data input records.

Hardware considerations must come first, since it is obvious that structures requiring random-access equipment cannot be implemented if the necessary hardware is unavailable. With the continuously declining price of direct-access disk drives, the cost of obtaining the necessary equipment is getting to be much less of a problem. However, the addition of more disk drives may not solve the problem. Depending on the central processing unit (CPU), there may be a limit as to the number of disk drives that can be added. The problem of configuring the hardware to accommodate the system design is primarily one of comparing the alternative from an economic point of view. For some extremely large files, which need only be accessed a few times to produce standardized reports, the use of magnetic tape is preferred, necessitating sequential access. Further, if output requirements are such that intensive demands are made on this file, but only for a short duration, the solution may lie in temporarily loading the file on disk during this demand period, but permanently storing the file on tape. Similarly, backup files are usually stored on tape rather than consuming valuable fixed direct-access-type storage. On the other hand, the declining price of memory and its availability in larger quantities on some computers certainly tempts one to load heavily accessed information into large

Table 4.3   Alternative File Specifications

File Type Consideration	Alternative*
1. Hardware	1. Sequential Access
	2. Random Access
2. File Organization	1. Simple-sequential
	2. Sorted-sequential
	3. Keyed-sequential
	4. Direct-linked
3. Search Technique	1. Serial
	2. Binary
4. Access Method	1. Sequential
	2. Random
5. Record Format	1. Fixed
	2. Variable
	3. Undefined

*Alternatives are locally ordered from the less sophisticated (the minimum) to the most sophisticated (the maximum).

Table 4.4   Alternatives Necessitated by Output Requirements

Output Requirements	Minimal Specifications		
	File Organization	Search Technique	Access Method
REPORTS			
Infrequent, and in no defined order	Simple-sequential	Serial	Sequential
Few, all in the same fixed order	Sorted-sequential	Serial	Sequential
Several in a variety of orderings	Keyed-sequential or Direct-linked	Serial	Random
QUERIES			
Few queries with slow response	Simple-sequential	Serial	Sequential
Few queries with faster response	Sorted-sequential	Serial	Sequential
Many queries with slow response	Direct-linked	Serial	Random
Many queries with faster response	Direct-linked or Keyed-sequential	Binary Serial	Random Random
Many queries with fastest response	Keyed-sequential	Binary	Random

arrays or structures, especially in virtual storage systems. Computer aided design (CAD) models, for example, must be immediately available for update and display during the actual design process; therefore, they should always reside in main storage.

For the remaining discussion in this section, let us assume that random-access hardware constraints are not a problem, or if so, that further economic analysis will be performed to determine if the acquisition of additional hardware is required. This will free us to explore the other alternatives, some of which are impossible if random access is unavailable.

The second consideration follows the general philosophy of design which requires output to be defined first. Thus, once free of hardware constraints, output requirements become paramount, specifically that subset of the output requirements which must be somehow satisfied by the file under design consideration. Table 4.4 presents some output requirements in terms of a qualitative overview of possibilities. These output requirements are given in terms of the categories of report and

Table 4.5   Alternatives Indicated by Input Requirements

Input and Maintenance Requirements	Maximal Recommendations		
	File Organization	Search Technique	Access Method
Relatively stable after setup	Direct-linked or Keyed-sequential	Binary	Random
Few periodic updates	Direct-linked or Keyed-sequential	Serial	Random
Frequent updates but not to key fields	Direct-linked or Keyed-sequential	Binary	Random
Frequent updates involving key fields	Direct-linked, Keyed-sequential, or Sorted-sequential	Serial	Random

query types which the file in question might be required to support. Any given file should be classified into the most stringent requirements which will be placed upon it. Obviously all possible output requirement categories are not listed; however, the ones given tend to cover the field. In addition, note that the specifications given are *minimal* specifications. That is, in addition to the one given, any of the specifications below that in Table 4.3 might also be appropriate.

The determination of the exact specification cannot be based solely upon the system output requirements. Closely integrated into this decision are the data input and maintenance considerations. Table 4.5 presents a guide to the specification of file type in terms of file input and maintenance requirements. Note that, unlike Table 4.4, these are *maximal* specifications. Obviously some compromise must be reached in obtaining the optimal balance between usability in response to output requirements and maintainability in response to input and file modifications. Tables 4.4 and 4.5, in conjunction with the alternatives stated in Table 4.3, provide a guide toward the specification of file type. These should provide for most design projects. However, recognize that alternatives and combinations of alternatives exist which have not been discussed because they are beyond the scope of this text. The designer is urged to develop a broad background in both data structures and data-base management systems in order to cope with this continuously advancing field. If this is not possible for the design project at hand, the designer should consult with someone who has such expertise.

A final consideration involves the record formats, which are primarily determined by the nature of the units of data which are collected. These were discussed in detail in Section 4.3 and summarized in Figure 4.10. In short, if the file record content is amenable to a fixed-format record without wasted space or loss of data, by all means use it. There also might be times when a small loss of insignificant data or wasted space is of such little concern that a variable-length record is not justified. However, when the record structure leads to variable repetitions of fields, and when both space and data completeness are at a premium, the variable-length record is mandated. Also, in cases where access demands will allow the special routines required to handle it, the undefined format might be considered.

References

Bradley, James. *File and Data Base Techniques*. New York: Holt, Rinehart and Winston, 1982.

Hubbard, George V. *Computer-Assisted Data Base Design*. Florence, Ky.: Van Nostrand Reinhold, 1981.

Knuth, Donald E. *The Art of Computer Programming*. (Several Volumes.) Reading, Mass.: Addison-Wesley Publishing Co., 1973.

Maurer, Hermann A., and Camille C. Price. *Data Structures and Programming Techniques*. Englewood Cliffs: Prentice-Hall, 1977.

Pfaltz, John L. *Computer Data Structures*. New York: McGraw-Hill Book Company, 1977.

Reingold, Edward M., and Wilfred J. Hansen. *Data Structures*. Boston: Little, Brown and Company, 1983.

Questions and Problems

1. What are: (1) single-process access, and (2) direct-access storage media? Why is a concentration upon these desirable for introductory purposes? Explain any limitations.

2. What are the two activities involved in documenting file specifications? What is the basis for the first of these? What is the primary emphasis of the first activity?

3. How do file content specifications depend upon the structure and organization of the file?

4. State the rules regarding redundancy.

5. Define physical record and logical record.

6. What are the primary differences between a keyed-sequential, a sorted-sequential, and a simple-sequential file?

7. Define what is meant by a key field and a key value. Why would a record have several key fields?

8. How many reads would it take, on the average, to find a record in a simple-sequential file, as opposed to the three other file organization types?

9. What ordering exists in the two components of an keyed-sequential file?

10. What is the advantage of using an empty/nonempty flag? What are some of the other methods for identifying empty records?

11. Why is index searching more efficient than serial searching the data records? Outline the process employed.

12. Give an example of an index which has key values that can be the subscript of an array.

13. Present an application which lends itself to the use of block anchors.

14. Define *inverted file,* and give an example of a completely inverted file.

15. Give the advantages and disadvantages of the three strategies for organizing overflow areas. How can these disadvantages be overcome if computer time is available?

16. What are the advantages and disadvantages of: (1) fixed-length record files, (2) variable-length record files, and (3) undefined-length record files?

17. Give the record layout, as in Figure 4.7, of a work-order file with records of the following characteristics:

   a. The work order is divided into four parts: a header, a craft code section, a materials section, and an equipment section.
   b. The header gives overall descriptive information (character string) and 7 other variables generally defining the type and location. It also tells the number of crafts, materials, and different pieces of equipment.
   c. The craft portion contains two variables pertaining to each craft and a third that indicates the number of craftsmen within that craft which are to be employed for that job. There is also a record for each craftsman that indicates time and activity requirements (5 variables).
   d. For each material required there is a record which includes a description, the amount, the inventory number, and its status (on hand, waiting receipt, waiting purchase).
   e. For each piece of equipment there is a record which includes a description, the equipment reference number, and its current location.

18. Set up the hierarchy tree (as in Figure 4.8) for the solution to problem 17.

19. Describe the two methods of formatting an undefined record file.

20. What major changes have occurred in job streams during the last decade, and what are the reasons for these changes?

21. How has hardware been a determining factor in file organization and access?

22. State clearly and definitively the difference between file-organization and file-access techniques. Why are the two so often confused?

23. Define each of the four basic file-organization techniques.

24. State and define the two basic file-access methods.

25. Which file-organization types and file-access methods are incompatible? Why?

26. On the average, how many records must be searched to find a target search key (assuming uniformly distributed record key and target values) under the following conditions: (1) simple-sequential, key record in file; (2) simple-sequential, key record not in file; (3) sorted-sequential, key record in file; and (4) sorted-sequential, key record not in file? (Note: key record = record containing target key value.)

27. Name the components of a structure diagram and illustrate them by drawing a structure diagram for a doubly linked list.

28. Define data structure and give some examples of various data structures.

29. Set up the indexes for the data in Table E4.1 such that the following records can be accessed directly in a keyed-sequential type organization:

    a. key 1 = 8, key 2 = 99, or key 3 = 19;
    b. key 1 = 25 or key 2 = 25;
    c. key 1 + key 2 < 20;
    d. key 1 + key 2 < 25 and key 3 < 30.

30. Draw the structure diagram for the following conditions: (a) key 1 ascending, (b) key 1 descending, (c) key 1 ascending and key 2 descending, and (d) concatenated (key 1 and key 2) descending and key 3 descending. (Note: the structure diagrams should be independent of any examples.)

Table E4.1   Data for Problems

Relative Record Position	Key 1	Key 2	Key 3	Data
0001	3	18	1	
0002	8	3	19	
0003	15	25	14	
0004	2	99	25	
0005	9	6	32	

31. Using the example given in Table E4.1, show the values of the link fields necessary to implement the structures of problem 30.

32. Given the structure of problem 30(c), give a stepwise procedure for inserting a record. Exemplify this using Table E4.1 and the results of problem 31 by inserting a record with values of key 1 = 10 and key 2 = 4.

33. Draw the structure diagram for key 1 doubly linked, ascending and descending. Exemplify this with Table E4.1 data. Show the stepwise procedure for inserting a record, and exemplify it by inserting a record with key 1 = 6 into the example.

34. State *all* ways (at least five) that the header record differs from a data record, both in content, format, and purpose. Give examples of parametric information that can be stored in the header record.

35. How does an application program know that it has arrived at the end of a linked list?

36. List the costs associated with adding one more link to a direct-linked structure. How does this cost compare with adding another index to a keyed-sequential file? In answering, state the different purposes that each would fulfill. Argue for and against these additions as would be done in formulating the decision.

37. Draw the structure diagram for a simply linked list. Also draw a spare record. Show graphically how the links are modified during the

Table E4.2   Example
Ordered Index

Relative Record Position	Key 1	Data Link
001	A	003
002	C	007
003	E	005
004	G	012
005	J	001
006	K	010
007	M	008
008	N	014
009	P	002
010	R	015
011	T	004
012	U	013
013	W	006
014	X	011
015	Y	009

following access operations: (a) fetch, (b) insert, (c) update nonkey data, (d) update key data field, and (e) delete.

38. Using the index given in Table E4.2, generate records in the original data file which it references. Let the records have the following format:

| Key 1 | Data |

Assign each its relative record position.

39. What sequence of records from E4.2 would be read to "find" the record with the key value of: (a) A, (b) J, (c) U, (d) X, and (e) Y?

40. Set up a low link and a high link as was done in Figure 4.22 to facilitate a binary search for Table E4.2.

41. Transform the results of problem 40 such that this index can be accessed sequentially by computation.

42. Add the following records to Table E4.2:

Key 1	Relative Record Location
F	016
B	017
I	018
S	019
Z	020
L	021
Q	022

Modify the linkage established in problem 40 accordingly.

43. Perform the transformation on the results of problem 42 as was required in problem 41. (NOTE: In order to do this, additional dummy elements must be added to make the total number of elements an even power of 2.)

44. *Case Study:* Multilevel Distribution System. Consider one aspect of the accounting system for a multilevel distribution system. These have become quite popular and several sound companies employ this approach (it is not a pyramid scheme, which is illegal). In this organization each distributor has the right to sign up distributors under him, and his distributors can do likewise. He receives commissions on his own sales as well as those in his part of the organization, according to the following schedule:

1. Personal Sales   25%
2. First-Level Distributors   10%
3. Second-Level Distributors   5%
4. Third-Level Distributors   2%
5. Fourth-Level Distributors   1%

Any given individual only receives commissions four-deep, i.e., he does not receive any commissions on distributors signed up by the fifth level under him. Figure E4.1 illustrates the organization level structure with an example for Distributor #10. Note that Distributor #10 signed up #11 and #12; #11 signed up #13, who in turn signed up #15 and #16, etc. Table E4.3 illustrates the concept of levels with regard to the discussion above. For example, Distributor #10 would receive a commission of 10% of #11's sales, 5% of #13's sales, 2% of #15's sales, 1% of #19's sales, but nothing from #23's sales or from anyone that #23 would sign up.

Orders come in from the field on a daily basis, as do requests to add new distributors. Orders for goods contain the distributor number

Table E4.3   Level Structure of Figure 4.1

Distributor Number	Subordinate's Distributor No.	Number of Levels Down
10	11	1
10	12	1
10	13	2
10	14	2
10	15	3
10	16	3
10	17	3
10	18	3
10	19	4
10	20	4
10	21	4
10	22	4
10	23	5
10	24	5
10	25	5
10	26	5
10	27	5
10	28	5
10	29	5
10	30	5

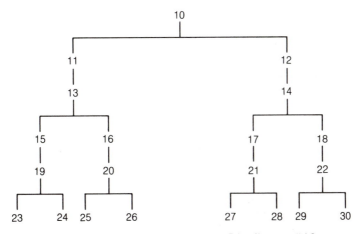

Figure E4.1 Example of Suborganization for Distributor #10

of the distributor who makes the direct sale. Any new distributor request will have the distributor number of his immediate sponsor.

The problem is to design a data structure which can be used to
determine, from a summary of the incoming orders for any given
month, the total amount of compensation due to each distributor. The
solution to this problem would be trivial if it were not for the following
requirement: Each distributor needs to know the dollar amount of his
income by source, i.e., by his subordinate's distributor number. Design
and specify the files, and their structures, that will produce these
monthly reports *efficiently*. Note that this specifically disallows the requirement of reading an entire file once for each distributor, since the
number of distributors could easily range in the hundreds of thousands.

Chapter  5

File Specifications

Chapter 4 caused an interruption to the design methodology initiated in Chapter 3. This was necessary in order to set forth some basic concepts which must be understood before the designer can effectively write design specifications for application system files. The purpose of this chapter is to continue the methodology of Chapter 3 by discussing the development of the documentation which specifies these files. Note that, in the stepwise procedure introduced in Chapter 1, this is the fourth step, namely, "specify file layouts and structures." This step follows the three steps which analyze: (1) the affected management system, (2) new system output requirements, and (3) input availability. Once the file-design step is completed, the design steps which follow will solidify the flow diagrams, specify input requirements, and develop program specifications. Thus, this fourth step in the design process is a critical pivotal step in the applications software system design.

The critical nature of file design can be better understood if it is recognized that the system files are the memory of the system. Virtually all of the information to be processed must go through the files in order to get back out to the system, analogous to the human brain. Obviously, if there is a problem within the file design, the impact upon the efficiency and integrity of the system could be enormous. Further, this step is critical because it is, in a sense, a point of no return. Prior to this step, changes in the system design could be accomplished merely by making the appropriate changes to the particular part of the design affected. Since subsequent steps involve synthesis, the file designs will have an impact in virtually every subsequent component of the design. This means that file changes made after this step is completed could be quite costly in terms of the redesign which they would necessitate.

Of course, if a deficient file design is not discovered until the software development is underway, the redevelopment consequences could be very expensive.

The following five activities will be performed in order to document the file specifications:

1. Specify file record layouts,
2. Develop variable specifications,
3. Develop organizational Warnier diagrams,
4. Develop structure diagrams, and
5. Verify file integrity.

Each of these will be discussed in a separate subsection below.

5.1 Specify File Record Layouts

The subject of the data dictionary, i.e., the area of the documentation dedicated to an orderly presentation of system specifications, was introduced in Section 2.3. This material should be reviewed before proceeding. This section will augment that introductory material by showing how some of the components of the data dictionary are constructed. We used the term *data dictionary* in the introductory material since this is the common expression of reference to the body of documentation used in specifying system details. The reader should recognize that the data dictionary is a subset of the Design Manual, which, after software development, becomes the Maintenance Manual. Since a definitive structure exists for the Design Manual, we will no longer reference this material as the data dictionary.

An example for specifying a file record layout was given in Figure 2.23.* This form provides an overall pattern, which is the output of this activity. Another example of the general form is presented in Figure 5.1. The following components of the form should be clearly understood by the programmer as well as the designer, who will use it to specify file content:

1. File reference number. Each data file should be assigned a reference number preceded by a "D" or some other unique letter which identifies the reference as being a data file.

2. File title. The file reference number followed by a short descriptive title (preferably no longer than two or three words). This is

---

*While standard pictorial layout forms could be substituted, they are not maintainable on text editors, they are nonstandard in size, they do not contain as much information, and, therefore, they are felt to be unnecessarily redundant.

D1.1 FARE FILE LAYOUT FORM - D1.1

FILE TITLE: Account File
COMMENTS: One record per account.

PAGE 1 OF 1
RECORD LENGTH: 434 bytes
FILE NAME: ie123db.acct.data

SEQ	ST POS	END POS	FLD SIZ	FLD *	REF NUM	FIELD DESCRIPTION	SOURCE
1	1	2	2	N		Fiscal Year	ADEM-1
2	3	8	6	N		Account Number	ADEM-3
3	9	10	2	A	1	Departmental Code	ADEM-2
4	11	11	1	N		Administrative Agency Code	ADEM-4
5	12	13	2	X		Fund Classification	ADEM-5
6	14	19	6	N		Creation Date, mmddyy	ADEM-6
7	20	26	7	N		Overhead Rate	ADEM-7
8	27	35	9	N	2	Exception Subcodes	ADEM-9
9	36	42	7	N		Fringe Benefits Rate	ADEM-10
10	43	49	7	N		Revised Budget - Subcode 100	ADEM-8
11	50	56	7	N		Expended this Month Subcode 100	ADEM-11
12	57	63	7	N		Expended this Quarter Subcode 100	ADEM-12
13	64	70	7	N		Year to Date Expended Subcode 100	ADEM-13
14	71	77	7	N		Total to Date Expended Subcode 100	ADEM-14
15	78	84	7	N		Current Encumbrances Subcode 100	ADEM-15
16	85	91	7	N		Total Encumbrances Subcode 100	ADEM-16
-23	92	140	49	N		SEQ 10-16 for Subcode 140	
-30	141	189	49	N		SEQ 10-16 for Subcode 150	
-37	190	238	49	N		SEQ 10-16 for Subcode 200	
-44	239	287	49	N		SEQ 10-16 for Subcode 600	
-51	288	336	49	N		SEQ 10-16 for Subcode 700	
-58	337	385	49	N		SEQ 10-16 for Subcode 800	
-65	386	434	49	N		SEQ 10-16 for Subcode 900	

*Field Characteristics: A = Alpha, B = Binary, N = Numeric,
SN = Signed Numeric, X = Alphanumeric.

Figure 5.1 Example of a File Layout Form

the reference name that should be consistently applied to the file in oral communications and written narratives; however, in narratives the title will usually be accompanied by the D reference number.

3. Page. If the layout form requires several pages indicate the page number and the total number of pages.

4. Comments. Any brief comments can be incorporated into the heading; these might include statements as to the sort/organization or give references to the location of such detailed specifications.

5. Record length. This can be in any unit, but bytes-per-record is the most common. This is an extremely handy bit of information to have available and therefore it belongs in the heading of the file layout specification. Variable-length records will be so indicated. (An example for variable-length records is given in Section 5.1.2.)

6. File name. The name by which the software references the file.

7. Sequence (SEQ).   For fixed-format, fixed-length records, this is the sequence of the variable fields in the file. Where groups of variables are repeated for variable record length, this will represent the sequence of "new and different" variables; that is, the record will be laid out as though there is but one occurrence of each different variable field (this is illustrated in conjunction with Figure 5.2, see page 206).

8. Start position (ST POS).   For fixed-format, fixed-length records, this is the relative start position of the field, generally measured in the number of bytes from the beginning of each record.

9. End position (END POS).   For fixed-format, fixed-length records, this is the relative end position of the field, generally measured in the number of bytes from the beginning of the record. The field includes the byte in this position.

10. Field size (FLD SIZ).   The field width, generally measured in the number of bytes in the field.

11. Field type (FLD).   As given by the key in the footnote, this will specify the field to be either purely alphabetic (A), binary (B), numeric (N), signed numeric (SN), or alphanumeric (X). In Figure 5.1 all field-type specifications are numeric, with the exception of variables 3 and 5. Additional specification types may be required for certain applications.

12. Reference number (REF NUM).   A reference to a paragraph of a narrative which accompanies the file layout form. The purpose of such a reference is to explicitly define the value that the variable field will contain for any given record. This column entry is completed by exception, that being only when the variable is not clearly defined by the data-entry or program specifications (see Source below). Thus, this column will not be completed during the initial file design; rather, it will be updated after the data-entry specifications are written. In Figure 5.1 it can be seen that reference paragraphs exist to further define variables 3 and 8. These paragraphs will be found with the narrative that accompanies the file layout form for D1.1.

13. Field description.   A brief description that conveys the meaning of the information stored in the field. This term should be used consistently in oral and written communication to reference the data element when descriptive references are required.

14. Source.   A reference to the data-entry mat (DEM), or the program specification number which generates the value stored in the respective variable location. Direct moves into the file from a data-entry mat will lead to a data-entry-mat specification, as exemplified in Figure 5.1 for the account data-entry mat (ADEM) items. The source column for variables whose values are calculated from input or from other file values will contain the program specification number of the program employed for these calculations. If several programs modify a given variable value, all program specification numbers will be listed. The logical determination of program specification numbers is given in Section 6.3. It should be clear that neither the data-entry nor the program

specifications have been written at this point, and therefore, this part of the file layout form cannot be completed now. (This will be performed during later steps of the methodology.)

The referencing of file variables was discussed in regard to the field description definition above. Quite often it is convenient to reference a variable by its location in the file as opposed to using the descriptive reference given in the Field Description. This is particularly true within program and data-entry specifications. The method for doing this was introduced in Section 2.3, above. It consists of using what was defined as the "file, seq" number. For example, the Overhead Rate in Figure 5.1 would be referenced as D1.1,7.

The file layout form, as exemplified by Figure 5.1, should be put on a computerized text editor so that modifications can be made as variables are added and rearranged within the various files. Variables which require definition not given by data-entry specifications will be referenced to accompanying narrative. However, the reference portions (i.e., Items 12 and 14) of the form will be deferred until the input and program specifications (Chapter 6) are developed. During this time it will also be helpful to defer the completion of the Start and End positions until all rearrangements and final field size assignments are made. Thus, the file layout form is a dynamic tool being updated through system design and possibly into development.

The file layout form presented and discussed above provides a pattern for the output of the activity of specifying file record layouts. Input for this activity is obtained directly from the specification of system output requirements, accomplished during step 2 of the design process. Since it is essential to have all the information in the files that will be used for system output, the definition of these outputs, which should now be available in the early sections of the Design Manual (see Sections 3.2.4 and 3.2.5), will provide all of the basic information required to complete an initial file layout.

If no tentative file layouts have been written, the procedure is quite simple. Go to the new system data flow diagrams, such as the one exemplified in Figure 3.13. Of course, at this point all file inputs and outputs will be completed on the DFD (unlike Figure 3.13). Using the *outputs* from each file, determine the file variable necessary to produce these outputs. In listing these variables try to maintain some logical ordering of the variables; this will aid in structuring the file later. However, concentrate upon making sure that all variables required for output production are available within at least one, and preferably only one, file.

Now that both the input and the output of this activity have been stated, the remainder is a matter of organization (ordering and categorizing). This is a major problem addressed in the following subsections. Three aspects of organization require consideration:

1. Organization of data elements into files,
2. Definition of record types, and
3. Ordering of variables within records.

Note that these activities do not include the formal "file organization" discussed in Section 4.4. Rather, at this point, the broader aspects of organizing the data elements into logically ordered subsets is under consideration. This is the first step in synthesizing the system, and it can be one of the most critical. Each of the three aspects will be discussed in a separate subsection below.

5.1.1 Organize Data Elements into Files

Consider the situation as we have depicted it above. From the output specifications the designer has a list of, possibly, hundreds of data elements that must be stored within the system. He also has a pile of blank file layout forms. Simply stated, the problem is to properly list the data elements on the layout forms in a logical order that has been organized into the optimal number and types of files. The complexity of this process depends upon the system itself as well as the experience of the designer. For some systems, and to some experienced designers, this might be a trivial exercise. For systems that are more complex, consider the following priority of organization:

1. By module,
2. By output-requirements classification,
3. By input/maintenance requirement, and
4. By special utility.

These considerations are not mutually exclusive, and their simultaneous consideration can lead to conflicting results. Thus, the ordering of their consideration follows a type of priority, as discussed in the following paragraphs.

1. *By module.* Modularization of the system was introduced in Chapter 1, and further discussed in conjunction with the definition of the system missions and projects in Section 3.1.3. Sometimes the modules of a system will be so independent of one another that the interaction between them is minimal. When this is the case, especially with respect to file creation and utilization, then separate files for each module might be justified. This could certainly lead to a more efficient modular development and implementation. Thus, data elements would be first classified according to their module, which should proceed directly, since this is the organization of the output reports. An example of such a modular system might be found in the production control area where the total system is broken into a work-order module and an

inventory-control module, among others. The advantages of having separate inventory-control and work-order files should be obvious, even though interactions between the two will occur.

2. *By output requirement classification.* This is the next-level breakdown within a given module. Each module will have been defined already in terms of a collection of common outputs (common with respect to the module definition). It might be possible to further organize these individual collections of reports such that the files can also be subdivided to facilitate their respective generation. To proceed one step further with the example given above, the work-order module in a system might generate work orders (first type of report) and exception listings of jobs with particular user-defined characteristics (second type). Since these reports might contain different elements, it might be advantageous to organize the data elements required for generation into different files if this does not produce excessive redundancy.

3. *By input/maintenance requirement.* This consideration, like that above, interacts heavily with the modularity of the system. If it seems clear that one type of data elements originates from one source, and that another, different set of data elements comes from another source, it might be preferable to maintain a separate file for each, rather than trying to integrate the two. This decision is much more dependent upon the purpose of the record, however, and this is the reason that this consideration is placed third. The "purpose of a record" is the objective that the record must satisfy. For example, a state traffic records system might have accident records and traffic citation records. This would seem to warrant two files from both the modular and the data input points of view. However, if it were required to correlate the accident circumstances with the historical violations of the drivers involved, then it might be necessary to either link these two files together (as might be done in a data-base management system) or to create a third file that had, within each record, the appropriate circumstances and driver history for each accident. Alternatively, it might be preferable to post the accident circumstances to the driver history (violations) record. These are complex decisions that must be made in light of the arguments for and against the variously proposed alternatives. However, the purpose of this example is to show that of primary concern is the integrity of the *output* for which the system is being designed, and that data entry and maintenance problems are secondary. Only when cost and feasibility enter strongly into the picture does the input/maintenance become a significant influencing factor. This is the reason that, although third in priority, this consideration cannot be ignored.

4. *By special utility.* Certain files are established for the primary reason of facilitating the mechanism of the system. For example, a separate, linked-list file which includes all of the key fields might be established for locating a given data record quickly. A pointer from the linked-list file to the data record would be employed to get more infor-

mation on the record, if required. Files of this nature do not fall under the same evaluation criteria as specified above. However, simultaneous with considerations of modularity, output, and input, considerations with regard to such utility files must be made.

We conclude this section with the realization that no stepwise methodology has been given for transforming the total list of data elements, as defined by the system outputs, into their respective files. A knowledge and enumeration of the possibilities, coupled with the knowledge of the new system structure and output requirements (as determined in Chapter 3), usually precipitates an obvious choice in this regard. Since file design is so critical to the success of the system, it is recommended that the designer formulate the alternatives and involve all of the design and development group in this decision.

5.1.2 Define Record Types

The decision as to the allocations of data elements into their respective files overlaps considerably with the definition of the record type. Both are resolved primarily in terms of the "purpose of the record." Here, record type refers to the alternatives discussed in Section 4.3, which also described the relative merits of each. The ultimate decision as to whether the record format will be fixed, variable, or undefined must be made in terms of the output requirements, coupled with the processing and input necessary to accomplish these requirements. Generally, if subsets of fields in a record recur at a variable rate, the variable-length record is in order. If this is not the case then a fixed record length is in order. Undefined-length record files are used in some data base applications, where ultimate flexibility must be maintained.

The example given in Figure 5.1 was for a fixed-length, fixed-format file. Notice that there was some repetition within the record, with the same data elements being repeated for each of the various subcodes, but the repetition is identical for all records. Here, referencing is used to repeat data-field descriptions in sequence numbers 10 through 16 (SEQ 10-16) without spelling out each. Not only does this save paper and clerical time, but it also improves clarity and eliminates errors which often accompany redundancy. A similar notion can be applied in referencing multiple-length records. Obviously, it is impossible to definitively state the record layout, as was done in Figure 5.1, because the number of repetitions of a field or grouping of fields is not fixed. However, the record format can be clearly defined by referencing.

To extend the example given in Figure 5.1, suppose that the record were to be formatted to allow a variable number of subcodes (as opposed to the fixed number of 8 subcodes in Figure 5.1). Further, suppose that the subcode did not have preassigned numbers (100,140,150,200, etc., in Figure 5.1), but that they were assigned dependent upon the

D1.1 FARE FILE LAYOUT FORM - D1.1

FILE TITLE: Account File   PAGE 1 OF 1
COMMENTS: One record per account.   RECORD LENGTH: Variable
                                                                    FILE NAME: ie123db.acct.data

SEQ	ST POS	END POS	FLD SIZ	FLD *	REF NUM	FIELD DESCRIPTION	SOURCE
1	1	2	2	N		Fiscal Year	ADEM-1
2	3	8	6	N		Account Number	ADEM-2
3	9	10	2	A	1	Departmental Code	ADEM-4
4	11	11	1	N		Administrative Agency Code	ADEM-3
5	12	13	2	X		Fund Classification	ADEM-5
6	14	19	6	N		Creation Date, mmddyy	ADEM-6
7	20	26	7	N		Overhead Rate	ADEM-7
8	27	35	9	N	2	Exception Subcodes	ADEM-8
9	36	42	7	N		Fringe Benefits Rate	ADEM-9
10	43	44	2	N		Number of Subcodes, N	P1.4
11	45	47	3	N		Subcode Number	ADEM-10
12	48	54	7	N		Revised Budget—1st Subcode	ADEM-11
13	55	61	7	N		Expended this Month—1st Subcode	ADEM-12
14	62	68	7	N		Expended this Quarter—1st Subcode	ADEM-13
15	69	75	7	N		Year to Date Expenditure—1st Subcode	ADEM-14
16	76	82	7	N		Total to Date Expenditure—1st Subcode	ADEM-16
17	83	89	7	N		Current Encumbrances—1st Subcode	ADEM-17
18	90	96	7	N		Total Encumbrances—1st Subcode	ADEM-18
−26	97	148	52	N		SEQ 11-18 for 2nd Subcode	
−34	149	200	52	N		SEQ 11-18 for 3rd Subcode	
⋮ 18 +8N				⋮ N		⋮ SEQ 11-18 for Nth Subcode	

*Field Characteristics: A = Alpha, B = Binary, N = Numeric,
SN = Signed Numeric, X = Alphanumeric.

Figure 5.2 Extension of Figure 5.1 to Variable-length Record Specification

particular account. Note that this is not an unreasonable generalization, since all subcodes may not apply to all accounts. Thus, to save space, it might be preferable to only utilize storage for those subcodes which apply. Figure 5.2 shows how Figure 5.1 would be modified to accommodate this change. To support the "variable" type of data-element specification, note the addition of two new fields. The first, SEQ 10, is the variable which specifies the number of subcodes for which the SEQ 11-18 will be required. The second, SEQ 11, is the value of the subcode which was identified in Figure 5.1 by *position*, but must now be explicitly defined within the variable grouping itself. As a final, subtle modification, note that the term *variable* is placed in the record-length specification. The extension of the concept exemplified in Figure 5.2 to multiple-variable occurrences of fields within variable-occurring field groupings should be easy to implement at this point.

5.1.3 Order Variables Within Records

From a systems point of view, the ordering of the variables within a record is of little significance. However, from the programmer's and, possibly, the user's points of view, proper ordering can be extremely critical. Unlike the other considerations, consistency with the input specifications might be the most important factor. If data are entered using a data-entry mat, considerable efficiency can be attained in development, debugging, and maintenance if the file to which these elements are written is in approximately the same order as the data-entry mat. This leads to a more systematic approach toward debugging, since the programmer does not have to perform a find operation every time the next data element is considered. While this might seem so obvious, the authors have observed such flagrant violations of this rule that we feel compelled to emphasize it. In one live commerical application the data collection forms, which were (mistakenly) developed after the data-entry-screen mats, were totally inconsistent with the screen requirements, thus causing the data-entry operator to have to find each item of data on the form. With a mere change in the form (or the screen format), the productivity of the data-entry operator could have been easily doubled. While this does not involve the ordering within the file itself, it points to the real and obvious problems that quickly accrue from a lack of planning. Consistency in ordering between the data-entry forms, the screen mats, queries, and/or menus, and the ordering of the data elements in the file, is extremely desirable.

This is not to say that the data-element ordering within the file is dictated by data entry. Indeed, there are times that minor rearrangements to accommodate screen formats need not be reflected in the file layout. It is preferable to think of the cause-effect mechanism being primarily motivated by a logical ordering per se. That is, classifications and subclassifications of the data elements should be formed such that those which have common characteristics fall into a common grouping. If similar elements appear in several subclassifications, then they should be similarly ordered (possibly according to some rationale) in each subclassification. This can best be illustrated by an example. Consider Table 5.1, which is a relatively unordered set of data elements which might be found in a traffic accident record. The real example had about 250 variables, but this subset will better serve to illustrate the points which have been made above. Suppose that the decision has been made to put these data elements into a file called the Accident File, as opposed to other files in the total traffic records system, which also might include a Driver's History File, Roadway Geometrics File, Skid Test File, Average Traffic Volume File, and possibly others. This initial decision would be performed according to the considerations given in Section 5.1.1.

Table 5.1   Example of Traffic Accident Record Data Elements

Original Variable Order

01	County	42	Milepost Indicator
02	City	43	Year of Accident
03	Time of Day	44	Est. Damage, Veh. #1
04	Month	45	Est. Damage, Veh. #2
05	Day of the Week	46	Action, Pedestrian
06	Number of Vehicles	47	Age, Pedestrian
07	Occupants, Vehicle #1	48	Sex, Pedestrian
08	Occupants, Vehicle #2	49	Occupation, Pedestrian
09	Accident Severity	50	Sobriety, Pedestrian
10	Action, Pedacycle	51	Clothing, Pedestrian
11	Lights, Improper	52	Age, Victim #1
12	Vehicle Unsafe Condition	53	Age, Victim #2
13	Driving in Wrong Lane	54	Age, Victim #3
14	Improper Signal	55	Age, Victim #4
15	Following Too Closely	56	Age, Victim #5
16	Defective Brakes	57	Sex, Victim #1
17	Exceeding Speed Limit	58	Sex, Victim #2
18	Disregard Sign or Signal	59	Sex, Victim #3
19	Pedestrian Drunk	60	Sex, Victim #4
20	Faulty Equipment	61	Sex, Victim #5
21	Walking Violation	62	Injury, Victim #1
22	Driving Under Influence	63	Injury, Victim #2
23	Est. Speed, Vehicle #1	64	Injury, Victim #3
24	Est. Speed, Vehicle #2	65	Injury, Victim #4
25	Weather	66	Injury, Victim #5
26	Defects in Road, #1	67	First Aid, Victim #1
27	Defects in Road, #2	68	First Aid, Victim #2
28	Age, Driver #1	69	First Aid, Victim #3
29	Age, Driver #2	70	First Aid, Victim #4
30	Sex, Driver #1	71	First Aid, Victim #5
31	Sex, Driver #2	72	Brakes, Vehicle #1
32	Sobriety, Driver #1	73	Lights, Vehicle #1
33	Sobriety, Driver #2	74	Steering, Vehicle #1
34	Action, Driver #1	75	Tires, Vehicle #1
35	Action, Driver #2	76	Brakes, Vehicle #2
36	Year, Vehicle #1	77	Lights, Vehicle #2
37	Year, Vehicle #2	78	Steering, Vehicle #2
38	Body Style, Vehicle #1	79	Tires, Vehicle #2
39	Body Style, Vehicle #2	80	Day of the Month
40	ON Road Code	81	Police Reporting Agency
41	AT Road Code		

Before ordering the variables, two factors need to be considered. The first involves the determination of record type by the methodology specified within Section 5.1.2, immediately above. Certainly, if records had to be maintained on every driver, victim, vehicle, and pedestrian involved in a traffic accident, a hierarchical variable-length record would be essential. However, suppose that the purpose of this file is to be able to produce quick problem-identification information where accuracy to one or two percent is quite sufficient. Since a third vehicle, a second pedestrian, or a sixth victim only occurs in a very small proportion of accidents, and since the causative factors are more accurately reflected by the first reported, as opposed to the last, the limitation of the record to two vehicles, one pedestrian, and five victims will be assumed to be justified. Note that all of the considerations of Sections 5.1.1 and 5.1.2 should be taken into account in making this decision, giving primary consideration to the *purpose* of the record.

Let us assume that a fixed-length record is in order, and that the variables in Table 5.1 are to be included. The second consideration which must precede the grouping and ordering of these variables is the categorization of the variables. Here again, the primary objective in terms of record use and output will dictate the variable grouping. Since the stated objective of our example record is to determine accident causes, the classification of the variables might follow the classification by causes. Here the practitioners should be consulted, since their expertise in this area will probably surpass that of the designer. For this example, traffic safety practitioners generally classify accident causes by the roadway, the driver, and the vehicle. Other categories might be added for general information, pedestrians, and victims. With this in mind, a more logical arrangement and grouping might be given by Table 5.2. Obviously this rearrangement may be further improved, depending upon specific system needs. This example brings to light additional advantages which can be visualized once a logical rearrangement is performed.

Table 5.3 presents some of the additional considerations which can be observed at this point. The first cut at posting variables to their respective categories usually uncovers some basic flaws in the categorization scheme itself. A variable does not clearly fit into any one category, or a number of variables are uniquely different from the others within that category. Usually a number of cross-categorizations exist, and no one set of classifications will solve all problems. For example, injured drivers might also be classified as "victims" for purposes of recording their physical characteristics. It would be conservative of space to set up a variable which identifies if a driver is a victim (i.e., injured) and, if so, which victim number applies. However, since this might limit processing ease and flexibility somewhat, total system needs must be considered. The point is that the logical ordering tends to bring these decisions clearly into focus.

Table 5.2   Example of Rearrangement of Records
                 New Ordering and Grouping

Roadway	**Driver**
01 County	13 Driving in Wrong Lane
02 City	14 Improper Signal
40 ON Road Code	15 Following Too Closely
41 AT Road Code	17 Exceeding Speed Limit
42 Milepost Indicator	18 Disregard Sign or Signal
26 Defects in Road, #1	22 Driving Under Influence
27 Defects in Road, #2	23 Est. Speed, Vehicle #1
General	24 Est. Speed, Vehicle #2
03 Time of Day	28 Age, Driver #1
05 Day of the Week	29 Age, Driver #2
04 Month	30 Sex, Driver #1
80 Day of the Month	31 Sex, Driver #2
43 Year of Accident	32 Sobriety, Driver #1
06 Number of Vehicles	33 Sobriety, Driver #2
25 Weather	34 Action, Driver #1
81 Police Reporting Agency	35 Action, Driver #2
Pedestrian/Bicyclist	47 Age, Pedestrian
10 Action, Pedacycle	48 Sex, Pedestrian
19 Pedestrian Drunk	49 Occupation, Pedestrian
21 Walking Violation	50 Sobriety, Pedestrian
46 Action, Pedestrian	51 Clothing, Pedestrian

The second consideration in Table 5.3 is that of variable consolidation. It might seem odd that Table 5.1 had a variable called "Pedestrian Drunk" (19) and at the same time, "Sobriety, Pedestrian" (50). However, this was taken from a real state accident form and data-entry system that had been in effect for over 10 years, processing well over 100,000 accident records per year. The only difference between the two variables was the coding scheme, one being binary (yes/no) while the other stated the combinations of drugs and alcohol involved. Clearly the two could have been combined to reduce storage and processing requirements. The variables related to driving under the influence also show potential for combinations, and considerable study should be applied to obtain the specific combination of these variables to satisfy user requirements.

We do not wish to leave the impression that Table 5.2 represents the ultimate rearrangement for this example. These are not five-minute decisions that can easily be illustrated by textbook examples. The redevelopment of the accident record, a small portion of which is illus-

Table 5.2   Example of Rearrangement of Records
                New Ordering and Grouping (continued)

Victim		Vehicle	
52	Age, Victim #1	07	Occupants, Vehicle #1
53	Age, Victim #2	08	Occupants, Vehicle #2
54	Age, Victim #3	11	Lights, Improper
55	Age, Victim #4	12	Vehicle Unsafe Condition
56	Age, Victim #5	16	Defective Brakes
57	Sex, Victim #1	20	Faulty Equipment
58	Sex, Victim #2	36	Year, Vehicle #1
59	Sex, Victim #3	37	Year, Vehicle #2
60	Sex, Victim #4	38	Body Style, Vehicle #1
61	Sex, Victim #5	39	Body Style, Vehicle #2
62	Injury, Victim #1	44	Est. Damage, Veh. #1
63	Injury, Victim #2	45	Est. Damage, Veh. #2
64	Injury, Victim #3	72	Brakes, Vehicle #1
65	Injury, Victim #4	73	Lights, Vehicle #1
66	Injury, Victim #5	74	Steering, Vehicle #1
67	First Aid, Victim #1	75	Tires, Vehicle #1
68	First Aid, Victim #2	76	Brakes, Vehicle #2
69	First Aid, Victim #3	77	Lights, Vehicle #2
70	First Aid, Victim #4	78	Steering, Vehicle #2
71	First Aid, Victim #5	79	Tires, Vehicle #2

Table 5.3   Additional Examples of Logical Rearrangement

Consideration	Example (See Table 5.2)
Revision of Categories	Pedestrian/bicyclist might be divided into separate categories.
Elimination of Variables	"Pedestrian Drunk" and "Sobriety of Pedestrian" would be candidates; also, driver variables related to alcohol (22, 32, and 33) might be combined.

trated above, took about six man-months in which hundreds of interactions with users were required. The result was a complete redefinition
of most of the variables which ultimately were stored in a hierarchical
file.

This subsection has concentrated upon the ordering of the variables
within the file. To place this in its broader perspective, this is merely
an adjustment (by reordering) of the basic file variable-assignment pro-

cess, the output of which was discussed in conjunction with Figures 5.1 through 5.3. Generally the grouping and ordering, i.e., the organizing of the data element will be performed simultaneously with the two activities described above, namely: (1) the separation of data elements into files, and (2) the definition of record types. The reiteration of these three activities ultimately results in a final file layout specification.

5.2 Develop Variable Specifications

Once the file record layouts have been completed, consideration should be given to developing the variable specifications. While some system design methodologies prescribe a data element specification for every variable in the file, this is rarely done in practice due to the labor involved. Further, the sheer volume of forms tends to make it difficult to get to the variable specifications which are needed. Since most of the variables within the files will not require additional specification, there are considerable benefits in developing variable specifications by exception.

The basic input to this activity is the file layout form, which should be completed to the extent possible. Minimally, this will include the field description for every variable in every file. This field description itself might be sufficient for specifying a variable. For example, "Creation Date, mmddyy" which is D1.1, 6 in Figures 5.1 and 5.2, is sufficient within itself to define the variable field, given that it is a numeric field. This should not be ambiguous, despite its brevity. However, it is recognized that in most cases the field description alone will be insufficient to define the value of the variable.

The designer must make a decision at this point as to whether or not additional narratives will be required for specifying each variable. The following two considerations are involved in this decision:

1. Will the value of the variable be directly determined by data entry? If so, reference to the item on the data-entry specifications (see Section 2.3.1) will adequately define the variable.
2. Will the value of the variable be defined within a program specification? If so, reference to the program specification (see Section 6.3) will adequately define the variable.

Both of these references are made under the source column of the file layout form, as was shown and explained in conjunction with Figure 2.31 (page 87).

The problem in making the above considerations is that neither the data-entry specifications nor the program specifications are yet in existence. Thus, it is necessary for the designer to anticipate, for each variable, whether or not these specifications will be adequate to define the variable. Experience has shown that they will be sufficient for the vast majority of variables. Exceptions can be anticipated in a few cases as follow:

1. Where the coding of the variable is going to differ significantly from that given on the data-entry mat, so that the data-entry specifications might be misleading to the programmer, or
2. Where the program specification is so technical that greater understanding of the variable value determination can be obtained by supplementary narrative documentation (including tables and figures).

Note that both of these situations require a judgment on the part of the designer.

The current activity, namely, "develop variable specifications," is placed chronologically after the file layout activity, since those variables which are regarded as difficult to understand should be documented at this point. In conjunction with the considerations given above, however, it should be recognized that this activity can be deferred for certain variables if, in fact, there is reason to believe that such specification will not be necessary for that variable. A general rule of thumb is: if the determination of the variable value is not intuitively clear to the designer, then additional documentation for clarification should be generated before proceeding. This may take the form of narrative, but the specification techniques given in Chapter 2 should also be employed as much as possible. What is recommended might more accurately be called "working documentation," and no attempt should be made to structure it into program specifications, since many other activities will be required before the program specifications can be efficiently developed. However, by cleaning up this working documentation which defines the few complex variables, the supplementary variable specifications will be available.

To finalize this activity, the designer need only reference the various documents developed to the appropriate file layout form. This is quite simple since, within the Design Manual, there will be a separate narrative section generally describing each of the files. Thus a sequential reference number (i.e., in the REF NUM column) within the particular file layout form tells the reader to go to the narrative for that file and look up the subsection which pertains to these variable references. Since the narratives for the files are not large (typically two or three pages), a sequential reference is quite adequate.

5.3 Develop Organizational Warnier Diagrams

During this activity the organizational Warnier diagrams (OWD) discussed in Section 2.1.5 will be developed. The technical aspects of the OWD should be reviewed, since it will be assumed to be known here.

Also, the background of Chapter 4, especially Section 4.7, is required. At this point, attention will be focused upon establishing the OWD for the files specified according to the activities presented in Section 5.1. The input to this activity consists of the file layouts, the output requirements, and, to some extent, a foreknowledge of the input requirements. The term *foreknowledge* is used since, although previous consideration has been given to input (Section 3.3), the full details of the input have not yet been specified (Section 6.2).

Figure 5.3 illustrates the inputs and outputs of this activity. It is necessary to synthesize these three inputs into the OWD specification for each file. While conflicts between the needs of each might sometimes arise, typically, the proper ordering for a file is not difficult to determine. The file layouts merely set the stage by specifying the subsets of data which are under consideration. The primary input to determining record order is the output requirements. The following considerations are associated with output requirements:

1. Order of output,
2. Access and search methods, and
3. Retrieval-time requirements.

Since these need simultaneous consideration, they will be discussed together in the remainder of this section.

As was thoroughly discussed in Chapter 4, there are tremendous time-saving advantages that can be obtained by either sorting or linking files to produce an ordered output. The objective of specifying an ordering is to take advantage of this time saving. It should be clear at this point that if output is required in a given order, the designer has the following options:

1. Maintain the records in random order and sequentially read as many times as it takes to produce the report,
2. Maintain the records in random order and sort before output,
3. Maintain the records in random order but link the records for the output order desired, or
4. Maintain the records in the physical ordering of the output required.

Figure 5.3 Activity Inputs and Output

The OWD is useful for specification of the file organization of any of the above, with the possible exception of the first, which should not be seriously considered except for small queries.

When an output sort-ordering is uniquely defined, the OWD can be developed directly from it. The OWD can also be used to establish the order of linkage in those cases where links are to be maintained, as opposed to sorting the records. When there are several output orderings required, one OWD will be required to specify each. This should be no problem as far as OWD development is concerned. However, it will be essential to spell out the following in the accompanying narrative: (1) the OWD which represents the order that is required for maintaining the file, (2) the data structure considerations (i.e., links and indexes) if output will be generated without re-sorting, and (3) the temporary data store requirements if sorting is to precede output.

Access- and search-method considerations introduced above become more important for query-type file outputs. Although there are a large variety of possibilities, discussed in Chapter 4, generally, the ordering can be specified in terms of an OWD. To illustrate this, consider what might be a difficult specification, that of the sequential binary representation exemplified by Figure 4.23. Suppose that we had a randomly arranged set of records with a unique key field. The objective would be to specify a sort that would result in an arrangement to facilitate a sequential binary search, an example of which is given in Figure 4.23. While these records do not appear to be "sorted," there is a logical sequence followed in obtaining this unique arrangement, and it is the specification of such that the OWD is all about. One solution to this problem is given in Figure 5.4. This OWD indicates that the first reordering is a sort of the Key Field (in a real application this would be defined) into ascending order. The second reordering depends upon the first, as indicated. The reader should verify that this indeed produces the arrangement exemplified by Figure 4.23.

The flexibility and clarity of the OWD specification enables the programmer to generate the code for the sort without taking the time to become familiar with the application. Certainly it would not hurt to involve the programmer, depending upon his level of competence, in the design procedure. However, the necessities of specialization often prevent this from occurring.

In summary, the specification of the OWD for the system files takes place during this activity according to the considerations given above and in Chapter 4. In simple cases where the ordering of the records is uniquely defined, this should pose no problem using the OWD approach originally given in Chapter 2. When several different output report orderings and queries are required, then the problem becomes one of trading off time and space resources such that an optimal system design with respect to cost results.

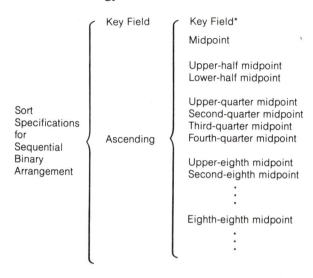

*If record does not exist it must be generated such that
the number of records is an even power of two.

Figure 5.4 Specification of "Sequential Binary" Arrangement

5.4 Develop Structure Diagrams

In Section 4.5.1 the concept of linking records together was discussed, and the means for communicating the structure of such a linkage was presented in terms of the structure diagram. It would seem reasonable that once the file-record layouts are specified and the ordering of the records within the files is determined, the next logical step is to specify any linkage which might exist. This section will briefly review this activity with minimal repetition of the basic concepts presented in Section 4.5.

Recall that the structure diagram consists of the following components:

1. An abbreviated record layout for the data records in the file,
2. A header record layout,
3. An example enumeration of the records within the file, the number of which is sufficient to illustrate the structure, and
4. Links which will illustrate the relationship between the records within or between files.

Unlike the specification for record layout and ordering, structure diagrams may involve several files simultaneously. When this is the case, the structure diagram should accompany the file documentation at the point where it is first required by the programmers. This diagram should then be referenced by the other files which are affected; generally, it should not be repeated.

Consider each of the four components of the structure diagrams exemplified in Figure 4.16. The record layout is presented to make the structure diagram understandable, and it can be created directly from the file-record layout specification (e.g., Figure 5.1). All links should be defined at this point; if not, the record layouts will require modification to include all link fields. The record-structure diagram need not include the names of nonkey data fields. These can be batched together and generically labeled as "data." However, link fields and key-data fields should be identified in the record layout of the structure diagram by the *same* field description as given in the file-specifications layout form. The layout in the structure diagram need not specify columns, as long as the field name is identical to that in the file specification, since it will only be used within the data structure diagram. It merely serves as a legend to identify the various fields which might be link or key fields.

The header record layout is necessary to identify the contents of the header record. This is essential programming information since, generally, the header will be read whenever the file is opened. In addition, since the header is the same size but is not a data record, the "data" fields of the header can be loaded with parameters. These may be special pointers to "get started" at particular entry points, or they may contain information about the overall file. In this regard, two points bear emphasis. First, if the designer wants particular information to be included in the header record, this should be named within the header-record layout and referenced to the backup documentation for the file. A separate header-record layout might be required for complete clarification and definition of these parameters. Second, if empty fields exist within the header record, then these should be made available to the programmers for their use. However, any use that they make of this excess space should be clarified, and the file documentation should be updated as described above.

The third component of the structure diagram discussed above is the example enumeration of the records. Since the record layout in the structure diagram labels the field by position, generally, additional labels are not required. The number of records shown need only be large enough to communicate the required linkage. Generally, the more complex the linkages, the greater number of records required to demonstrate them. Links should be drawn, as in Figure 4.16, from the center of the link field to the record in the position which its value would indicate. Note, however, that Figure 4.16 does not show the reason for the links in terms of key fields. Thus, unless the programmers are given supplementary narratives to explain the reason for the linkage, they will not know how to maintain them. The objective of the structure diagram is to provide an example which the programmer can use to establish and maintain the data structure. This is the reason that it is necessary to provide examples of key-field values, as was stated above.

Figure 5.5 presents an example with key-field values included. In this example there are two key fields used for linkage, Due Date and Project Number. Three linkages are defined: (1) DD, forward by Due Date; (2) PN, forward by Project Number; and (3) PNB, backward by Project Number. While all three could have been displayed on the same figure, the resultant diagram would be quite confusing to read, and it would offer no advantages other than paper savings. By displaying each linkage on a separate diagram, the specification is made quite clear. The presence of three link fields in the record structure portion of the diagram, as well as the accompanying narrative, will indicate that the set of (in this case, three) diagrams are to be considered as a unit.

Two other features of the example structure diagrams bear mentioning. The first is the use of the link to "point" at the relevant key field of the right node record. Technically, the link can point to any part of the record (as was the convention applied in Chapter 4), since the value of the link field is just the location of this entire record. However, why not focus attention upon the key value which determines the value of the link field? As long as both the writer and the reader understand the significance and limitation of this convention, it can certainly be used to advantage. The second aspect of the example, which bears mentioning, is the use of a reference from the "parameters" section of the header record to a record layout form. This form would be practically identical to that given in Figure 5.1 for specifying record layout.

This section has presented the considerations involved in developing the data structure diagrams to complete the file specification process. An example was presented to illustrate the format and content of the structure diagram. However, the methodology for determining which linkages should and should not be incorporated was presented only in a general way. For additional technical aspects, the reader is referred to Chapter 4, as well as to the many excellent books in the literature on the subject of data structures.

5.5 Verify File Integrity

The integrity of a file is the degree to which it can be called upon to accurately provide the data required of it. The subject of integrity can be subdivided into the following areas:

1. Assurance that required data elements are included in the file,
2. Protection against data alteration, and
3. Backup and recovery.

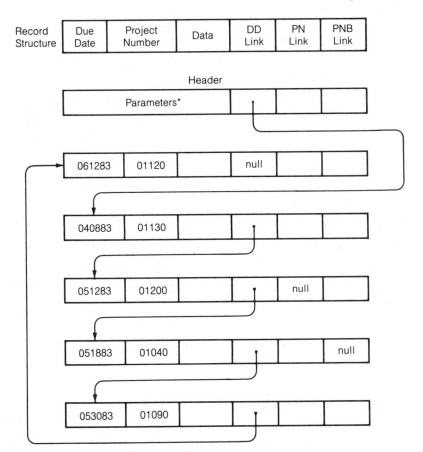

Figure 5.5a Example of a Structure Diagram (Due Date Linkage)

These will be covered in separate subsections below.

5.5.1 Verify File Content Integrity

This is a fairly time-consuming activity which is often ignored. However, the time required to perform this verification is a small fraction of the time which will be required if a variable has been inadvertently omitted from the file. The objective of this activity is to assure that all required output can be produced from the data stored in the files. This is accomplished by posting the reference number of each variable back to the output report specifications, which were completed during step 2 (Chapter 3). This has already been done in the example given in Section 3.2.6.4, and the reasoning for it was discussed in Section 2.3.3. At the end of each variable specification which is obtained from a file, a

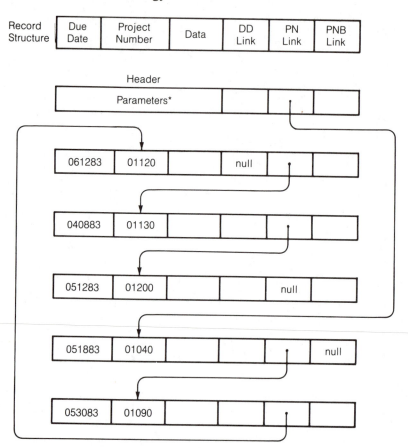

*Give the reference to the complete header record layout.

Figure 5.5b Example of a Structure Diagram (Project Number Linkage)

number appears in parentheses. This number gives the file and the sequence number for the source of the information presented; for example, D2.1,3 is variable sequence 3 in file D2.1. Where calculations are required, the output specifications should reference the source variables to be used in this same way. Note that, while Section 3.2.6.4 showed this activity to be completed, in reality this cannot be completed until the record layouts are finished, i.e., now.

The verification step requires a total enumeration of all output specifications, item by item. This step is often omitted by designers in a hurry to get the software development underway. However, it is the only way to assure that the data necessary for every report are available within the system. Thus, it is a very beneficial exercise, and in the long run it proves very cost-effective. In terms of file content integrity, it is by far the most important activity, and therefore it should not be neglected.

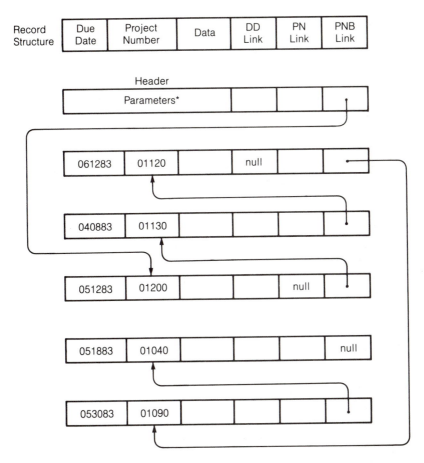

*Give the reference to the complete header record layout.

Figure 5.5c Example of a Structure Diagram (Project Number Back
Linkage)

5.5.2 Assure File Protection

File protection assurance is not nearly as simple as the "check-list"
verification of file content. Methods and procedures to protect files
against inadvertent destruction of data must be documented in the
narratives which accompany the file specifications. Many of these are
system dependent, such as the availability of password protection. The
objective here is to prescribe some general considerations along this
line. To define file protection, note that catastrophic destruction of
hardware is not under consideration; this will be considered in the next
subsection. Of primary importance here is the inadvertent modification
of data by a user or maintainer of the system. The worst case is where
such alterations are made without the knowledge of the person who
does it.

During this activity it is essential that the design specifications be augmented to consider such inadvertent data modification. Input and modification of files should be performed only through controlled user programs that protect the files in every way possible. If information is classified, or if it should only be modified by a select group of users, then password protection should be incorporated either at the system or the application program level. If at all possible, files should not be capable of being "edited." That is, users should only be able to access the files under the control of the programs within the system. Otherwise there is no way to control the procedures by which data can be modified. This limits the flexibility of the user and necessitates additional software development for updating. However, these additional costs are justified since data integrity is one of the most critical parts of any system.

Many of the considerations given above involve software and input specifications. These will be discussed in further detail in Chapter 6. However, the subject of file protection interacts heavily with that of file design, and therefore it should be given some consideration at this point.

5.5.3 Provide for Backup and Recovery

The subject of backup and recovery is more a consequence of catastrophic storage media failures, as opposed to the inadvertent data modifications discussed above. Rarely is backup and recovery used to address minor data flaws, which can be more easily fixed through direct modifications. While catastrophic hardware failures are quite rare, their mere possibility mandates some protection against them. Further, the probability of an erroneous deletion of an entire file is not small on a system-wide basis. While every precaution should be taken to prevent this from happening, it is impossible to totally protect the files from everyone. Thus, some backup and recovery provisions are required.

This activity requires special attention at this point, since, generally, backup files have been excluded from the design consideration. For example, the data flow diagrams generally exclude backup files. These tend to clutter and obscure the design specifications. Yet they are quite important, and each file specification should be accompanied by provisions for backup and recovery. These specifications will vary by: (1) frequency of backup, (2) media requirements, and (3) recovery considerations. Given that the data within a system is quite volatile and extremely valuable, there might be a requirement to back it up daily, or even hourly; whereas, slowly changing data, which can easily be re-entered and updated, might only need weekly or monthly backups.

The media chosen for backup will be a function of the backup frequency requirements. However, cost and media resource availability generally play a larger role in determining the media for backup. Frequent backups would necessitate a quick method of backup. While the same media (e.g., disk to disk) might be warranted, security mandates that the same device or account not be used. Generally, complete dumps of all files to tape are performed routinely by most data-processing centers. It is not wise to count upon this for system backup. Especially in smaller hardware systems, provisions should be made by the applications software for backup.

Finally, for every backup operation there must be a means of re-covery. A mirror image of the backup is desirable in this regard. Usually such recovery will totally wipe out the current data file being recovered. Thus, the provisions for recovery should warn the user that such is the case, and that an interim update to bring the backup to current will be required. These procedures must be specified to the user like all other user requirements. Similarly, program specifications for backup and recovery software will follow the general guidelines for other program specifications given in Chapter 6.

5.6 Closure on File Specifications

The objective of this chapter has been to provide the methodology for writing file specifications, given that the basic considerations of Chapter 4 are understood. Two chapters have been dedicated to the subject of files, the first to design and the second to specification. Yet, "Specify File Layouts and Structures" is only one step in the design process. By far, this is the most important step, since errors in the other parts of the design specification are so much more easily rectified than those in the file specifications. Given that the files have been adequately defined and specified, the next three steps in the design process should proceed with considerably fewer difficulties than any of the preceding steps.

Questions and Problems

1. Name three reasons that file specification is the most critical step discussed so far.

2. Name the five activities within the file specification step.

3. Complete a detailed file layout form for the files specified in the case study of problem 44 in Chapter 4.

4. How does the designer know which data elements will go into the files at this point?

5. What three aspects must be considered in order to perform the "specify file record layouts" activity?

6. How do each of the following influence the assignment of variables to files: (1) modular breakdown of the system, (2) system output

requirements, (3) input and maintenance requirements, and (4) use of special utility files?

7. What inputs and primary considerations are available after step 3 on which to base the decision as to the type of record?

8. Explain why the ordering of variables within a given file is of little significance to the hardware requirements. Why, then, does such become so significant to the programmer and the user?

9. How does the record type influence the arrangement of the variables within the record?

10. Name three advantages of arranging the data elements within a record by category.

11. What is meant by, and what is the value of, defining variables by exception?

12. How are the variables within the file layout form defined by exception (three ways)?

13. What considerations determine the development of the organizational Warnier diagrams (OWD) for a given file?

14. How are OWDs and structure diagrams related? What information is conveyed by each which is not conveyed by the other?

15. State the purpose of each of the four components of the structure diagram.

16. Contrast the purpose and content of the file-specification layout and the data-record layout in the structure diagram.

17. What does the content of a link field represent? What is the significance in drawing a link to the key field?

18. What three areas are covered by the term "file integrity"? How is each addressed in the system design?

Chapter

The Process of Synthesis

After two chapters which concentrated upon file design, the reader may be lost with respect to the stepwise sequence of the design process. To re-establish the overall perspective, consider Table 1.3 in Chapter 1. Chapter 3 was dedicated to those steps which were primarily analytical in nature. The first and most important synthesis step, that of specifying the files, was covered in two chapters, namely, 4 and 5. While this step was clearly a synthesis of variables into records and records into files, it was also called a "pivotal" step between analysis and synthesis. This is true primarily because the files themselves are actually components of the system which will be further synthesized in the remaining steps. Note that as we allowed some synthesis to be performed within the analytical steps, some analysis will be performed in the remaining steps which are primarily dedicated to putting the final system together.

The three remaining design steps which follow file specification are:

1. Solidify flow diagrams,
2. Specify input requirements, and
3. Develop program specifications.

Once these three steps are completed, all of the documentation required to develop the system software will be in existence. In addition, all documentation needed to operate the system will also exist. Thus, the highway will be paved and waiting for the software development to begin. The conversion of the design into a plan for the software development will be covered in the next chapter. Prior to that, a major subsection will be devoted to each of the steps given above.

6.1 Solidify Flow Diagrams

This step is primarily one of verification and validation. As such, there is not much that can be said on a conceptual basis. However, this step clearly deserves its independent status. While much data-flow analysis has been performed already, it is clear that this work must be considered preliminary. For, until the files are totally specified, the information to solidify the flow diagrams itself is tentative. Files may have been created, combined, or restructured during the file design step to result in a completely different set of interfacing programs. However, only now that the files and their data structures have been defined can the final flow diagrams be generated.

The term *flow diagrams* has been applied above as opposed to using the term *data flow diagrams* (DFD), since there are several types of flow diagrams that are in need of either creation or updating at this point. These are as follow:

1. Management system Warnier DSTs and DRTs,
2. Warnier flow diagrams,
3. Data flow diagrams, and
4. File-program cross-reference.

Each of these will be discussed as a separate activity in the subsections below.

6.1.1 Solidify Management System Warnier Diagrams (MSWD)

The management system Warnier diagram (MSWD) provided an essential tool to aid in first analyzing the affected management system (AMS), and then transforming this into a definition of the general design of the new system. The former was performed in the very first step, while the transformation took place during step 2. As a result, the perception of the potential new system accomplishments in terms of its outputs and capabilities were documented. The MSWD for the new system should now be updated to reflect the actual file designs. This will be performed by a direct update of the new system MSWD by revising the Data Retrieval Transactions (DRTs) and the Data Storage Transactions (DSTs) appropriately.

6.1.2 Develop Warnier Flow Diagrams

Warnier flow diagrams (WFD) were introduced in Section 2.2.2 primarily as the interface between the data flow diagram and the rest of the system documentation. It can now be seen that the WFD is derived directly from the management system Warnier diagram by simplification and substitution. Thus the WFD becomes an intermediate step in data-flow specification, as well as a useful roadmap of the new system design specifications.

A comparison of Figure 3.1 with Figure 2.28 (pages 100 and 80 respectively) can illustrate how the various elements in the MSWD are transitioned into their counterparts in the Warnier flow diagram. Generally the transition process takes the following form:

1. Replace the extreme left-hand descriptor with the name of the new software system and the term *System Software*.
2. Replace the missions with the modules of the new software system; generally there will be a correspondence although it may not be one-to-one.
3. Replace the projects and activities, including decision-action or mandate procedure groupings, with the corresponding computer programs which will perform or aid in these operations.
4. Use additional levels of analysis if the program structure requires, such that all programs are enumerated.
5. Convert DRTs to links from the respective program to one or more files. The arrowhead pointing to a file "D number" indicates that a file record is to be written or updated. If several files are written by the same program, use a Warnier bracket and list all files to the right (see Figure 2.28 for a generalized example of this symbology).
6. Convert DSTs to links from the respective file or files to the program which reads or queries it. In the case of a query, follow the arrowhead by the letter Q (example: in Figure 2.28, Program 2.2 queries file D5).

Obviously the above procedure is an oversimplification of the process involved. Most experienced designers could, after the study that has been required to this point, produce the Warnier flow diagram with little consideration of the MSWD. The procedure given above illustrates the dependence between these two models. The Warnier flow diagram is just an abstract of the mechanized portions of the MSWD. As such, it provides a condensed illustration of the system design; hence, its value for design purposes.

6.1.3 Solidify Data Flow Diagram

The input to this activity is the data flow diagram (DFD) which resulted when the DFD for the AMS was transformed into the DFD for the new system (Section 3.2.2). Recall that this was done specifically for the purpose of defining output requirements. Since that time the data files have been specified and possibly redefined. For this reason a verification and possible modification is required. During this process the following changes within the data flow diagram will be considered:

1. Addition or combination of files to reflect the final file specifications,
2. Addition or deletion of process nodes to accommodate file changes, and
3. Addition or deletion of data flow links to show the updated data flows between the new or modified nodes.

The output of this activity will be the final system data flow diagram which will form the basis of all subsequent design steps.

6.1.4 Develop File-program Cross-reference

Now that the interactions between the finalized files and processes have been defined, it is possible to create a file-program cross-reference table. This was introduced and an example was given in Chapter 2, Figure 2.32 (page 89). The data files are listed across the top of the table and the programs are listed in the first column. The elements of the table indicate whether the interaction is nonexistent (blank), or if it is input only (I), output only (O), both input and output (B), or query (Q). Certain programs which are used only for maintenance may not be included in the data flow diagrams. For example, a program might be required to create a file and initialize the header record. Such programs may be included as maintenance programs, as exemplified by M1, M2, and M3 in Figure 2.32.

6.2 Specify Input Requirements

The previous step, which finalized the data flows based upon the file specifications, provides essential information for specifying the input requirements. Combined with the file specifications themselves, the designer is now ready to document that part of the system which will be most apparent to the user. It is therefore imperative that a user orientation be assumed at this point. The output documentation for this step will become a major part of the User Guide. However, consistency in the naming of variables will enable the programmer to use this document as a reference in designing all menus, queries, and data-entry mats.

Since the user is the primary target of the documentation resulting from this step, the ordering of the input specifications should be logically directed toward him. In a menu-driven system the organization most easily followed will be ordered as given below:

1. A brief explanation of the supervisory menu (this material might also be appropriate for the system overview part of the documentation),
2. A separate section of the User Guide for each of the options of the supervisory menu,
3. Within each of these sections, separate subsections for each submenu option if another menu is generated, and
4. For nonmenu queries or data-entry mat, an explanation of the input requirements in identically the same order as they will appear to the user.

Figure 6.1 gives a general prototype for a table of contents to set the stage for this documentation. It helps to visualize the target document in this way before delving into the detailed requirements of each component. (For a complete example, see the case study in Part 2.) For this prototype the display of the system supervisory menu is used in the Overview to explain the modules of the system and generally state the objectives of each. The User Guide goes into detail explaining the requirements of the user in a step-by-step manner. Note that the ordering proceeds as though the user began at the top of the supervisory menu and systematically executed each option. However, prior to executing the next option within any given menu, all menu options within pre-

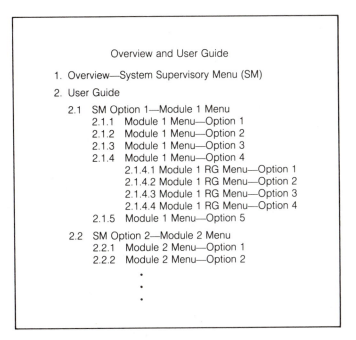

Figure 6.1 Prototype of Input Specification Table of Contents

viously occurring menus would be executed. For example, in Figure 6.1, the Module 1 Report Generator Menu falls within Option 4 of the Module 1 Menu, so all of its options will be explained before going on to Module 1, Option 5, and the Module 1 Menu is totally cleared before going on to the supervisory menu Option for Module 2.

Figure 6.1 could be extended to additional levels of menus, but this presentation should be sufficient to illustrate the structure of the target documentation. This logical ordering is imperative for ease of reference for both the user and the programmer. At this point the contents of each section of the input specifications bears further definition. In order to describe the input types, they will be classified from the most to the least structured, as follows: (1) menus, (2) mats, and (3) queries. The input specifications for each of these will be described in separate subsections. A final subsection is then devoted to the activity of verifying that all necessary data for system output is being generated.

6.2.1 Menus

Technically, menus are not generally used to input data; they are used to give the user control over the system. Please pause to recognize this subtle difference, since the distinction is of great practical significance. When a user selects an option on a menu, no data is being sent to a file (generally speaking), or to a program to be used in calculation. Rather, the information that is sent merely tells the controlling program which one of a number of different options to execute. Indeed, before menus became popular, users often had a set of commands to enter onto the terminal, each of which would execute the option desired directly. No one would have felt that one of these commands was data to the application software*. It is for this reason that we have excluded menu interactions from the data flow diagrams. The reader should verify that the addition of all (and for consistency it would have to be *all*) menu options to a data flow diagram renders it virtually unreadable. Thus, we make a clear distinction between data entries and menu entries.

This distinction, which strongly influences the data flow diagram, should be transparent to the user. In fact, both are inputs to the system. It is for this reason that these are discussed together in the input specification section of the documentation. As discussed above, the documentation sequence followed is that which the user would encounter if each option in the system were systematically executed. Therefore, menus, mats, and queries will be discussed in the order in which they occur. However, menus play a more predominant role in that they dic-

---

*Indeed, it is data to the operating system, but consistency compels us to either ignore this or else consider all interactions with the operating system as data flows. We choose the former for obvious reasons.

```
        WORK ORDER REVIEW AND CONTROL SYSTEM (WORCS)
               WORCS SUPERVISORY MENU (SM)

            1 - INPUT NEW WORK ORDER(S)
            2 - UPDATE EXISTING WORK ORDER(S)
            3 - WORCS REPORTS MENU
            4 - SCHEDULE WORK ORDERS
            5 - PRINT SCHEDULED WORK ORDERS
            6 - PRINT SELECTED WORK ORDER(S)
            7 - DELETE WORK ORDER(S)
            8 - BACK UP OR RECOVER WORK-ORDER FILE
            9 - TERMINATE/LOGOFF

        ENTER FUNCTION =>
```

Figure 6.2 Example of a Supervisory Menu (SM)

tate the structure of the input specification ordering. This can easily be seen from Figure 6.1 in that the word *menu* appears in each title. (Note that in practice this would generally be replaced by the actual name of the option.)

The procedure for the menu specification activity is as follows:

1. Draw the menu under consideration *exactly* as it is to appear on the terminal for the user,
2. In narrative form describe all of the general considerations of the menu that cannot be classified within any one of the options, and
3. In separate subsections, each numbered according to the respective menu option, describe the user interaction in terms of subsequent menus, data-entry mats, or queries.

Figure 6.2 gives an example of a supervisory menu which might be applied to a work-order generation system. Since this is the supervisory menu (i.e., that which appears when the system is brought up), it might be documented in very broad terms within the system overview. This depends heavily upon the total system scope. For example, the work-order system might be a component or a module of a total management information system, including work order, inventory control, preventive maintenance, purchase order issuance, etc. In this case the *system supervisory menu* would be used for illustrative purposes in writing the system overview directed primarily to upper management. One of the options within that menu would be the work-order system, which would be described broadly in terms of its overall purposes and features. Figure 6.2 would then be used within the subsection of the User Guide which describes the user requirements when the work-order module is selected from the system supervisory menu. In turn, there will be a sub-subsection for each of the options on this menu. Further, since Option 3 in Figure 6.2 opens the door to a menu of a number of reports, this menu will be illustrated for further discussion within the third of these sub-subsections. If each report input requirement is brief, further subsectioning need not be required.

The documentation of report-generator menus might require further clarification. The user needs to know the report that is being ordered. Usually, a descriptive name of the report on the menu is sufficient, along with its corresponding explanation in the User Guide. The question may arise as to whether example outputs are required in the User Guide. First note that if this is done it will be integrated into the *input* specifications. Such is generally not the place for output report examples. Recall that the output reports were documented in detail during step 2, "Analyze New System Output Requirements" (see Chapter 3), which was an essential step prior to the subsequent design steps. The output reports were documented in the highest level of detail, and this became part of the Design/Maintenance Manual. In the interest of minimizing redundancy, if an output report needs to be illustrated to the user, these output specifications and prototypes should be referenced. In most cases such referencing will not be required. However, systems which have a very large number of different reports might require an index of output reports, including an example of each, documented within a separate section of the User Guide.

To illustrate the documentation of report-generator menus with an example, consider Option 3 of the menu given in Figure 6.2, and suppose that it, in turn, leads to Figure 6.3. Each option of this menu would require further explanation, possibly in separate subsections. However, detailed output reports need not be given, and they might only need to be referenced for the first few cases. Efficiency can be obtained here by the ordering within the menus. For example, in Figure 6.2 the input and update of the work orders precedes the "print reports" option, as it does in practice. From a documentation standpoint, this arrangement is quite beneficial since the explanation of the input and update functions will explicitly define the terms used in Figure 6.3. Thus, after reading the previous sections of the User Guide, the reader will understand that in selecting Option 6 of Figure 6.3, a listing of all jobs in the system which are waiting for equipment in order to get started will be

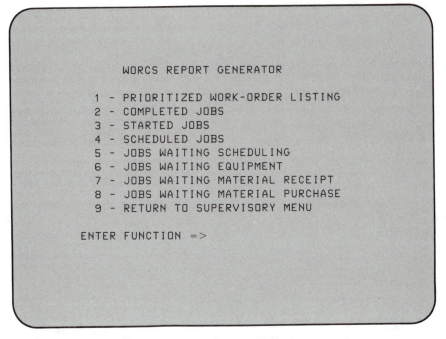

```
              WORCS REPORT GENERATOR

        1 - PRIORITIZED WORK-ORDER LISTING
        2 - COMPLETED JOBS
        3 - STARTED JOBS
        4 - SCHEDULED JOBS
        5 - JOBS WAITING SCHEDULING
        6 - JOBS WAITING EQUIPMENT
        7 - JOBS WAITING MATERIAL RECEIPT
        8 - JOBS WAITING MATERIAL PURCHASE
        9 - RETURN TO SUPERVISORY MENU

      ENTER FUNCTION =>
```

Figure 6.3 Example of a Report-generator Menu

obtained. The particular format and additional information which may be reported about each job is relatively unimportant, since this will become quite self-explanatory as soon as the option is selected.

Complete examples of menu input specifications are given in Part 2. It will be noticed that in describing the documentation of menus we have been skipping around some other necessary aspects of data entry. The next two subsections should fill in these gaps.

6.2.2 Mats

Menus were discussed above first for two reasons: (1) the input specifications are organized primarily by the order of the menu options, and (2) menus are the most structured form of input. While quite powerful in leading the user to understand and effectively interact with the system, menus are limited by the number of options that can be included on the screen at one time. Therefore, most forms of data entry cannot be performed by the use of menus; thus, menus are generally reserved for system control.

A mat is to a computer terminal screen as a form is to paper. Thus, a mat is a data collection form superimposed upon the CRT screen. Data collection is sometimes performed on paper prior to being entered

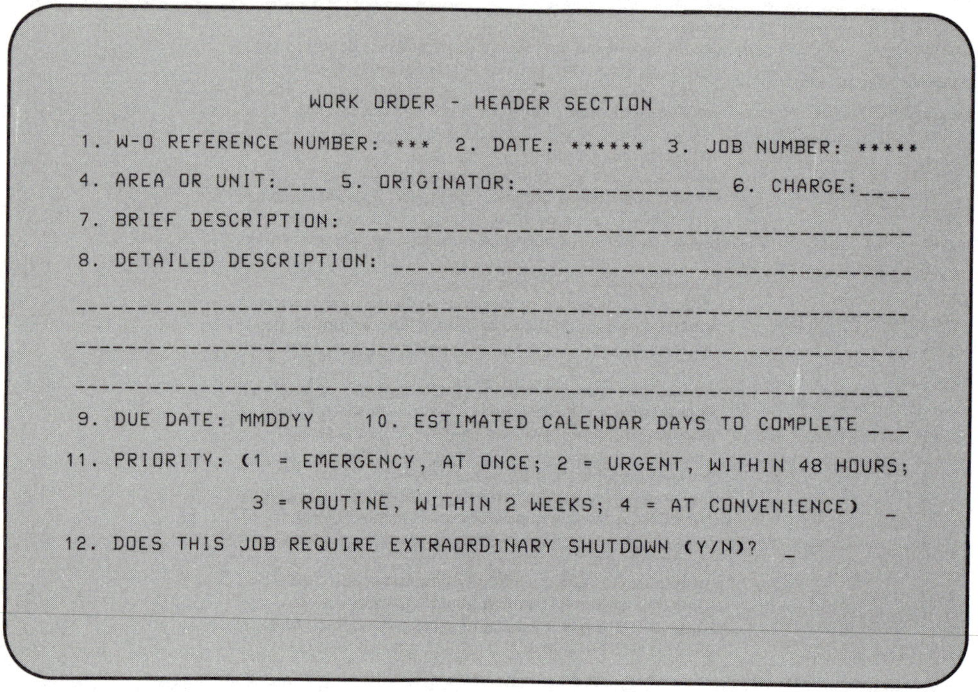

```
                    WORK ORDER - HEADER SECTION
   1. W-O REFERENCE NUMBER: ***  2. DATE: ******  3. JOB NUMBER: *****
   4. AREA OR UNIT:____  5. ORIGINATOR:_____  6. CHARGE:____
   7. BRIEF DESCRIPTION: _____
   8. DETAILED DESCRIPTION: _____

   ------------------------------------------------------------------

   ------------------------------------------------------------------

   ------------------------------------------------------------------
   9. DUE DATE: MMDDYY    10. ESTIMATED CALENDAR DAYS TO COMPLETE ___
   11. PRIORITY: (1 = EMERGENCY, AT ONCE; 2 = URGENT, WITHIN 48 HOURS;
                    3 = ROUTINE, WITHIN 2 WEEKS; 4 = AT CONVENIENCE)  _
   12. DOES THIS JOB REQUIRE EXTRAORDINARY SHUTDOWN (Y/N)?  _
```

* Indicates computer-generated information.

NOTE: A query to make corrections or proceed will appear immediately after the response for Item 12 is entered. The actual mat presented will not be numbered since cursor control during data entry makes this unnecessary. Numbers are used here to assist in the explanation.

Figure 6.4 Example of a Work-order Mat—Header Section .

on the screen. If this is the case, then the two formats should be as close to each other as possible. As obvious as this rule seems, amazingly enough it is probably violated at least as much as it is followed. This can be remedied either by making the paper form coincide with the mat or, which is usually easier, reformatting the mat to look like the paper form.

Ideally, in many systems the data creator should be the user. In such a case there is no reason why the paper step cannot be bypassed and the data entered directly on the input mat. The work-order example introduced in the previous section is such a system. The user would select Option 1 in Figure 6.2 and the data-entry mat given in Figure 6.4 would appear, possibly after some intermediate queries. The cursor will be positioned on the screen to accommodate the first item to be entered. Once entered, the cursor moves to the next field, etc. Mats should be designed in an order which is logical and easy to follow. The mat layout in the documentation should be *exactly* as it will appear on the screen, including the number of spaces provided for each field.

With a little experience, the user will be able to input most data, just by responding to the data-entry mat. However, backup documentation is still required for the following purposes: (1) as an educational tool to get the user started and to provide an understanding of the system data-entry requirements, (2) to provide additional information on those data items which are impossible to thoroughly explain because of space limitations, (3) to provide the programmer with the exact layout of the mat, and (4) to provide the programmer with a knowledge of the data-entry program input specifications.

In order to accomplish these purposes, an itemized list of descriptions should accompany each data-entry mat. Generally, narrative is sufficient for describing the entry requirements. At times, if judgmental coding is required, tables may be included. As a rule, however, it is best to minimize and, if possible, eliminate all need for backup documentation referencing by the user. Thus, if at all possible, programs and mats should be developed to eliminate the second purpose for backup documentation given above. Note, for example, that Item 11 in Figure 6.4 presents the alternatives right on the screen, making backup documentation referencing unnecessary.

To serve the third and fourth purposes given, the backup documentation must enable the programmer to identify the variables being supplied by the user and determine where they are to go in the file. If some are not to be moved directly into file storage, some reference as to the program which will perform their manipulation will be required. The main, unambiguous point of reference here is the file, sequence (file,seq) number. Secondarily, references are made to the variable by name, especially in the backup narratives. If possible, the name assigned in the documentation of the mat should be identical to the field description given in the file layout for those variables which will be moved directly into the file. Since program specifications will be taken up in Section 6.3 below, it is sufficient to state that any subsequent reference to a variable in the program specification must use the identical name as that given. In addition, the file, sequence reference number should be used to eliminate all ambiguities.

The format for the backup documentation is as follows:

1. Item reference number,
2. Item name,
3. Narrative explanation, and
4. File, sequence reference.

The item reference number corresponds only to the number in the mat within the documentation. As indicated by the note at the bottom of Figure 6.4, the item numbers given are for local documentation purposes. They need not appear on the actual mat itself, and generally they have little correspondence to their ultimate sequence number in the file (for those variables which go directly into a file). To try to force such an exact correspondence is unwise; however, the ordering should be as consistent as possible.

Sometimes the item name will not appear directly on the data-entry mat. For example, Item 12 in Figure 6.4 does not show consistency between the item name and the request for information. Since the data-entry mat is to be as user friendly as possible, the forcing of perfect consistency here is not recommended. The mat should respond to the paper data-entry forms and/or the needs of the data-entry operator. Programmers should recognize that the mat is not the place to look to identify the item name; rather, this is stated in the file layout form.

The narrative description portion has as its objective to clarify anything that is not totally clear from the mat itself. It provides a concise definition of the source of the data as well as its ultimate disposition. The disposition is in terms of the "file, sequence" reference which is found in parentheses at the end of the narrative. Since the file layouts have been determined at this point, the assignment of the file and sequence numbers should be no problem. However, their inclusion is not trivial. Not only does it tell the programmer the destination of the data item (both for initial programming and maintenance), it also provides a verification that the data required by the file is being created. At the same time that this is done, the "source" column can be completed on the file layout form (see Section 6.2.4 below).

As an example of possible backup documentation to Figure 6.4, consider the following:

1. Reference number.   This will be assigned by the system. This is a "rotating" three-digit number used for quick and easy reference to specific work orders within the system. All subsequent referencing to particular work orders will use this reference number (D1.1,1).

2. Date.   This is also provided by the system from the system calendar (D1.1,2).

3. Job number.   This number is computer generated in the following way: (a) the first two digits are the year, and (b) the last four digits count the number of work orders for that year. The third character will enumerate the alphabet after counting through 9. This will enable 37,999 work orders to be counted in one year before rolling over a duplicate number (D1.1,3).

4. Area or unit.   A four-character area or unit designation of the location in which the work is to be performed. This must, of necessity, be a short, coded descriptor since it must share one

line of the CRT screen with other data during subsequent processing. Thus, actual locations should be assigned meaningful abbreviations. If there are many of these locations, a code list can be maintained for easy reference (D1.1,4).

5. Originator.   The name of the authorized individual who originated the work order from the field (D1.1,5).

6. Charge code.   The company-assigned departmental or organizational code of the unit that will be charged for the work (omit if not applicable) (D1.1,6).

7. Brief description.   A brief title (40 characters) which describes the job. This will appear on the screen during interactive scheduling so it *must* be descriptive enough to differentiate between jobs (D1.1,7).

8. Detailed description.   This is a detailed description of the work to be performed. It will be printed out on the work order. A maximum of 300 characters is provided (D1.2,1-4).

9. Due date.   The date that the job is expected to be completed in order to maintain the normal operational flow (D1.1,8).

10. Estimated calendar days to complete.   The estimated number of calendar days from the initiation of *work* until the job will be completed. This figure does not have to be perfectly accurate; a good estimate is sufficient. It will be used to warn the user when the due date may not be met because of a late-start condition. Include weekends and holidays in this estimate (D1.1,9).

11. Priority.   The priority assigned to the job according to the codes listed (D1.1,10).

12. Shutdown.   Most maintenance requires some minor interruption of normal activities; this is not what is called a shutdown. Only extraordinary shutdowns that will necessitate special scheduling will be given a "Y." All such jobs will be grouped together and considered separately during interactive scheduling (D1.1,11).

A few comments on this example are in order. Note that an attempt was made to satisfy all four objectives described above. In particular, a brief explanation of each entry informs a new user of the meaning of the terms employed (see especially item 5). Those items, such as item 10, which require interpretation are given such. As importantly, however, some descriptive information is included as much for the programmer as for the user (such as for items 1-3). Finally, the file, sequence references are given exclusively for the programmer, and a note in the introductory narrative should be made to this effect so that it in no way interferes with other users.

6.2.3 Queries

A second user-oriented method of data input is the user query (also called *prompt*). Here the software is programmed to request some information from the user and then to immediately read this information from the terminal. As with mat entries, the information input should be edited carefully for all possible errors that can be detected by the software. Several queries may appear one after the other; however, the mat or menu method of data entry should be employed as opposed to an excessive number of queries.

The specification of queries is best accomplished within the narrative in the sequence in which they occur. This is illustrated in the case study given in Part 2. When specifying queries, be sure that the query itself is spelled out and separated, by spacing, from the surrounding narrative. The query in the documentation should be exactly as it is expected to be seen on the screen. In this way the programmer can know exactly how to word the query by reading the User Guide. Considerable thought should be given to the design of queries to assure that they are: (1) unambiguous, (2) concise, and (3) easy to answer. Queries should be worded and reworded by several designers simultaneously in order to achieve these objectives. Several examples will be given below to illustrate query-design considerations.

In the example given above for work-order scheduling, Option 5 on the supervisory menu (see Figure 6.2) prints out the scheduled work orders. Suppose that we wish to give the user the option to modify the issuance date so that, for example, a work order generated today could be issued some time in the future. Let us further assume that the computer system has the capability of generating the current date, as most do. Immediately upon the user selection of Option 5 from the supervisory menu, the following query might appear:

ENTER ISSUANCE DATE (mmddyy) => _ _ _ _ _ _

If it has not been described already, the User Guide should explain exactly the meaning of the term "issuance date." This is true of any terminology in the query which may not be generally understood. Note that this query has also supplied the format (mmddyy) so that the user need not have to remember or inquire as to the entry format. A preferable method of stating the format, which is supported on many microcomputer systems would appear as follows:

ENTER ISSUANCE DATE => mmddyy

The advantage of this method of prescribing the format is that it is right before the user, the cursor being under (or on) the first character to be entered.

As stated above, the program specifications should assure that any software-detectable errors are trapped and corrected before proceeding. For example, a month entry greater than 12, could be trapped and a quick error message to this effect displayed on the screen. The software should then return immediately to the query to give the user another chance to re-enter. Further, single-digit month or day entries might be immediately reformatted and presented so that the user is reassured that the right value was entered (if this is not possible then this too would need to be detected for correction). In no case should a program abnormally terminate (abend) due to an incorrect user entry. Provisions should be made for all possibilities. However, the User Guide is *not* the place to document all of these exceptions. This will bog users down in unnecessary and (to them) irrelevant detail. A statement may be made to the effect that "any erroneous entry will be intercepted, giving the user an opportunity to re-enter an acceptable value." If the details of the entry-edit procedures need to be spelled out, this can be done within the program specifications.

The use of a default in a query can greatly facilitate user acceptance. In the example given above, it might be known that 95% of the time the current date will be entered for the issuance date. Since the current date is retrievable, why should the user have to re-enter it? One way to handle this would be to use a first query as a condition, such as:

```
IS ISSUANCE DATE SAME AS CURRENT DATE (Y/N) => _
```

Depending upon the user's entry of Y or N, the user will be further queried to enter the issuance date. Since Y and N are easier to enter than the date, this will save considerable time, especially if all output reports have such a dating option. However, there may be an even better way, as follows:

```
ENTER ISSUANCE DATE (default: 052484) => mmddyy
```

Here the documentation will inform the user of which key (usually the ENTER) will automatically produce the default, which is the software-generated date. Users very easily get used to quickly hitting the enter (or other default) key to obtain the default value. The example given immediately above is preferable since it eliminates one of the queries and at the same time presents the value of the default to the user.

If improperly applied, defaults can be counterproductive. For example, some systems, probably out of designer or programmer laziness, have been set up to check for only one value, allowing any other entry to default to the other alternative. As an example, if a (Y/N) response is required, the program would check for a Y entry, and if not Y, it would take anything else as a default to N. Such can be extremely aggravating to the user, especially if Y was the intended target and the wrong key was hit. If the options are Y/N, or any other subset of responses, the software should check to see that one of the acceptable response options has been selected and give the user another chance if any other key is hit.

At times it is good to augment a query with additional information. For example, suppose that the user has just been through a fairly complex work-order scheduling operation, after which a list of those jobs which were scheduled is presented, and the user is asked to verify if they are correct or not. The following query might be applied:

```
IS THIS SCHEDULE CORRECT (Y/N)? _
    Y - Return to Supervisory Menu;
        Save Schedule
    N - Return to Scheduling.
```

Technically, this is a menu, and, generally, if the number of discrete options requiring some explanation (as opposed to single word queries) exceeds two, the menu format is preferable.

Sometimes the sequence of queries becomes quite complex. This occurs where a subsequent set of queries depends upon the first. To illustrate a simple case of conditional execution, Figure 6.5 presents the documentation for Option 6 of Figure 6.2. In this case the middle query will be omitted if the answer to the first query is N. Note how simple

---

F6 - PRINT SELECTED WORK ORDERS

     The purpose of this function is to provide a direct method for printing a specific work order. This might be required if, for example, a work order got lost. By using this function, one or more work orders can be viewed on the screen or sent to the printer. The queries are quite simple:

```
DO YOU WANT A LISTING OF ALL WORK ORDERS (Y/N)? => _
```

If the response is N, then the following appears:

```
ENTER WORK ORDER REFERENCE NUMBER => _
```

In either case, this will be followed by:

```
DO YOU ALSO WANT OUTPUT SENT TO THE PRINTER (Y/N)? => _
```

The system will now produce the output requested and then return to the Supervisory Menu.

Figure 6.5 Example of Query Documentation For Figure 6.2, Option 6

this documentation is, assuming that all terminology local to the work-order system has been defined. (If not, such terms as *Work-order Reference Number* would have to be defined here.) It provides both the user and the programmer with the information needed to proceed. However, in more complex cases of conditional query execution the narrative might need to refer to an interactive process specification Warnier diagram as discussed in Section 6.3.

6.2.4 Verify File Data Sources

There is a natural tendency to attempt to order the design activities in the same sequence as the system itself. If this were the case, the input requirements would come first. However, we have placed this step sixth in the design methodology, after the output and the files are completely specified, and it is to be performed at the next to the last design step, which is the development of the program specifications. This reflects one of the major tenets of this book: that of fitting the design to the output requirements of the organization.

The input specification documentation activities given above were stated totally in terms of producing a User Guide. This document, which guides the user in a logical manner through the system, must be in existence *before* the first line of code is written. This is true because it provides the programmer with the overall structure of the system as well as the exact wording of all menus, mats, and queries. The bypassing of this step improperly places design responsibility upon the programmer. On the other hand, if the input specifications have been fully developed based upon the successful completion (documentation) of all previous steps, the program specifications can now be written.

One final activity is essential prior to proceeding to the development of the program specifications. This is the verification that all data requirements of the files have been met by the data entry. This can be performed quite easily by systematically going through the data-entry specifications and posting the data-entry mat and item-number reference to the "source" column, as was described in Section 2.3.1. The data-entry mat should be given a short acronym for this reference (example: Order Data-Entry Mat → ODEM). The item number can then be appended to this (example: ODEM-5 is Item 5 of the ODEM). Figures 5.1 and 5.2 gave examples of file layout forms which were so annotated. However, it was stated at that time that the posting of these source references could not be performed until the data-entry specifications were completed. It was recommended in Section 6.2.2 above that this posting could be done at the same time that the file, sequence numbers were added at the end of each data item description, since one is a reference to destination while the other is to source. This is the most convenient time to perform the cross-referencing, and it assures that the two references are consistent.

All data items which are not moved directly from a data-entry mat into a file will lack a source column reference. These data elements require modification (preprocessing) before their values are determined. Their source references are not adequately described by a data-entry mat reference. Rather, a program specification number is required in these cases. It is essential before concluding the data-entry specification activity to assure that all information required by the files is being generated by data entry. A review should therefore be made of all variables which do not currently have a source reference. All such variables should be obtainable, through numeric computation, from the others.

Finally, a list of all the variables which do not have a data-entry mat source reference will be made at this time. Ultimately these will receive a program specification number as a source reference. However, at this point the program specifications have not yet been developed or numbered. This list will provide a check for the program specifications, since all of these variables must be defined within at least one of them. As the program specifications are written, the number of the program specification which defines a variable will be posted to the source column of the file layout form for that variable. Following this, a final review of the source column for all variables will verify that *all* of the data requirements for the system have been met.

6.3 Develop Program Specifications

At this point all of the information necessary for writing the program specifications exists. Because of this, some designers have erred in thinking that the program specifications are redundant and therefore unnecessary. Further, the writing of the program specifications is a time-consuming process which can only be partially delegated to technicians. It is no wonder, then, that in many shops program specifications never get written. Yet, the writing of the program specifications is essential in that it serves general purposes that cannot be accomplished by any other mechanism. Among these are: (1) to furnish a final synthesis step which validates all previous design steps, (2) to provide an unambiguous work statement for the programmers, and (3) to provide the documentation for all subsequent system maintenance and modifications. For these reasons we state what we believe should be a rule in any good system development effort: Never begin software development on any system until all of the program specifications have been written.

The amount of work that is employed in many current program specification efforts may be quite frightening to many designers. To view such as wasted effort is largely justified since much of it is unnecessary. However, the reason that such rigor would be required would be that the foundation work had not been done. Specifically, two basic shortcomings could cause this problem: (1) the system programs have not been adequately analyzed in terms of their inputs, outputs, and relationships with the files and each other; and (2) inadequate documentation exists to ease the program specification effort by referencing. An attempt is then made to make up for these two shortcomings by increasing the complexity of the program specifications.

As opposed to this rather grim picture, let us assume that the six previous steps in the design effort are satisfactorily completed. This leaves the designers with the following inputs to be exploited during the program specification step:

1. The system flow diagrams,
2. The system input specifications,
3. The system file specifications, and
4. The system output report specifications.

In essence, the program specifications either transform Item 2 above into Item 3, or else they transform Item 3 into Item 4. It is the flow diagrams which initially define the program that performs each transformation. By referencing the four items listed above, the writing of most program specifications can become quite routine and simple.

The techniques to be employed in this process were given in Sections 2.1.3–5. They included the use of: (1) process specification Warnier diagrams (PSWDs), (2) decision Warnier diagrams (DWDs), and (3) organizational Warnier diagrams (OWDs). A knowledge of the basic concepts presented in Section 2.1 will be assumed in the discussion below, and the reader is encouraged to review this material before proceeding. The question that we wish to address at this point is: given that we have the four "inputs" to this decision step presented above, and given that we know the techniques presented in Section 2.1, how do we now proceed to generate the program specifications? We suggest that the designer start at the beginning and *attack*, one program at a time, until each is specified. In this case the "beginning" is the first thing that the user sees when the system is brought up—the supervisory menu. And the ordering involved will follow much the same pattern as did the input specifications; namely, a systematic, orderly traversal of the system. Since menu programs are slightly different in their approach than processing programs, these two will be discussed in separate subsections below.

6.3.1 Menu Program Specification

A menu program is a program which puts a menu on the screen and, after reading the user's response to it, executes another program based upon that response. The discussion above with respect to the four items of input to the process of program specification do not apply to menu programs. Menus generally do not appear on the data flow diagram (DFD), since they do not directly impact upon data flows. Rather, their function is primarily that of control. Since the menu programs are not included in the DFD, they cannot be numbered and referenced by it. While these considerations might seem to be restrictive, they present no problem in practice, as shown below.

Let us first take up the problem of numbering. Each process in the DFD is given a number. In referencing this process, we have preceded the number by a "P" to indicate that it was a process, e.g., P1, P2, P3. Generally, each module will be assigned a primary number. It is convenient, if a module is exploded to several processes within the DFD, to add a decimal in order to number the processes, e.g., P1.1, P1.2, P1.3. In turn, programs within processes, subprocesses, or sub-subprocesses can be further assigned a number. Thus, if P1.1 requires three programs to accomplish its function, they will be numbered P1.1.1, P1.1.2, and P1.1.3. Thus we have a unique numbering system which ties each program specification to its process node in the DFD. These program specifications can be ordered in the documentation by these assigned numbers, which form a quick index to any given program specification. As a table of contents to the program specifications, the description of each program will be listed in sequence by program number.

To accommodate menu programs into this scheme, their program numbers can be preceded by a process number of zero. Thus, the supervisory menu can be P0.1. Each subsequent *menu* called by the supervisory menu will be P0.1.1, P0.1.2, etc. If any of these executes a further menu program, a further extension can be made. Figure 6.6 illustrates the case in which there are four options in a supervisory menu, each of which invokes another menu. While Figure 6.6 uses the program specification Warnier diagram (PSWD) methods explained in Section 2.1.3, there are a number of features of application to menu programs that bear mentioning, as follow:

1. The general format of major descriptor (on left of Figure 6.6) is: program name, program reference number, calling program reference number. In the supervisory menu case only, there is no calling program; therefore, the term "from logon procedures" is substituted.
2. The input to the procedure is just the user selection; note that the Overview and User Guide document is referenced so the programmer can see the actual menu in the documentation.
3. The logical statement references programs that will be executed or "chained to." Since each begins with a zero, it is known that each is another menu.
4. After the logical statement there is a reference to the program specification Warnier diagram (PSWD) which specifies the pro-

.INPUT—User selection from
            Supervisory Menu—O & UG PART 1

Chain to P0.1.1—PSWD 0.1.1
.Condition—User selects 1
            Descriptor*

            (+)

Chain to P0.1.2—PSWD 0.1.2
.Condition—User selects 2
            Descriptor*

System   (+)
Supervisory
Menu   Chain to P0.1.3—PSWD 0.1.3
Program   .Condition—User selects 3
P0.1   Descriptor*

from   (+)
Logon
Procedures Chain to P0.1.4—PSWD 0.1.4
            .Condition—User selects 4
                        Descriptor*

            (+)

Logoff User
.Condition—User selects 5
            Logoff/Terminate

.OUTPUT—None

PSWD 0.1 System Supervisory Menu

*In actual application this will be the menu descriptor corresponding to each option.

Figure 6.6 Prototype Program Specification for Supervisory Menu Program

gram mentioned. Generally the PSWD number corresponds exactly to the P number, although additional digits are required
where several PSWDs are required for one process node on the
data flow diagram. However, it is always possible to tell the node
number from the PSWD number.

5. Under each logical statement is the condition under which that
statement will be executed. This condition also includes a statement of the actual menu descriptor for the option (in the Figure
6.6 prototype this is indicated by the word *Descriptor*).

6. The output from this program, as well as most menu programs,
is nonexistent. The program does nothing other than invoke
other programs. Therefore the output is stated to be "none."

Once the system supervisory menu program is specified the sequence of the program specification is defined. In Figure 6.6, four programs have been referenced within P0.1, and each of these will require
a program specification. To illustrate this by extending the prototype
example, Figure 6.7 presents a possible specification for P0.1.1, the pro-

Module 1
Supervisory
Menu
Program
P0.1.1
from
P0.1
{
.INPUT—User selection from Module 1
Supervisory Menu—O & UG 2.1.1

Chain to P1.1
.Condition—User selects 1
Descriptor*

(+)
Chain to P1.2 (Option: NEW)
.Condition—User selects 2
Descriptor*

(+)
Chain to P1.2 (Option: OLD)
.Condition—User selects 3
Descriptor*

(+)
Chain to P0.1.1.1
.Condition—User selects 4
Descriptor*

.(+)
Chain to P1.3
.Condition—User selects 5
Descriptor*

(+)
Chain to P1.4
.Condition—User selects 6
Descriptor*

(+)
Chain to P1.5
.Condition—User selects 7
Descriptor*

(+)
Chain to P0.1
.Condition—User selects 8

.OUTPUT—OLD/NEW to P1.2

PSWD 0.1.1 Module 1 Supervisory Menu

*In actual application this will be the menu descriptor
corresponding to each option.

Figure 6.7 Specification for P0.1.1 from Figure 6.6

gram invoked by the first option. To keep this example as general as possible, we will call this program the Module 1 Supervisory Menu Program; we are assuming that there are several of these module supervisory menus which are called up by the system supervisory menu program. Note the following differences which further illustrate the features of this method of specification:

1. The major descriptor references back to the calling program P0.1. It is essential that the reader always be able to determine the calling program(s). When several programs could invoke the same program this reference will need to be generalized; however, it is usually unique.

2. The input to this program is not, in this case, from the calling program. The distinction between the invoking of a program and the provision of data to a program must be clearly maintained. Thus, the input does not reference the calling program. Note that it does reference the particular section of the Overview and User Guide where the menu is given.

3. Options 1 through 3, and 5 through 7 invoke processing programs as opposed to menu programs. Note that the PSWD reference can be omitted since it has the same number as the program. The program numbers given here refer to the process nodes of the data flow diagram. Note that Options 2 and 3 within this example invoke the same program, P1.2. In this case data must be passed to P1.2 so that it can determine which option was selected. This data can be generated by the calling program, and it is specified in the parentheses following the logical statement. Thus a variable called *Option* will be set equal to "NEW" if Option 2 is selected, while it may be set equal to "OLD" if Option 3 is selected. This will appear as one of the inputs to P1.2, and portions of P1.2 will be executed conditionally depending upon this value.

4. Option 4 invokes another menu program, P0.1.1.1, so identified by a zero in the first-digit position. This might be, for example, a report-generator menu for this module, by which the user can select the report desired.

5. Option 8 returns control to the supervisory menu.

Figure 6.7 has defined the need for the development of program specifications for six additional programs. There might be a tendency for the designer to become overwhelmed while performing this synthesis. We urge that a systematic approach be adopted by developing the program specifications consistently in the order in which they are numbered. That is, develop P0.1 first, followed by P0.1.1, P0.1.1.1, etc.; then P0.2, etc.; then P1.1, P1.1.1, P1.1.1.1, etc.; followed by P1.1.2, P1.1.2.1, etc.; and, ultimately, P1.2, etc.; until Module 1 is cleared; then proceed to Module 2. This systematic development of the specifications will generally guarantee that a program specification is not written before the programs which call it are specified. Further, it provides a systematic approach to solving what seems collectively to be an extremely complex problem. However, when one program specification is written at a time, and when it is written at the point when all calling and input programs to it have already been specified, then the specification task becomes greatly simplified.

The program specifications in the documentation, which will become a part of the Design/Maintenance Manual, will also be ordered in this numerical sequence. The development of the program specifications in the same order in which they appear in the documentation brings with it a great clerical savings as a side benefit. An attempt has been made to order the presentation here in this way. At this point, additional consideration can be given to the specification of processing programs.

6.3.2 Process Program Specification

The source materials used to write the program specifications were presented in the beginning of this section. To reiterate, they are: (1) flow diagrams, (2) input specifications, (3) file specifications, and (4) output report specifications. We can now add to these the menu program specifications, which will provide both order to the development of the process program specifications and, in certain cases, input data to be referenced by them.

The format for the process program specifications is identical to that for menu program specification. However, while the menu program specifications did little more than invoke new programs, the process program specifications must explicitly state to the programmer exactly how data is to be transformed from input to output. This would indeed be a formidable task were it not for the resources at the designer's disposal due to the prior completed steps in the design process. Through referencing, the program specifications can be made very brief. As a rule, the designer should never repeat anything that already appears in the system documentation. The objective is to make the program specifications as brief as possible without omitting any unnecessary information.

Since the overall structure of the program specifications were given in Chapter 2, the emphasis upon brevity can best be illustrated at this point by an example. Let us consider Figure 6.7 as a basis and suppose that P1.1 within that specification is a data-entry program. Assume that the descriptor within the menu is:

    ACCOUNT CREATION OR MODIFICATION

which indicates to the user that this option enables either the creation of a new account or the modification of the details of an existing one. The reader should not be concerned with the details of the system under consideration; rather, assume that the input, file, and output specifications are available (as the programmer would). Figure 6.8 gives a possible PSWD for P1.1. Let us traverse it systematically and see that it contains all that the programmer needs to write the software to perform this process. This will be done in the following:

1. The major descriptor on the left gives the program name and number, and identifies which program invokes it.
2. The input section references the Overview and User Guide (O & UG) Section 2.1.1, which contains the input mat and a description of all variables which are to be input to this program.
3. The first statement defines the variable "Option." This requires a query from the user which is totally spelled out in O & UG 2.1.1. Its value will either be an existing account number or an indication that a new account is to be created.
4. The second step is a read into a temporary record in memory. If the user specifies an existing account number, this will be read

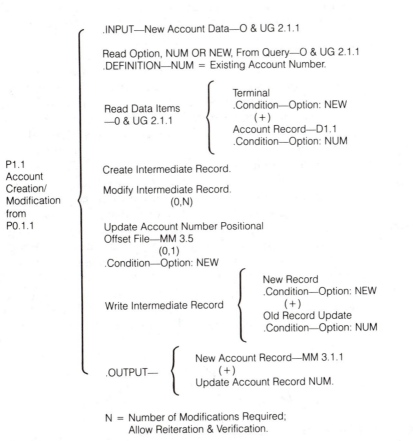

.INPUT—New Account Data—O & UG 2.1.1

Read Option, NUM OR NEW, From Query—O & UG 2.1.1
.DEFINITION—NUM = Existing Account Number.

Read Data Items —0 & UG 2.1.1
{
Terminal
.Condition—Option: NEW
(+)
Account Record—D1.1
.Condition—Option: NUM
}

P1.1 Account Creation/ Modification from P0.1.1

Create Intermediate Record.

Modify Intermediate Record.
(0,N)

Update Account Number Positional Offset File—MM 3.5
(0,1)
.Condition—Option: NEW

Write Intermediate Record
{
New Record
.Condition—Option: NEW
(+)
Old Record Update
.Condition—Option: NUM
}

.OUTPUT—
{
New Account Record—MM 3.1.1
(+)
Update Account Record NUM.
}

N = Number of Modifications Required;
Allow Reiteration & Verification.

Figure 6.8 PSWD 1.1 Account Creation/Modification Program

from D1.1. If the user indicates that a new record is to be created, then the data-entry mat specified in O & UG 2.2.1 will be used to input a new record from the terminal.

5. After the intermediate record is created, the user is given as many chances to modify the record as required.

6. The next step is executed only if this is a new record. This is the procedure that generates the new account number as well as the file offset index. It is documented in the Maintenance Manual (MM) Section 3.5.

7. The next step is to write out the intermediate record either to a new record position or as an update to an existing account.

8. The program output is stated to be either a new record or the update of the old record. In either case the record layout is given in the Maintenance Manual Section 3.1.1. Since there are no new variables created or variables modified in this specification, the source column of the file layout form will totally specify all variable moves. Computations and conditional executions would be specified using the basic principles of the Warnier diagram.

The above specification is adequate to provide any programmer sufficient guidance in developing the software necessary for this process. It is simple only because all inputs and outputs have already been specified in the documentation. Yet, as simple as it is, it still provides an essential, unambiguous specification of the programming requirements. It pulls together all of the references needed. Generally, these will fit on one page, and an attempt should be made to keep each program specification to one page, if possible. This can generally be done by the use of subprogram specifications (see Case Study, Part 2).

Program specifications can become much more complex than Figure 6.7 when there are a large number of transformations and/or decision steps. The use of decision and organizational Warnier diagrams greatly facilitates this process. These techniques are further exemplified in the case study given in Part 2. Although the writing of the program specifications completes the design steps, two things remain to be done prior to effective software development. The first is the assembly of the documentation into its final form, and the second is the documentation of the software development plan. These will be covered in the next chapter.

Questions and Problems

1. Explain why the last three steps are primarily synthesis steps. What elements of analysis might be performed in these steps?

2. Name the four types of flow diagrams that need to be finalized, and state the step and activity in which they were first created.

3. Since the Warnier flow diagram is just a condensation of the management system Warnier diagram (MSWD), what advantage does it have over the MSWD?

4. What are the primary and secondary uses of the file-program cross-reference tables, and who are the users of these?

5. What steps are taken to assure that the program input requirements satisfy both the needs of the user and those of the programmer?

6. Why does the ordering of the input specifications not follow the overall organization of the system?

7. State the similarities and differences between a menu, a mat, and a query.

8. Why are menu interactions not shown on data flow diagrams?

9. Why do menus dictate the overall order of the input report specifications as opposed to mats and queries?

10. What problems would be encountered if output report examples were included in the documentation of the input required in response to a selection from a report-generator menu?

11. If menus are the most efficient means of obtaining information from the user, why are they not used for all data entry?

12. Define what is meant by a data-entry mat.

13. When would hard-copy data collection forms be required as opposed to their elimination?

14. State and discuss the four purposes for data-entry mat backup documentation.

15. How does the programmer identify those variables which are moved directly from the data-entry mat to the file? If computations of the data-entry variables are required prior to storage, how would these variables be referenced?

16. Why might there be inconsistency between the item name (used in the file layout and the item backup documentation), and the description on the data-entry mat? Explain in terms of the purposes of each.

17. State the basic difference between a menu and a query. When is each appropriate?

18. Name the three criteria for an effective query.

19. What are the advantages and dangers of incorporating defaults into queries?

20. State how the following variables might be error-checked to assure that erroneous entries will not be further processed:

   a. Day of the month,
   b. Integer between zero and 10, inclusive,
   c. Numeric value between $-10,000.00$ and $+10,000.00$.

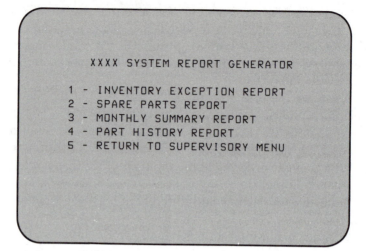

```
            XXXX SYSTEM REPORT GENERATOR

        1 -  INVENTORY EXCEPTION REPORT
        2 -  SPARE PARTS REPORT
        3 -  MONTHLY SUMMARY REPORT
        4 -  PART HISTORY REPORT
        5 -  RETURN TO SUPERVISORY MENU
```

Figure E6.1 Sample Menu for Problem 27

21. Why is it essential to complete the User Guide before the software is developed?

22. What are the three purposes of the program specifications?

23. What insufficiencies in prior documentation efforts might lead to the necessity of considerable additional work in program specification?

24. What is the key to the simplification of the program specifications? Discuss in terms of those documents referenced.

25. What three techniques are employed for program specification?

26. In what order are the program specifications developed?

27. Assume that Option 4 in Figure 6.7 leads to the report-generator menu given in Figure E6.1. Write the program specification for this menu program. Assume that the report-generator programs are numbered 1.7, 1.8, 1.9, and 1.10 for the four reports, respectively.

28. Write a program specification Warnier diagram with the following characteristics:

a. It is for a subroutine with a descriptor "Update Indirect Costs," which is called from program P1.2; its reference number is P1.2.6.

b. Its input is the file D1.1, the dollar amount of the transaction (DA), the transaction subcode (TS), and account number (AN); these last three variables are queried from the user.

c. The subroutine will either perform item d or item e below, but not both. It will perform item d if the TS is equal to the eighth variable within D1.1 for record AN. Otherwise it will perform item e.

d. Terminate the procedure without making an update.
e. Update overhead items by adding DA times the seventh variable within D1.1 for record AN to variables 60, 61, 62, and 63 within this same record.
f. The output is the update to D1.1, items 60-63.

29. Write the program specifications for P1.2.2 and P1.2.3, both of which are subroutines called within program P1.2. P1.2.2 will be exercised if the "expenditure type" being handled by the routine is type code 1 (non-purchase-order expenditure), while P1.2.3 will be employed with a code 2 type, which is a purchase-order expenditure. The input to both routines is the same:

a. Transaction data-entry mat (TDEM) information documented in the Overview and User Guide (O & UG) Section 2.1.2,
b. The account record of the account specified by the user, from file D1.1, and
c. The OFFSET which *has been determined* by a previous routine. This is a function of the account record specified, and it tells the beginning point of this account record in D1.1.

The procedure for routine P1.2.2 is:

a. Add the dollar amount of the entry from TDEM Item 10, to D1.1, variable sequence numbers 11, 12, 13, and 14; this will be the position of the variable in the file plus the OFFSET.
b. Determine if the balance is negative and print the appropriate message as given in the O & UG 2.1.2.
c. Print a query in order to verify as specified in O & UG 2.1.2.

The procedure for routine P1.2.3 is identical, with the exception that for item a, variable sequence 15 and 16 are modified, rather than 11-14.

Chapter

Design and Development Plans

Toward the end of Section 1.2 some discussion was given differentiating between planning tools and communication tools. One main theme of the first six chapters of this book was the need for stimulating effective communication from management to the designers so that they could understand the true needs of the organization. Responsibility was placed upon the designers to generate effective written communication (documentation) for the programmers and users of the system so that the software could be developed to meet those needs. Thus, the emphasis has been upon communication tools rather than upon planning tools.

Now that the communication tools are in the reader's background, we can turn our attention to the application of some planning tools for the design and development of the software under design consideration. These will be presented in terms of applications of some of the simple planning techniques, without which the control of the design and development processes can be very easily lost. To understand this, consider the basic elements of control. It should be intuitively obvious that to maintain control over any system, the following three elements must be present: (1) a goal, analyzed into a measurable set of objectives; (2) measurements to determine the degree to which the goal is being met; and (3) a correction capability to modify the process when the measurements indicate that there is a deviation from the accomplishment of the goal. Visualize any system subject to control (e.g., a thermostatically controlled heating system, a driver-automobile system, etc.) and identify the goal, measurement, and correction capabilities. Then remove any of these elements and verify that the result is a total lack of control.

The objective of this chapter is to show how the communication tools presented in the first six chapters can be further developed into documents useful for establishing control over the software development process. Consideration will also be given to the development of a plan for the design process itself. Without a plan for accomplishing both the design and the development, the goal and measurement elements of project control are disabled, and success becomes largely a matter of chance. Thus the plans provide the roadmaps to both of these complex efforts.

It is not the objective of this chapter to present a comprehensive review of planning tools. Very basic tools, such as Gantt charting and the critical path method (CPM), will be used to illustrate how a plan can be developed as a function of the design documentation. Readers who are interested in the more sophisticated planning tools and techniques should consult the references at the end of this chapter.

Planning will be divided into two categories addressed in separate sections below, as follow: (1) the design plan, and (2) the development plan. The use of both of these will depend heavily upon an understanding of the components of the system software documentation described in Chapters 3, 5, and 6 above. Since all planning is formulated in terms of this documentation, these sections will provide a review of the methodology previously presented. Prior to these, an introductory section on planning will be given.

7.1 Introduction to Software Project Planning

There are certain concepts and techniques of planning that can best be presented apart from their applications. These will apply to large design and development projects in any area. Their presentation provides the means by which project goals and measurements can be established. In this way, two of the three necessary elements of control are established.

7.1.1 Milestones and Deliverables

A *milestone* is a clearly defined event which indicates that a measurable level of progress on a project has been attained. A *deliverable* is either a document or a demonstrably functioning software module (or both) which provide evidence that a milestone is passed. Thus, the two terms might be used interchangeably to designate two aspects of defining the goals of a project.

Table 7.1   Example of System Software Documentation
Table of Contents

Overview and User Guide (O & UG)

1.0 System Overview

2.0 System Functions
    2.i   Module i Functions
    2.i.j Option j of Module i

3.0 Technical Operations

Design/Maintenance Manual (MM)

1.0 Introduction

2.0 Output Report/Capability Specifications
    2.i   Module i Output Reports
    2.i.j Report j Documentation of Module i

3.0 File Design
    3.i Design of File Group i
        (i = module or other category type)

4.0 Process Specifications
    4.1 Flow Diagrams
    4.2 Program Specifications

5.0 Documented Program Listings

Appendix: Management System Warnier Diagrams

For planning and control purposes, it is essential that milestones and their accompanying deliverables be established. Note the difference between this definition of objectives and that discussed in Section 3.1.2. There, the goals and objectives of the *organization* were under consideration; now the concern is the *project*. This might be either the design project or the development project. In either case, the objectives, as defined by project milestones and deliverables, are clearly distinguishable from those of the organization (although the two should be consistent).

It has been stated that one reason for the early success of NASA in putting a man on the moon is that everyone could clearly see the goal (at least during most clear nights). Milestones and deliverables cannot be abstractly specified, such as "Documentation of System Goals." Rather, they must be clearly defined in terms of their content and format, preferably in terms of an example. It is for this reason that Part II of this book gives a complete example of the end product of the design project. The design team must have a clear target as to their final product; experienced design teams rarely "start from scratch," but rather, they transform existing example documentation into that which is required for the new application.

In order to reference the target documentation in the remaining sections of this chapter, a general table of contents for the system software documentation is given in Table 7.1. This organization should be modified to fit the specific needs of the software design project.

7.1.2 Warnier and CPM Techniques

The theory behind the use of Warnier diagrams for analyzing the activity steps necessary for project plan specification was given above in Section 2.1. The particular application to be made here is given by the general structure displayed in Figure 7.1. This application of the Warnier technique is called a planning Warnier diagram (PWD). Here the *project* may be either the design project (Section 7.2) or the development project (Section 7.3). Large projects of either type may require an analysis into *phases*. Each phase will be further broken into those *activities* which are required for its completion. Associated with each activity are the milestone and deliverables which measure its successful completion.

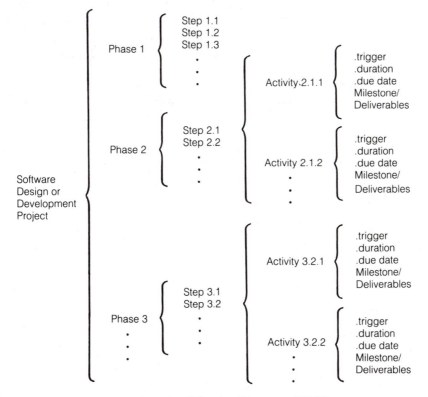

Figure 7.1 Generalized Planning Warnier Diagram (PWD)

The PWD is the most convenient way to get started in establishing a plan. Further levels, such as subactivities, may be added if required. The top-down ordering indicates the general order in which the activities will be performed. Contingency plans can be formulated using the OR symbol, or by using the label descriptor .condition for conditional execution. Milestones and/or deliverables are specified right on the diagram, enabling the project manager to determine if the project is meeting its target objectives. Further, associated with each activity is a *trigger*, a *duration*, and a *due date*. The activity trigger is generally specified in terms of other activity completions, which must take place before the activity can begin. The duration is the number of time units (days, weeks, or months) that the activity is expected to require for completion. The due date is just the expected date at which the milestone will be passed and/or the deliverables for the activity received in completed form. The numbering system used for phases, steps, and activities enables ready reference both within the diagram and to backup documentation, which might be required to thoroughly define each activity.

While the PWD is a powerful and flexible tool for the analysis necessary to generate a plan, it is limited because of its straight-line list structure. Because of the concurrent performance of activities, and their possible nonsequential interdependence, the PWD does not enable the project manager to effectively visualize the relationship between activities. Another planning tool, the critical path method (CPM), aids in solving this problem. CPM is a graphic technique for representing the activities in a plan. The PWD provides the basic inputs to CPM, namely, the specification of the activities, their interdependencies, and their durations. Table 7.2 gives a general example of the CPM data requirements. The first is an estimate of each activity duration in some com-

Table 7.2   Example of CPM Data Requirements

Activity	Activity Duration	Precedent Activities
2.1.1	2	---
2.1.2	1	2.1.1
2.1.3	5	2.1.1
2.2.1	3	2.1.2
2.2.2	2	2.1.2
2.2.3	1	2.2.1
2.2.4	4	2.2.2
3.1.1	5	2.2.3, 2.2.4
3.1.2	3	2.1.3
3.1.3	1	3.1.2

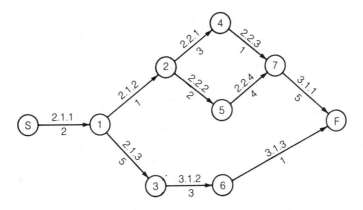

Figure 7.2 Graph of CPM Example

mon time unit, usually days. The second is the activity or activities which must be completed before the activity in the first column can be started. This can be obtained from the ".trigger" entries of the PWD (see Figure 7.1).

To understand the value of CPM, it is advantageous to present this method graphically. The CPM graph has links and nodes. The links correspond to activities which take time for their completion. Nodes correspond to one or more milestones, which either have or have not been accomplished. Thus, a node represents an instantaneous point in time when all of the input links have been completed and their milestones have been realized.

Figure 7.2 presents the graph for the generalized CPM example given in Table 7.1. Conventionally, the links are labeled above, to identify the activity, and below, to identify the time duration. Similarly, the nodes are numbered for ease of reference, S being the start node and F being the finish node. Each node number could be further used to create a list of milestones and deliverables for project control.

The full value of Figure 7.2 can be appreciated only if it is first recognized that the CPM graph is not drawn on a time scale. Such charts, called *Gantt charts*, are quite valuable, and they will be considered in the next subsection. However, a failure to recognize from the outset that CPM charts are not drawn to time scale can result in great confusion. Thus, in Figure 7.2, the link for activity 3.1.3, between nodes 6 and F is of duration 1 day, while the link for activity 3.1.1, between nodes 7 and F, is of duration five days. Yet the link 6-F is actually longer than the link 7-F. This is necessary in order to connect the nodes, and therefore, the link length has no significance.

With this in mind it is now possible to see the value of CPM. This is accomplished by determining the *critical path*, which is the path from S to F which has the longest total time duration. In Figure 7.2 the critical path is S-1-2-5-7-F, which has a total time of $2+1+2+4+5 = 14$ days. No other path through the network takes longer to accomplish.

It should be clear that the only way to shorten project duration is to shorten one or more of the activities on the critical path. For example, in Figure 7.2, the reallocation of personnel to activity 2.1.3 to shorten it from five days to three days will have no effect upon total project completion, and such reallocation might be quite wasteful. On the other hand, it might be judicious to take some personnel away from activity 2.1.3 (as well as other noncritical activities) and reallocate these resources to the critical activities. If flexibility in personnel allocation exists (and it usually does in software development projects), then the optimal allocation is the one in which all paths are critical (i.e., their completion times are identical). It should be recognized that each time a reallocation is made, it could cause a different path to become critical.

The advantage of CPM over the PWD representation in allocating limited resources to activities should be clear. While CPM is not the best tool for the initial analysis since it does not reflect the overall system structure, its clarity in revealing activity interactions makes it an indispensable tool for the planner. Much more sophisticated tools exist which extend the basic concepts of CPM presented here. The most popular, called project evaluation review technique (PERT), is heavily documented in the literature. Further, packages are available for most computer systems for performing CPM and PERT analyses. The value

ACTIVITY TIME PROGRESS CHART

ACTIVITY		DAY FROM PROJECT START														
NUMBER	DESCRIPTION	1	2	3	4	5	6	7	8	9	10	11	12	13	14	15
2.1.1		S	S													
2.1.2				S												
2.1.3				S	S	S	S	S								
2.2.1						S	S	S								
2.2.2						S	S									
2.2.3									S							
2.2.4								S	S	S	S					
3.1.1												S	S	S	S	S
3.1.2											S	S	S			
3.1.3													S			

LEGEND:
S = scheduled duration; last S is scheduled completion.
C = actual completion.

Figure 7.3 Gantt Chart for Table 7.1

of these depends upon the number of *concurrent* activities which are to be performed, as opposed to the total number of activities in the project. That is, even though a project might have hundreds of activities, if they must be performed sequentially (one right after the other), then sophisticated CPM/PERT techniques will not be of great assistance. However, if there are many activities which can and should be performed simultaneously, then the use of these techniques becomes quite valuable. And, due to the complexity and the number of computations for such projects, computerized CPM/PERT packages are recommended highly.

This subsection has presented an introduction to the use of PWD and CPM tools for project planning, mainly to provide the background for their use in the following sections. While the concepts presented here will certainly provide a basis for project management and control, the full discussion of these tools is beyond the scope of any text primarily concerned with software design. Those who have heavy project-management interests should consult the references at the end of the chapter.

7.1.3 Gantt and Project Cost Control Charts

To review the development of the planning tools presented above, the process would begin by translating either the design or the software development activities into the planning Warnier diagram (PWD) format using the basic rules of Warnier diagram analysis. These numbered activity descriptors will be fully documented in sequentially numbered narratives as referenced backup to the PWD. From these, time estimates and precedence relationships will be established for each activity. These will provide all information required for the establishment of a CPM synthesis of the project. Node numbers will provide a ready reference to the milestones which must be passed before subsequent activities can be initiated. With these tools in place the project director has the capability to assign tasks and reallocate resources to maintain project efficiency. Measurements against estimated activity times will enable some degree of control to be exercised.

At this point a large variety of additional project-control tools could be introduced which undoubtedly would aid the project manager. Concentration here will be on two which are most basic to overall project control: (1) Gantt charts, and (2) project cost control charts. The first of these provides a pictorial representation of project progress by activity on a time scale. Figure 7.3 gives an example Gantt chart for the data in Table 7.1. Note that this is a planning tool and, as such, there is no necessity to show an activity as beginning at its earliest start time. The exceptions are the critical path activities which must start at the earliest possible moment if the project is not to be delayed. For example, Activity 3.1.2 could be programmed any time after Activity

2.1.3, since it is not on the critical path. However, to schedule Activity 2.2.4 any later than it currently appears would result in a delay in the project.

There are many variations on the basic Gantt chart. Traditionally, manually drawn rectangles have been used to indicate the planned schedule, and these were manually colored in to reflect the proportion of project completion. The use of letters facilitates the maintenance of these charts on standard word-processing equipment. Also, the activity description (same as the PWD descriptor) might be included to aid the project manager in identifying the activity. These variations depend highly upon the extent of application. In actual practice, calendar dates, as well as work days, should be included to give the project manager a clear indication whether activity due dates are not being met so that corrective action can be taken. Times for large projects may be given in units of weeks, or even months.

Clearly the time scale on the Gantt chart has an advantage over CPM for measuring the consumption of time resources (despite its weakness in determining critical activities). For this reason both are required for project control, and time and manpower allocation. An additional tool is essential, however, to measure a second, sometimes even more critical, resource—money. An example project cost control chart is given in Figure 7.4. Time is plotted on the horizontal scale, generally in months, in synchrony with accounting reports. Expenditures on the vertical scale should be allowed to extend 10% to 20% above the projected budgeted amount. For planning purposes, a continuous line should be drawn from the zero point to the projected budget over the life of the project, according to the expected cash flow. Rarely will this be a straight line since most projects have irregular expenditures, especially concentrated toward the beginning of each new phase.

The actual expenditures are plotted on the project cost control chart as they are reported back to the project manager. For closer control, a weekly, or even a daily, update may be required. However, for most long-term projects a monthly update is sufficient. By plotting the actual points and connecting them with straight lines, a clear distinction is maintained between the actual and anticipated time lines, as shown in the example in Figure 7.4 where three months of actual expenditures have been entered. Note further that this example indicates a deviation in the third month of well over 10% of anticipated expenditures. This is cause for alarm. The project manager should take immediate action to identify the source, and, if necessary, correct the inconsistency.

This section began with the concept of control. The planning process provides the first two elements of control, the goals (in terms of milestones and deliverables) and the means of measurement. Table 7.3 summarizes this section by presenting the steps involved in the planning process (the plan for the plan). This is a systematic method of

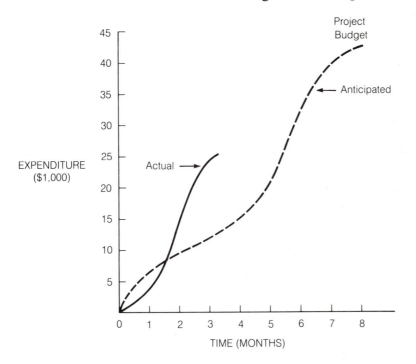

Note: Actual line is for three months of project reporting.

Figure 7.4 Example of a Project Cost Control Chart

proceeding from the definition of project activities to the establishment of the control tools. Once these tools are in place, project success will depend upon the third element of control—*correction capability*. This capability is largely within the realm of personnel management, requiring such attributes as motivation and communication.

The remainder of this chapter will reiterate the process presented in Table 7.3 as applied first to the design project itself (Section 7.2), and then to the development project (Section 7.3).

7.2 Documentation of the Design Plan

This section combines the design methodology chapters (Chapters 3, 5, and 6) with the planning techniques presented immediately above. One subtheme from the very first chapter of this book was that design tools and planning tools should not be confused; thus, planning is a separate activity from design. Now that both have been given independent consideration, the interaction between the two can be clarified. Indeed, the design activities provide all of the input required for the development of the design plan.

Table 7.3   Summary of Steps Involved in Project Planning

Step	Document Generated
1. Perform Warnier analysis of the planning process by Phase, Step, and Activity to the Milestone/Deliverables level.	Planning Warnier Diagram (PWD)
2. Describe each activity in sequential narrative form, referenced to the PWD.	PWD Backup Narratives
3. Determine and incorporate into the PWD .trigger, .duration, and .due date for each activity.	Updated PWD
4. Use activity durations and precedence dependencies to perform CPM. Number nodes and assign Milestone/Deliverables.	CPM Data Requirements CPM Network Critical Path Activities Milestone/Deliverables by Node
5. Use CPM data requirements and due dates to establish Gantt chart.	Project Gantt Chart
6. Use project budget, and activity time and manpower estimates to project cash flow.	Project Cost Control Chart

7.2.1   Generalized Design Plan PWD

Table 7.3 indicated that the first step in project planning was to analyze the project into phases, steps, and activities, as given in Figure 7.1. In turn, each of the activities would be assigned a trigger, duration, and due date, and then be analyzed further into their respective milestones and deliverables. Figure 7.1 presented a very general planning Warnier diagram (PWD) which could apply to any type of project. The project which is now of concern is the software design project. This project has already been analyzed into steps and activities in Chapters 3, 5, and 6. The presentation of these in terms of the PWD is given in Figure 7.5. Note that this is still very general in that it is not applied to a particular project. Descriptors, as well as the analysis, might be modified to reflect the needs of a particular project. The "Phase" level of analysis has also been omitted. In large modular design projects this would be included to reflect the organization of the design effort (not necessarily consistent with the modularity of the system).

Figure 7.5 begins the analysis of the software design project by considering the steps required for its accomplishment. Then, each step

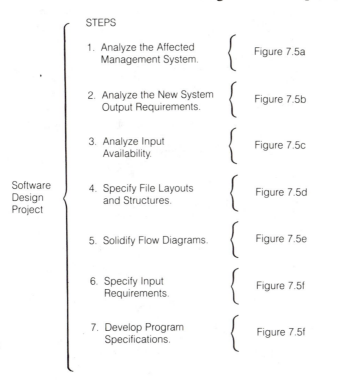

STEPS

1. Analyze the Affected Management System. ⎰ Figure 7.5a

2. Analyze the New System Output Requirements. ⎰ Figure 7.5b

3. Analyze Input Availability. ⎰ Figure 7.5c

Software Design Project

4. Specify File Layouts and Structures. ⎰ Figure 7.5d

5. Solidify Flow Diagrams. ⎰ Figure 7.5e

6. Specify Input Requirements. ⎰ Figure 7.5f

7. Develop Program Specifications. ⎰ Figure 7.5f

Figure 7.5 PWD for Generalized Software Design Project

is further analyzed by milestones and/or deliverables—tangible objectives whose degree of accomplishment can be measured. Within this final level of analysis the trigger, duration, and due date are assigned. While the trigger can be assigned in this generalized model, the duration and due date cannot. In a real application this information would be completed for purposes of project control, CPM, and Gantt chart development.

The meanings of the step and activity descriptors should cause no problem in interpretation since these have been thoroughly discussed in Chapters 3, 5, and 6. See the descriptor section title in these chapters for more information. The milestone/deliverable descriptor may not be as clear. With the exception of step 1, most of the milestone/deliverable descriptors make reference to sections of the Overview and User Guide (O & UG) or the Design/Maintenance Manual (MM), an example table of contents of which was given in Table 7.1. This is significant. The last thing that a project manager should want is a large number of isolated, disorganized documents floating around. Thus, each activity output which is part of the Overview and User Guide (O & UG) or the Design/Maintenance Manual (MM) should be immediately inserted, via word

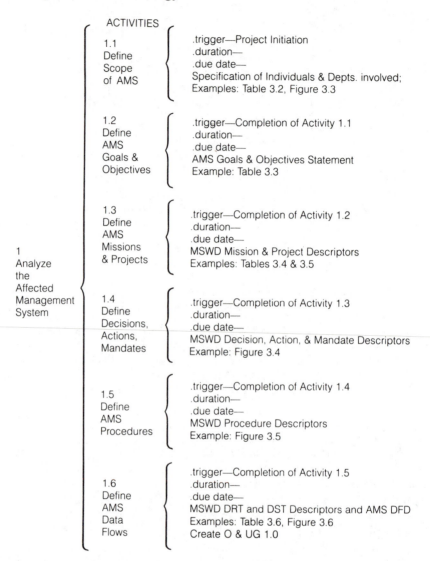

ACTIVITIES

1.1
Define
Scope
of AMS

.trigger—Project Initiation
.duration—
.due date—
Specification of Individuals & Depts. involved;
Examples: Table 3.2, Figure 3.3

1.2
Define
AMS
Goals &
Objectives

.trigger—Completion of Activity 1.1
.duration—
.due date—
AMS Goals & Objectives Statement
Example: Table 3.3

1.3
Define
AMS
Missions
& Projects

.trigger—Completion of Activity 1.2
.duration—
.due date—
MSWD Mission & Project Descriptors
Examples: Tables 3.4 & 3.5

1
Analyze
the
Affected
Management
System

1.4
Define
Decisions,
Actions,
Mandates

.trigger—Completion of Activity 1.3
.duration—
.due date—
MSWD Decision, Action, & Mandate Descriptors
Example: Figure 3.4

1.5
Define
AMS
Procedures

.trigger—Completion of Activity 1.4
.duration—
.due date—
MSWD Procedure Descriptors
Example: Figure 3.5

1.6
Define
AMS
Data
Flows

.trigger—Completion of Activity 1.5
.duration—
.due date—
MSWD DRT and DST Descriptors and AMS DFD
Examples: Table 3.6, Figure 3.6
Create O & UG 1.0

Figure 7.5a PWD for Generalized Software Design Project (continued)

processor, into these documents. The table of contents used in Table 7.1 is consistent with that of the Case Study in Part 2, which should be reviewed for a complete set of examples.

Some of the milestone/deliverables do not get included in the software system design documentation. Most notable of these are the products of step 1. These outputs document the existing system before the design, which we have referenced as the affected management system

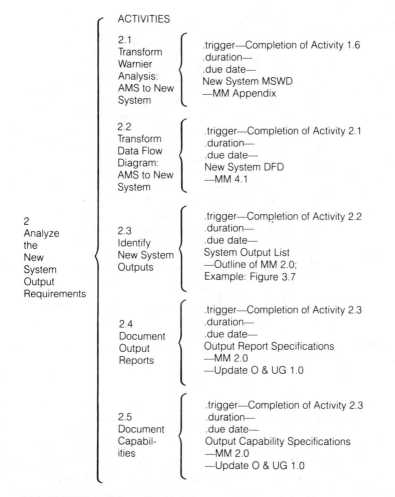

ACTIVITIES

2.1
Transform
Warnier
Analysis:
AMS to New
System

.trigger—Completion of Activity 1.6
.duration—
.due date—
New System MSWD
—MM Appendix

2.2
Transform
Data Flow
Diagram:
AMS to New
System

.trigger—Completion of Activity 2.1
.duration—
.due date—
New System DFD
—MM 4.1

2
Analyze
the
New
System
Output
Requirements

2.3
Identify
New System
Outputs

.trigger—Completion of Activity 2.2
.duration—
.due date—
System Output List
—Outline of MM 2.0;
Example: Figure 3.7

2.4
Document
Output
Reports

.trigger—Completion of Activity 2.3
.duration—
.due date—
Output Report Specifications
—MM 2.0
—Update O & UG 1.0

2.5
Document
Capabil-
ities

.trigger—Completion of Activity 2.3
.duration—
.due date—
Output Capability Specifications
—MM 2.0
—Update O & UG 1.0

Figure 7.5b PWD for Generalized Software Design Project (continued)

(AMS). These are utility documents which are ultimately transformed, during activity 2.1, into their counterparts for the new design. It is recommended that these be compiled for future reference into a separate set of documentation entitled "Pre-Conversion System." The output of activity 2.1 may either be included as an appendix of the Maintenance Manual or compiled separately.

Most of the other activity products go directly into the system documentation (i.e., the O & UG or the MM), or they serve to make modifications of this documentation. This is facilitated by the ease with which documentation can be modified and grow "from the middle" due to the use of word processing equipment.

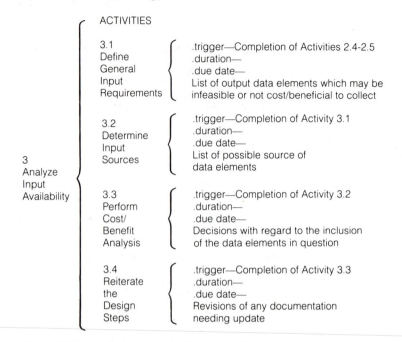

ACTIVITIES

3.1 Define General Input Requirements	.trigger—Completion of Activities 2.4-2.5 .duration— .due date— List of output data elements which may be infeasible or not cost/beneficial to collect
3.2 Determine Input Sources	.trigger—Completion of Activity 3.1 .duration— .due date— List of possible source of data elements
3.3 Perform Cost/ Benefit Analysis	.trigger—Completion of Activity 3.2 .duration— .due date— Decisions with regard to the inclusion of the data elements in question
3.4 Reiterate the Design Steps	.trigger—Completion of Activity 3.3 .duration— .due date— Revisions of any documentation needing update

3 Analyze Input Availability

Figure 7.5c PWD for Generalized Software Design Project (continued)

7.2.2 Generalized Design CPM

Once the analysis of a project is performed as given in Figure 7.5, all information is available to establish the CPM network. This is given in Figure 7.6. For simplification, activities 1.1-1.6, 2.1-2.3, and 3.1-3.4 have been combined on the network because their dependency is sequential. Thus, for CPM purposes their duration could be combined in determining the critical path. Activity times appear as descriptors under the links. The values given assume a thirty-week project duration. These values will be different for each application, and considerable thought should be given to the allocation of available manpower such that these values can be effectively estimated.

Table 7.4 presents a backup list of deliverables by node. The value of reorganizing the deliverables by CPM node should be obvious. It reorders them according to their criticality for subsequent dependent activities. It also provides a checklist to assure against the premature initiation of an activity.

Given that the application of Figure 7.6 to an actual design project would result in estimates of activity duration times, the critical path through the network could be calculated. It should be recognized that Figure 7.6 is the simplest network for a design project. Most ongoing

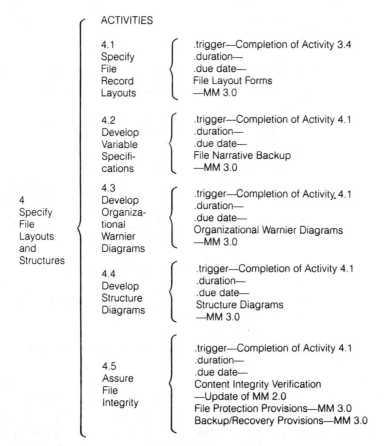

ACTIVITIES

4.1
Specify
File
Record
Layouts

.trigger—Completion of Activity 3.4
.duration—
.due date—
File Layout Forms
—MM 3.0

4.2
Develop
Variable
Specifi-
cations

.trigger—Completion of Activity 4.1
.duration—
.due date—
File Narrative Backup
—MM 3.0

4.3
Develop
Organiza-
tional
Warnier
Diagrams

.trigger—Completion of Activity 4.1
.duration—
.due date—
Organizational Warnier Diagrams
—MM 3.0

4.4
Develop
Structure
Diagrams

.trigger—Completion of Activity 4.1
.duration—
.due date—
Structure Diagrams
—MM 3.0

4.5
Assure
File
Integrity

.trigger—Completion of Activity 4.1
.duration—
.due date—
Content Integrity Verification
—Update of MM 2.0
File Protection Provisions—MM 3.0
Backup/Recovery Provisions—MM 3.0

4
Specify
File
Layouts
and
Structures

Figure 7.5d PWD for Generalized Software Design Project (continued)

software system designs are integrated into larger efforts which might include interactions with other systems or other phases of the system under design consideration. When this is the case, resources are being dynamically reallocated on an ongoing basis. Several parallel networks, like that given in Figure 7.6, will exist at different stages of completion, and resources will be shifted to maintain the efficiency of the total development effort. It is in this multi-project environment that CPM and the more sophisticated planning tools become indispensable.

7.2.3 Generalized Design Gantt Chart

The design project Gantt chart follows directly from the CPM network. Recall that the basic difference between these two models is that the Gantt chart is on a time scale. Thus, while CPM is used to reallocate

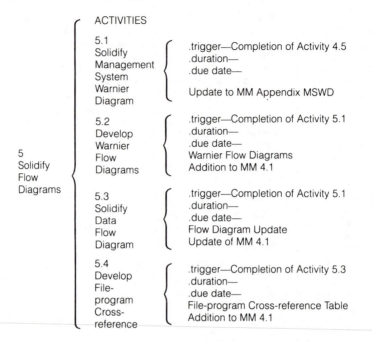

Figure 7.5e PWD for Generalized Software Design Project (continued)

and balance the network, the Gantt chart is used more effectively for project management. The Gantt chart associated with Figure 7.6 is given in Figure 7.7. The time scale in this case is given in weeks from the start of the project (1 week in Figure 7.7 = 5 days in Figure 7.6). Once the project start date is determined, actual dates should be assigned to aid in project control. An "S" (for Scheduled) is used within the chart to indicate the week or portion thereof which is scheduled to be employed in each activity. Once the project is underway, other letters can be superimposed upon the chart to indicate progress. For example, "A" may be used for the actual start and "C" for the completion. The overprint capability of most text editors can facilitate this updating without obscuring the original chart.

This example illustrates the need for selecting the proper time scale. Past experience and time records should be used to estimate activity duration times, usually in person-days. This must be converted into real time durations, by the planners, by determining the number of people to be assigned to each activity task. This will vary between design projects. For simplicity, the days given in Figure 7.6 were in increments of 5 days/week, which enabled an easy conversion to equivalent weeks in Figure 7.7. However, this was done merely to facilitate the example. In reality, the estimates should be as accurate as possible. Note that the choice of units on the Gantt chart depends heavily on the

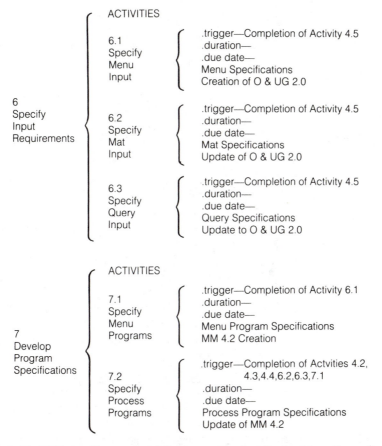

ACTIVITIES

6
Specify
Input
Requirements

6.1
Specify
Menu
Input

.trigger—Completion of Activity 4.5
.duration—
.due date—
Menu Specifications
Creation of O & UG 2.0

6.2
Specify
Mat
Input

.trigger—Completion of Activity 4.5
.duration—
.due date—
Mat Specifications
Update of O & UG 2.0

6.3
Specify
Query
Input

.trigger—Completion of Activity 4.5
.duration—
.due date—
Query Specifications
Update to O & UG 2.0

ACTIVITIES

7
Develop
Program
Specifications

7.1
Specify
Menu
Programs

.trigger—Completion of Activity 6.1
.duration—
.due date—
Menu Program Specifications
MM 4.2 Creation

7.2
Specify
Process
Programs

.trigger—Completion of Actvities 4.2,
        4.3,4.4,6.2,6.3,7.1
.duration—
.due date—
Process Program Specifications
Update of MM 4.2

Figure 7.5f PWD for Generalized Software Design Project (continued)

total project duration time, which often is determined, sometimes quite arbitrarily, before activity times are estimated. Obviously, days would be a poor choice for a project that was to span several years, while months would be a poor choice for a project that was expected to be completed in a few weeks. Recall that the objective of the Gantt chart was to provide measurements for control purposes. The time scale should be selected such that this purpose is attained. In a thirty-week project, precision to the day is unnecessary.

Since some activities which are performed in sequence are quite short with regard to others, they might be combined on the Gantt chart as they were in the CPM network. Figure 7.7 shows the difficulty of depicting activities 1.1-1.6 in terms of weeks when, in fact, some of these are performed in a few days. Some overlap must be shown, although this was not allowed according to the original analysis in Figure 7.5.

The Gantt chart can also be used as a tool for smoothing the personnel allocations. In Figure 7.6 activities 4.2, 4.3, and 4.4 are allowed

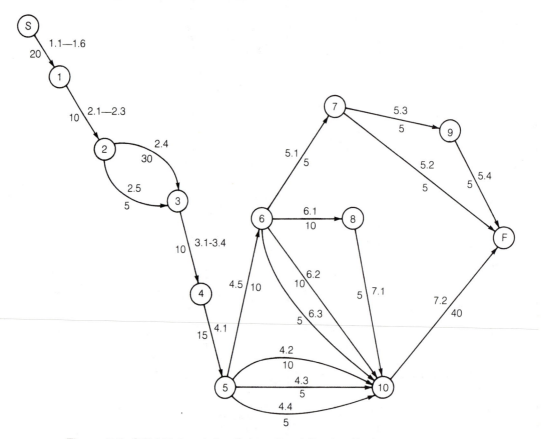

Figure 7.6 CPM Network for Generalized Design Project

to be performed simultaneously for CPM purposes. However, since these are not on the critical path, there is no advantage to be gained in total project duration from doing this. The Gantt chart shows them being performed one after the other, which will probably result in a more efficient allocation of resources.

Once the CPM and Gantt charts are constructed, a verification of the planning Warnier diagram should be made to assure that the .duration and .due-date specifications are accurate. The initial estimates must be considered as tentative prior to the CPM and Gantt analyses.

7.2.4 Closure on Design Planning

This subsection has presented some basic planning tools as they apply to the software design process. In summary, it took the project planning

ACTIVITY NUMBER	DESCRIPTION	*	TIME FROM PROJECT START (WEEKS) 0000000000111111111122222222223 1234567890123456789012345678901234567890
1.1	Define Scope of AMS	*	S
1.2	Define AMS Goals & Objectives	*	S
1.3	Define AMS Missions & Projects	*	S
1.4	Define Decisions, Actions, & Mandates	*	SS
1.5	Define AMS Procedures	*	SS
1.6	Define AMS Data Flows	*	SS
2.1	Transform Warnier Analysis	*	S
2.2	Transform DFD	*	SS
2.3	Identify Outputs	*	S
2.4	Document Output Reports	*	SSSSSS
2.5	Document Capabilities		S
3.1	Define General Input Requirements	*	S
3.2	Determine Input Sources	*	SS
3.3	Perform Cost/Benefit Analysis	*	S
3.4	Reiterate the Design Steps	*	S
4.1	Specify File Record Layouts	*	SSS
4.2	Develop Variable Specifications		SS
4.3	Develop OWD		S
4.4	Develop Structure Diagrams		S
4.5	Assure File Integrity	*	SS
5.1	Solidify MSWD		S
5.2	Develop Warnier Flow Diagrams		S
5.3	Solidify DFD		S
5.4	Develop File-Program Cross-Ref.		S
6.1	Specify Menu Input	*	SS
6.2	Specify Mat Input	*	SS
6.3	Specify Query Input		S
7.1	Specify Menu Programs	*	S
7.2	Specify Process Programs	*	SSSSSSSS

* Indicates Critical Activity

Figure 7.7 Generalized Design Project Gantt Chart

steps given in Table 7.3 and applied them to the design activities detailed in Chapters 3, 5, and 6. It is recommended that these very general steps and activities be further adapted to the specific needs of the software design effort under consideration. These will adequately provide for the control measurements which, along with sound motivational techniques, should lead to a successful design.

7.3 Documentation of the Development Plan

Section 7.2 dealt with the documentation of the *design* plan. Let us assume that this plan has now been developed and executed. The result

Table 7.4   Deliverables/Milestones by Node for
Generalized Design Project

CPM Node (Fig. 7.6)	Precedent Activities	Deliverable Description	Documentation Reference
1	1.1	Specification of Individuals and Departments	
	1.2	AMS Goals and Objectives Statement	
	1.3	MSWD Mission and Project Descriptors	
	1.4	MSWD Decision, Action, and Mandate Descriptors	
	1.5	MSWD Procedure Descriptors	
	1.6	MSWD DRT and DST Descriptors	O & UG 1.0
2	2.1	New System MSWD	MM Appendix
	2.2	New System DFD	MM 4.1
	2.3	System Output List	Outline MM 2.0
3	2.4	Output Report Specifications	MM 2.0
	2.5	Output Capability Specifications	Update O & UG 1.0
4	3.1	List of Questionable Output Data Elements	
	3.2	List of Possible Sources	Update:
	3.3	Cost/Benefit Decisions	MM 2.0
	3.4	Revision of Documentation	MM 4.1

will be a completed set of software design specifications in terms of an
Overview and User Guide (O & UG) and a Design/Maintenance Manual
(MM), examples of which are given in the case study in Part 2. The
development plan proceeds directly from the design specifications, since
software development is the process of coding, testing, and debugging
the software. The purpose of this section is to demonstrate how the
project planning steps of Table 7.3 can be applied to the product of the
design effort in order to facilitate the development effort. In this regard,
it is essential that the reader discern the difference between: (1) the
design plan, (2) the design itself, (3) the development plan, and (4) the
development process. Confusion with regard to these separate entities
can introduce inefficiency into the total process. For example, programmers generally do not need planning information to perform their coding; they need the program specifications and the reference documen-

Table 7.4   Deliverables/Milestones by Node for
Generalized Design Project (continued)

CPM Node (Fig. 7.6)	Precedent Activities	Deliverable Description	Documentation Reference
5	4.1	File Layout Forms	MM 3.0
6	4.5	Content Integrity Verification	Update: MM 2.0
	4.5	File Protection Provisions	MM 3.0
	4.5	Backup/Recovery Provisions	MM 3.0
7	5.1	MSWD Update	MM Appendix
8	6.1	Menu Specifications	O & UG 2.0
9	5.3	Flow Diagram Update	MM 4.1
10	4.2	File Narrative Backup	MM 3.0
	4.3	Organizational Warnier Diagrams	MM 3.0
	4.4	Structure Diagrams	MM 3.0
	6.2	Mat Specifications	O & UG 2.0
	6.3	Query Specifications	O & UG 2.0
	7.1	Menu Program Specifications	MM 4.2
F	5.2	Warnier Flow Diagram	MM 4.1
	5.4	File-program Cross-reference Table	MM 4.1
	7.2	Process Program Specifications	MM 4.2

tation that gives these specifications meaning. It is the development project manager who needs the planning information such that the project can be directed and controlled.

The establishment of the development plan must come after the design is completed. Obviously, consideration must be given as to the time and cost of the development during the early design activities. However, this is design as opposed to planning. Without having a completed design, any plans would be tentative.

Given that the software design exists, the procedure for developing the plan is quite straightforward. No generalized development plans can be given since this is totally dependent upon the nature of the software under consideration. However, the planner can begin with the ordered program specification Warnier diagrams, which were developed as the last activity of the design process (Section 6.3). These are essentially the same as the planning Warnier diagram described for

Table 7.5   Example PSWD Planning Data

PSWD Number	Module	Program Name	Precedent PSWD Nos.	Activity Duration	Activity Due Date
0.1	—	System Supervisory Menu	---	1	
0.1.1	1	Supervisory Menu	0.1	1	
0.1.1.1	1	Report-generator Menu	0.1.1	1	
0.1.2	2	Supervisory Menu	0.1	1	
0.1.2.1	2	Report-generator Menu	0.1.2	1	
0.1.3	3	Supervisory Menu	0.1	1	
1.1	1	Data Entry	0.1.1	10	
1.2	1	Transaction Update	1.1	15	
1.3	1	Delete Transaction	1.1	5	
1.4	1	Report Generator 1	1.1	10	
1.5	1	Report Generator 2	1.1	12	
1.6	1	Report Generator 3	1.3	15	
2.1	2	Data Entry/Mod. 1	0.1.2	8	
2.2	2	Data Entry/Mod. 2	0.1.2	6	
2.3	2	File Update Routine	2.1, 2.2	11	
2.4	2	Report Generator 1	2.1	4	
2.5	2	Report Generator 2	2.2	3	
2.6	2	Report Generator 3	2.1, 2.2	8	
3.1	3	Data Entry/ Modification	0.1.3	6	
3.2	3	Report Generator	3.1	7	

project planning in Figure 7.1, with the exception that there is no .trigger, .duration, .due date, or milestone/deliverable specification. In this application it is not recommended that these be added to the PSWDs since this information is not of interest to the programmer either for development or future maintenance. Neither is it recommended that a separate PWD be devised for the programming activities, since these would be largely redundant. Rather, this information may be incorporated into a tabular format, as exemplified in Table 7.5.

Table 7.5 presents the necessary data for a generalized three-module system. Typically, systems requiring the design and development planning discussed here would have several times the number of programs, but these serve adequately to exemplify the data requirements. The program names are described in general terms, which should be replaced with the actual descriptive name in practice. Further, the activity due dates have been omitted until the CPM and Gantt chart analyses are performed. Despite these generalizations and simplifications, this example typifies the transition from program specification to development project planning which would be encountered in practice. Essentially, the steps give in Table 7.3 simplify to the following:

1. From the program specification Warnier diagrams obtain each program to be developed as a separate activity (smaller sub-references may not require listing).
2. Determine the precedent programs for each listed. These would be programs which have to be developed first (e.g., data entry before report generator).
3. Determine the amount of time required for the development of each program.
4. Using the data from Steps 2 and 3, formulate the CPM analysis, Gantt chart, and cost control chart.

Note in Table 7.5 that the order of numbering of the PSWDs generally follows the order of development. Exceptions occurred because of the grouping of certain types of programs, such as the menu programs. Recall that the general ordering of the PSWDs started with the menus, followed by the input, processing, and output. This "common sense" arrangement from start to finish was mandated by the requirement that the Overview and User Guide be logical and readable to the users. However, recall further that the design steps were performed in just the opposite direction (output—file design—input—process specification). Software development is generally consistent with the design documentation as opposed to the design activities. This is consistent with the modular concept of *implementation* (as opposed to design), discussed in Chapter 1. It is also necessitated within each module by the requirement to build the files (data entry) before they are processed.

References

Antill, J.M., and R.W. Woodhead. *Critical Path Methods in Construction Practice.* New York: John Wiley & Sons, 1982.

Bruce, P., and S.M. Penderson. *The Software Development Project.* New York: John Wiley & Sons, 1982.

Sisk, H.L., and J.C. Williams. *Management & Organization.* Cincinatti: South-Western Publishing Co., 1981.

Questions and Problems

1. Why is it important to differentiate between communication tools and planning tools? State the basic difference. Which is applied first in the design and development process?

2. What are the three essential elements of control? How are these exemplified in the following systems: (1) car/driver, (2) thermostatically controlled air conditioning, (3) software design group, and (4) software development team.

3. State how the design plan differs from the development plan.

4. Define and differentiate *milestones* and *deliverables*. Why are these terms used concurrently?

5. What is the difference between the goals and objectives in project control and those discussed in Chapter 3?

6. What is meant by target documentation, and why is it so important to have a clearly defined target?

7. Name two ways that the planning Warnier diagram (PWD) can handle contingency planning.

8. Define, in the PWD context, the meaning of the nonanalytical entries: trigger, duration, and due date.

9. State the basic advantage of CPM over the PWD. Why is the PWD developed first, and is it even necessary?

10. Given Table E7.1, set up the CPM network and determine the critical path.

11. Using the Solution to problem 10, determine the effect of the following reallocations on total completion time:

   a. Shorten Activity 8 from 5 to 3 at the expense of lengthening Activity 6 from 5 to 10.

   b. Shorten Activity 5 from 4 to 2 at the expense of lengthening Activity 9 from 1 to 3.

   c. Shorten Activity 8 from 5 to 4 at the expense of lengthening Activities 7 and 5 by two units each.

12. What is the problem in attempting to draw a CPM network to a time scale?

13. Draw a Gantt chart for the data given in Table E7.1. What advantage does the Gantt chart have over a CPM network for project control purposes?

14. When is the best time to assign activity due dates? Why?

15. What resources are ignored by the Gantt chart? How could a similar chart record the expenditure of these resources?

16. How could the project cost control chart be modified to record the expenditure of resources other than dollars?

Table E7.1   Data for Problem 10

Activity	Activity Duration	Precedent Activities
1	3	—
2	2	—
3	2	1
4	2	1
5	4	2
6	5	3
7	4	4
8	5	4
9	1	5
10	5	8, 9

17. Summarize the steps in proceeding from design activities to the generation of tools necessary to control the design project.

18. Summarize the steps in proceeding from the completed design documentation to the generation of the tools necessary to control the development project.

PART 2

Case Study

MANAGEMENT ANALYSIS
TO SUPPORT TRANSACTIONS
IN EDUCATION AND RESEARCH

(M A S T E R)

Overview and User Guide
Maintenance Manual
Commentary on the MASTER System

MANAGEMENT ANALYSIS
TO SUPPORT TRANSACTIONS
IN EDUCATION AND RESEARCH

(M A S T E R)

Overview and User Guide

Designed and Developed by:
Scientific Computer Applications, Inc.
P.O Box 1391
Auburn, Alabama 36830

MASTER OVERVIEW AND USER GUIDE

Table of Contents

Part 1   MASTER System Overview

MASTER is a computer-based system designed to enable the academic department heads and the Office of Administrative Services (OAS) personnel to effectively manage and control their respective budgets, proposals, and contracts. It is a totally interactive system designed to combine the experience of management with the efficiency of computer technology. The time-consuming tasks of filing and processing inherent. in any management information system are assigned to the computer, leaving the decision makers free to more effectively utilize the experience and skills so essential to good management.

The MASTER system was designed with ease of operation as one of its main objectives. It guides the user through each function with menus, or lists of choices, and it queries for all information needed by the system. No prior knowledge of computers or data processing is needed and no supplementary documentation is required once a minimal amount of experience is obtained.

The management and control capabilities of MASTER are not confined to the manipulation of paperwork. MASTER enables the user to perform functions which would be impossible to perform with a manual system. These include the on-line updating of budget expenditures, enabling the most recent budgetary figures to be reviewed at the departmental level at any time. A similar capability exists with regard to proposal and contract status updates. Any of these functions may be selected from the menus which provide a logical guide to all system functions.

To illustrate this concept, consider the MASTER system supervisory menu (SM) given in Figure 1, which appears on the CRT when MASTER is requested. The user merely enters the number corresponding to the module which is desired. While an entire section will be devoted to each of these modules (in Part 2), the following summary demonstrates the MASTER capabilities:

1. Budget Management and Control (BMC). This module deals with all functions of MASTER related to departmental and OAS budgets, with the exception of Extension project budgets. When this module is selected by the user, the BMC module supervisory menu will appear, enabling all of the BMC data-entry or report-generation options to be selected. This will include the capability to perform updates to all accounts as transactions are received by OAS. Reports out of this module may be obtained at the CRT terminal or on a printer. These include the following: (1) Subsidiary Ledger Summary (SLS), a summary of each account, including expenditures, encumbrances, and balances available; (2) Budget Subcode Analysis (BSA), a summary of obligations and expenditures according to funding sources by de-

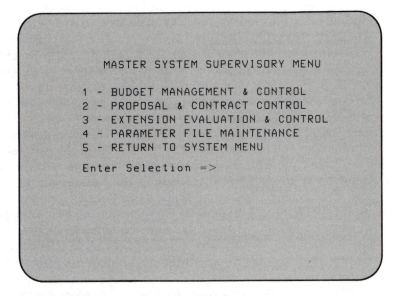

```
          MASTER SYSTEM SUPERVISORY MENU

      1 - BUDGET MANAGEMENT & CONTROL
      2 - PROPOSAL & CONTRACT CONTROL
      3 - EXTENSION EVALUATION & CONTROL
      4 - PARAMETER FILE MAINTENANCE
      5 - RETURN TO SYSTEM MENU

      Enter Selection =>
```

Figure 1 MASTER Supervisory Menu (SM)

partment; (3) Departmental Account Statement (DAS), a further breakdown of accounts by budget control code for given accounts, accompanied by a detailed list of purchase orders to date; (4) Departmental Account Statement by Transaction (DAST), which lists all transactions processed against a given account within a user-specified time span; (5) Budget Expenditure Comparison (BEC), a comparison of estimated-with-actual expenditures based upon the expenditure rate of previous years; and (6) Budget Projection Report (BPR), which projects current-month and current-to-date expenditures over the future months of the contract.

2. Proposal and Contract Control (PCC). This module includes all functions related to proposals and contracts, with the exception of account expenditures which are handled by BMC. When the user selects this option, the PCC module supervisory menu will appear, enabling the performance of any of the data-entry, maintenance, and report-generation functions associated with proposals and contracts. Output reports, most of which may be obtained either on the CRT or on hard copy output, include the following: (1) Extramural Proposals and New Funding (EPNF), which provides a comparison of proposals submitted and new awards received against their counterparts for the previous years; (2) Summary of Active Projects (SAP), a listing of vital information on each project, including expenses-to-date and unobligated balances; (3) Research Activities Report (RAR), in-

cluding basic information on each project formatted for the OAS Research Activities Report; (4) Percent Support List (PSL), which indicates the percent support of each project from extramural sources along with other basic information; (5) Proposal Status Report (PSR), which presents the current status of each proposal; and (6) Contract Status Report (CSR), which presents all nonfinancial information on each contract.

3. Extension Evaluation and Control (EEC).   This module includes all functions related to extension projects. When the user selects this option, the EEC supervisory menu will appear, giving the opportunity to perform any required data entry, file maintenance, or report generation. Outputs obtained from the EEC module include: (1) Extension Budget Control Report (EBC), which presents the estimated and current actual costs for each line budget item within any given project; and (2) Extension Project Summary (EPS), a summary over all of the projects currently active.

4. Parameter File Maintenance.   This option enables file maintenance to be performed on those files which do not fall under any of the other functional modules. This would include departmental information and account code descriptions.

5. Return to System Menu.   This function provides a systematic method of getting out of MASTER in order to either logoff or utilize other software systems.

The value of any computerized system can only be measured in terms of the output produced. It is not the quantity or the variety of outputs that is important. In fact, too much information can be worse than not enough. The system must save time and money to be cost justified. The two questions that must be answered are: (1) What new information does the system provide that is *essential?* and (2) How does this system help reduce the time to do what *must* be done? MASTER does not invent new and complex methods for performing simple tasks. Rather, MASTER proceeds compatibly with the established manner to store, manipulate, and present information as consistently with current practice as possible.

The philosophy under which MASTER was developed is to free the department heads and their staffs from many of the clerical tasks that have been required to maintain budgetary control. The on-line nature of MASTER provides the environment in which this can be accomplished. The remainder of this manual details the MASTER system functions and design.

Part 2   MASTER System Functions

This part of the manual details the functions of MASTER organized according to the MASTER supervisory menu (SM) given in Figure 1. It provides reference material to those unfamiliar with MASTER operational procedures, and it enables potential users to know what to expect from MASTER.

2.1 Budget Management and Control (BMC) Module

When the BMC module is selected from the MASTER supervisory menu (Figure 1), a second menu will appear as given in Figure 2. Each of these options will be discussed in separate subsections below. Note that these subsections are numbered corresponding to the BMC supervisory menu option (example: Subsection 2.1.3 is for menu option 3).

2.1.1 Account Creation/Modification

This entry on the BMC supervisory menu indicates to the system that a new account is to be created, i.e., a new record will be created in the account file (D1.1). It is also used to update existing account information with the exception of those items which must be updated by transactions (see Section 2.1.2). Only OAS logon codes will be able to enter

```
                       MASTER
           BUDGET MANAGEMENT AND CONTROL MODULE
               BMC SUPERVISORY MENU

          1 - ACCOUNT CREATION/MODIFICATION
          2 - ACCOUNT TRANSACTION
          3 - TRANSACTION MODIFICATION
          4 - GENERATE REPORTS
          5 - DELETE TRANSACTIONS
          6 - DELETE ACCOUNTS
          7 - PERFORM END-PERIOD UPDATES
          8 - RETURN TO MASTER SUPERVISORY MENU

          ENTER SELECTION =>
```

Figure 2 Budget Management and Control Supervisory Menu

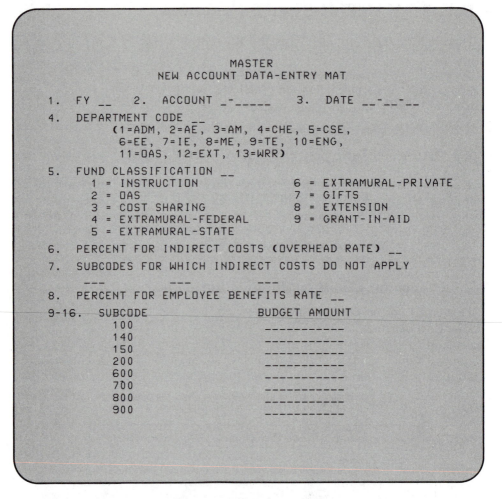

Figure 3 New Account Data-entry Mat

new accounts. When this option is selected, the following query will appear:

```
ENTER ACCOUNT NUMBER (OR RETURN
FOR NEW ACCOUNT) =>_____
```

After this, the mat given in Figure 3 will appear in order to provide the necessary information for account creation or modification. The following items are included in the new account data-entry mat:

1. FY. Fiscal year. This item is output by the system for general information; it is not supplied by the user (protected field). It is obtained from the general parameter file, D4.3,1 and included in the account file records to identify the fiscal year in which the account was created (D1.1,1).

2. Account number.   This is a single digit followed by a five-digit number (D1.1,2).
3. Date of entry.   This is the current date obtained from the system calendar (D1.1,3). This is not supplied by the user (protected field). It is written to the account file for future reference.
4. Department code.   Select and enter the applicable number from the choice given (D1.1,4).
5. Fund classification.   Select and enter one of the choices given. Very close edits will be performed to assure that the fund classification is consistent with the account number (D1.1,5).
6. Overhead rate.   This is the percentage rate that will be applied to expenditures (exceptions given in Item 9) to calculate indirect costs (D1.1,6).
7. Subcodes for exception.   If there are subcodes for which the overhead rate entered in Item 6 should not be applied, they will be entered here (D1.1,7). Up to three subcodes may be entered. A "Return" in lieu of a numeric entry will result in preparation for entry of the next data item.
8. Percent for employee benefit rate.   This is the percent of salaries that is charged to fringe benefits (D1.1,8).
9-16. Budget amount.   Enter the amount in dollars for each budget subcode (D1.1,9-16).

Once the mat is completed, the user will be given a chance to review and modify any of the entries. When acceptance is indicated, the new account record will be created and ready for subsequent transactions using BMC supervisory menu function 2.

2.1.2 Account Transaction

This option from the BMC supervisory menu is used for all transaction types. When entered, a mat will appear as given in Figure 4. Data will be entered according to the following detailed descriptions:

1. FY. Fiscal year.   This item is output by the system for general information; it is not supplied by the user (protected field) (D1.2,1). It is necessary to provide a unique transaction reference number, which is a concatenation of the FY and the Transaction Ref., Item 2. This figure is obtained from the general parameter file (D4.3,1).
2. Transaction reference.   This is a sequential-transaction reference number which starts with 1 for the first transaction of any given fiscal year (D1.2,2). This is not supplied by the user (protected field). It is generated by the system and maintained current in the general parameter file (D4.3,2). This number must be written manually on the purchase order in order to facilitate future transaction modification.

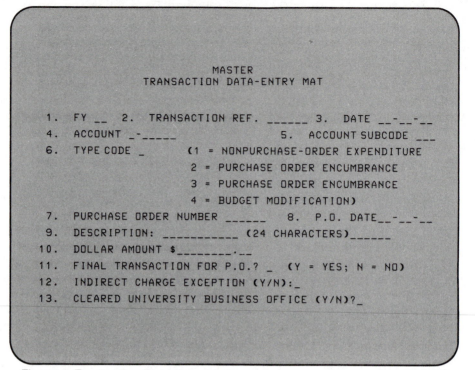

```
                                MASTER
                    TRANSACTION DATA-ENTRY MAT


    1.   FY __   2.   TRANSACTION REF. _____   3.   DATE __-__-__
    4.   ACCOUNT _-_____                 5.   ACCOUNT SUBCODE ___
    6.   TYPE CODE _      (1 = NONPURCHASE-ORDER EXPENDITURE
                           2 = PURCHASE ORDER ENCUMBRANCE
                           3 = PURCHASE ORDER ENCUMBRANCE
                           4 = BUDGET MODIFICATION)
    7.   PURCHASE ORDER NUMBER _____    8.   P.O. DATE__-__-__
    9.   DESCRIPTION: _____ (24 CHARACTERS)_____
   10.   DOLLAR AMOUNT $_____.__
   11.   FINAL TRANSACTION FOR P.O.? _  (Y = YES; N = NO)
   12.   INDIRECT CHARGE EXCEPTION (Y/N):_
   13.   CLEARED UNIVERSITY BUSINESS OFFICE (Y/N)?_
```

Figure 4 Transaction Data-entry Mat

3. Date entered. This is the current date obtained from the system calendar (D1.2,3). This is not supplied by the user (protected field). It is, however, written to the transaction file (D1.2) for future reference.
4. Account. The code of the account which is being modified by this transaction will be entered (D1.2,4). A verification will be made at this point to assure that the account number exists; if not, an error message will be returned.
5. Account subcode. The subcode for the entry transaction. When this is entered, the system will go to the account subcode parameter file (D4.2,1) and verify that the account subcode exists; then it will present the account subcode description for visual verification by the user (D1.2,5).
6. Type code. The entry type as given in Figure 4 (D1.2,6).
7. Purchase-order number. This field will be skipped if the entry code (Item 6) is a 1 or 4. Otherwise, a valid purchase order number will be accepted. Generally this number will not be available at the time that a transaction is first entered; it will be updated later using BMC SM Option 3 (D1.2,7).
8. Purchase-order date. The date on the purchase order. This field will be skipped if the entry code (Item 6) is a 1 or 4 (D1.2,8).

9. Description. A 24-character description of the transaction (D1.2,9).
10. Dollar amount. The dollar amount to be used in modifying account balances (D1.2,10). The dollar amount for a budget modification may be positive or negative (use minus sign for negative entries). These are the changes to be added or deducted from the existing balance.
11. Final transaction for purchase order? Enter Y if this is the final transaction for this purchase order; otherwise, enter N (D1.2,11). This field will be skipped if the entry code (Item 6) is a 1, 2, or 4. A check will be performed if N is entered to assure that there is a positive balance in the purchase order.
12. Indirect charge exception. Enter E if indirect costs are not to be charged against this expenditure; otherwise, enter nothing (D1.2,13).
13. Cleared university business office (Y/N)? Enter Y if the transaction has cleared the university business office; N if it has not (D1.2,14).

When the mat given above is completed by the user, a query will appear to allow for verification. If the user detects an error, any of the items may be changed. If not, the transaction record will be written to the transaction file (D1.2), and the account file (D1.1) will be updated appropriately. If the system detects that an account subcode is negative the following warning will appear:

```
BALANCE FOR SUBCODE XXX IS NEGATIVE
BY $_____
ENTER A TO ABORT THIS TRANSACTION OR
        S TO SAVE THIS TRANSACTION.
```

Corrections of errors after this point will require an additional transaction. The following note will appear at this point if this is a purchase-order encumbrance transaction (i.e., Item 6 = 2):

```
NOTE: BE SURE TO WRITE THE FY AND TRANSAC-
TION REFERENCE NUMBER (ITEMS 1 AND 2 ABOVE)
ON THE PURCHASE ORDER NOW IN ORDER TO FACIL-
ITATE FUTURE REFERENCING OF THIS TRANSACTION
FOR UPDATE. PRESS ENTER AFTER THIS IS DONE.
```

After the completion of a first transaction, a second transaction data-entry mat (Figure 4) will appear. To facilitate entries with common information, a set of defaults may be used on the second and all subsequent transactions during one session. The defaults will be the figures previously entered during the previous transaction. To obtain the default value, enter a RETURN in lieu of a character entry for any item on the mat. The figure previously entered will appear. Corrections can be made after the completion of the entire mat. When the user indicates that all transactions have been made, the system will return to the BMCM supervisory menu.

2.1.3 Transaction Modification

Either for reasons of error correction or update, there will be occasions when it is desirable to go back and modify a transaction. Budgetary information will not be accessible for update so that an audit trail of all budgetary transactions can be maintained. However, all other data items given in Figure 4 may be updated. This is of particular value for adding previously unavailable purchase-order numbers to their corresponding transactions. This is the reason that the transaction reference number was required to be manually entered on the purchase order prior to its processing through University Accounting.

When this option is selected the following query will result:

```
ENTER TRANSACTION REFERENCE NUMBER => FY_____
```

Upon this entry, the system will produce the transaction information requested for update in the same format as given in Figure 4. Updates can be made in the same manner as the original entries were made.

2.1.4 Generate (BMCM) Reports

This option indicates that the user wishes to generate one or more of the BMCM reports. Before doing this, all transactions which are desired to be included on the reports must be input. Thus, if possible, it is recommended that all of the available transactions be input prior to report generation.

The possible BMCM reports will appear on a menu when this option is selected, as given in Figure 5. Depending upon the report selected, the user will be queried for the information necessary to generate the report. The following subsections detail these queries for each report. Note that since options exist on most reports for either terminal or hard-copy output, the queries for this will not be repetitively documented below. Also, due to the screen width constraints of the CRT terminals, certain outputs require the user to specify the output items to be viewed. This will be done by optional scrolling right and left as well as down the page. See Section 2.1 of the MASTER Maintenance Manual for a detailed specification of each report.

2.1.4.1 Subsidiary Ledger Summary (SLS) Queries

The following will appear after Option 1 is selected from the BMCM RG Menu:

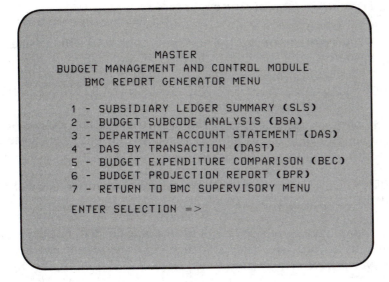

```
                        MASTER
          BUDGET MANAGEMENT AND CONTROL MODULE
             BMC REPORT GENERATOR MENU

          1 - SUBSIDIARY LEDGER SUMMARY (SLS)
          2 - BUDGET SUBCODE ANALYSIS (BSA)
          3 - DEPARTMENT ACCOUNT STATEMENT (DAS)
          4 - DAS BY TRANSACTION (DAST)
          5 - BUDGET EXPENDITURE COMPARISON (BEC)
          6 - BUDGET PROJECTION REPORT (BPR)
          7 - RETURN TO BMC SUPERVISORY MENU

          ENTER SELECTION =>
```

Figure 5 BMCM Report-generator Menu

```
ENTER DEPARTMENT CODE OR ALL
ENTER => ___
```

An entry of ALL will produce a listing by account numbers irrespective of department. Upon entry of a department code, the SLS report for that department will be output to the screen or hard copy, depending upon the user response to:

```
DO YOU WANT SCREEN (S) or PRINTED (P) OUTPUT? => _
```

Following this entry the SLS report will be generated as requested.

2.1.4.2 Budget Subcode Analysis (BSA) Queries

The following will appear after Option 2 is selected from the BMCM RG Menu:

```
ENTER DEPARTMENT CODE => ___
```

Upon response to this query, the BSA report will be generated for this department.

2.1.4.3 Departmental Account Statement (DAS) Queries

The following will appear after Option 3 is selected from the BMCM RG Menu:

```
ENTER ACCOUNT NUMBER => _____
```

Upon response to this query, the first part of the DAS will be generated (i.e., expenditures by subcode). After this the following will appear:

```
DO YOU WANT A PURCHASE ORDER LISTING (Y/N)? _
```

If Y is entered a purchase order listing will be output, otherwise the system will return to the BMCM RG Menu. An affirmative response will lead to the following query:

```
DO YOU WANT PAST MONTH ZERO-BALANCE PURCHASE
ORDERS INCLUDED (Y/N)? _
```

If Y is entered, all purchase orders for the account will be output. If N is entered, all purchase orders without current month activity (i.e., those which have zero balances) will be excluded from the listing.

2.1.4.4 Dast Queries

The following will appear after Option 4 is selected from the BMCM RG Menu:

```
ENTER ACCOUNT NUMBER => _____
```

Following this the user will be queried for the time period for which output is required, as follows:

```
ENTER T FOR TOTAL CONTRACT PERIOD OR
ENTER THE TIME SPAN FOR WHICH OUTPUT
IS DESIRED (MM-DD-YY)
=> mm-dd-yy TO mm-dd-yy
```

A RETURN entry in lieu of the second date will default to the current date. If the first date entered precedes the contract start date for this account, an appropriate warning will be issued and the report will be generated for the entire contract period. At this point the user will be queried for the transaction type for which output is required, as follows:

```
ENTER TRANSACTION TYPE:
1 = Nonpurchase-order expenditure
2 = Purchase-order encumbrance
3 = Purchase-order payment
4 = Budget modification
5 = Types 2 and 3 combined
6 = All transactions (default)
ENTER TYPE => _
```

This entry will restrict the listing to the type specified.

2.1.4.5 Budget Expenditure Comparison (BEC) Queries

The following will appear after Option 5 is selected from the BMCM RG Menu:

```
ENTER DEPARTMENT ABBREVIATION => ___
```

Following this, the BEC report will be generated for that department. An entry of ALL will result in an output containing all departments.

2.1.4.6 Budget Projection Report (BPR) Queries

The following will appear after Option 6 is selected from the BMCM RG Menu:

```
ENTER ACCOUNT NUMBER => _____
```

Following this, the user will be queried for the time period over which a projection is to be made, as follows:

```
ENTER THE NUMBER OF MONTHS OVER WHICH
THE PROJECTION IS TO BE MADE (DEFAULT -
REMAINDER OF CONTRACT) ENTER => _
```

A RETURN entry without a numeric specification will result in the default to the remainder of the contract.

2.1.4.7 Return to BMC Supervisory Menu

This option has no queries; it merely returns control to the BMC supervisory menu (Figure 2).

2.1.5 Delete Transaction

The fourth option on the BMC supervisory menu (Figure 1) is for deleting transactions. This option must be used with care to prevent current transactions from becoming lost to further processing. However, this is a necessary maintenance function to be performed by the user to eliminate those transactions from the system which are of no further reporting value. Two backup systems are used to assure that information is not lost due to transaction deletion. First, when an account is purged, all transactions in that account are listed (in the same format as the DAST report) on hard copy. Second, no transaction will be allowed to be purged as long as its account still exists.

All of the transactions for a given account will be deleted simultaneously. Thus, the query following Option 4 of the BMC supervisory menu is:

```
ENTER ACCOUNT NUMBER FOR TRANSACTIONS TO BE
PURGED => _____
```

This account number will be checked against those in the account file (D1.1), and if the account exists the message will be printed out:

```
ACCOUNT STILL EXISTS, NO TRANSACTIONS PURGED.
```

If, in fact, the account is no longer to be kept by the system, Option 6 of the BMC supervisory menu will be utilized (see next section). If the account has already been purged then the transactions will be purged.

2.1.6 Delete Accounts

This option of the BMC supervisory menu (Figure 1) will be used to eliminate an account from the account file (D1.1). Obviously, if any further activity or any further reporting is required for an account, it should not be deleted. However, when accounts are of no further informative value, they must be deleted from the system to allow room for others. As a backup, the DAS and DAST reports will automatically be generated whenever an account is deleted. This will provide a hard copy archive for all accounts.

When this option is selected, the following query will appear:

```
ENTER ACCOUNT NUMBER OF ACCOUNT TO BE
DELETED => _____
```

This account number will be echoed back, and a chance to correct any improper entry will be given prior to account deletion.

2.1.7 Perform End Period Updates

The account file (D1.1) keeps a record of expenditures to date during the month, quarter, and fiscal year. Therefore, at the end of these periods, these variables must be "zeroed-out" to start a new record. Option 7 of the BMC SM (Figure 2) will enable this to be done. The system will automatically generate a monthly reminder for this. The reminder will appear on the CRT, for any OAS logon code, until it has been cleared for that month. In order to be sure that all reports for the previous period have been produced prior to performing this function, the following reminder will appear:

```
THE FOLLOWING REPORTS SHOULD BE RUN PRIOR
TO ANY END-OF-MONTH CLOSEOUTS:
```

REPORT	MENU	OPTION
SUBSIDIARY LEDGER SUMMARY	BMC RG	1
DEPARTMENTAL ACCOUNT STATEMENT	BMC RG	3
BUDGET EXPENDITURE COMPARISON	BMC RG	5
BUDGET PROJECTION REPORT	BMC RG	6

RUN THE FOLLOWING BEFORE END-OF-YEAR CLOSEOUTS:

REPORT	MENU	OPTION
BUDGET SUBCODE ANALYSIS	BMC RG	2
EXTRAMURAL PROPOSALS		
AND NEW FUNDING	PCC RG	1
SUMMARY OF ACTIVE PROJECTS	PCC RG	2
RESEARCH ACTIVITIES REPORT	PCC RG	3

CHECK TO BE SURE THAT ALL OF THE AP-
PROPRIATE REPORTS ARE IN HAND BEFORE CON-
TINUING WITH THE END-OF-PERIOD
CLOSEOUT. ENTER "CONTINUE" TO PRO-
CEED; ALL OTHER ENTRIES WILL ABORT => _____

After this, the following query will appear:

    ENTER PERIOD (M=MONTH, Q=QUARTER, Y=FISCAL
    YEAR) => _

Note that Q will automatically produce the M update, and Y will produce both the M and the Q updates. A verification query will follow the
Y entry to assure that no mistakes have been made in requesting the
end-of-year closeout.

2.1.8 Return to Master Supervisory Menu

This option enables the system to return to the MASTER supervisory
menu (Figure 1), from which any of the other MASTER modules may
be executed.

2.2 Proposal and Contract Control Module

When the PCC Module is selected from the MASTER supervisory menu
(Figure 1), a second menu will appear as given in Figure 6. Each of
these options will be discussed in separate subsections below. Note that
these subsections are numbered corresponding to the PCC supervisory
menu option (example: Subsection 2.2.3 is for menu option 3).

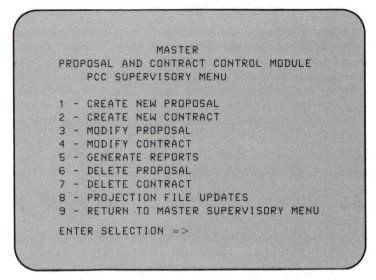

Figure 6 Proposal and Contract Control Supervisory Menu

2.2.1 Create New Proposal

This entry on the PCC supervisory menu indicates to the system that a new proposal is to be created, i.e., a new record will be created in the proposal file (D2.1). Only OAS logon codes will be able to enter new proposals. When selected, the mat given in Figure 7 will appear. These items will not be numbered on the data-entry mat. They are numbered here for reference to the following descriptions for each item:

1. FY. Fiscal year. This item is output by the system for general information; it is not supplied by the user (protected field). It is necessary to provide a unique proposal reference number, which is a concatenation of the FY and the Proposal Reference, Item 2. This figure is obtained from the general parameter file, D4.3 (D2.1,1).

2. Proposal reference. This is a sequential transaction reference number which starts with 1 for the first proposal of any given fiscal year. This is not supplied by the user (protected field). It is generated by the system and maintained in the general parameter file, D4.3 (D2.1,2).

3. Proposal number. This is the number which OAS assigns to each proposal (D2.1,3).

4,5. Principal investigators. Space limitation on output restricts this field to two entries of 20 characters each. Enter the principal investigator on these two lines exactly as they should be output (e.g., do not break a single name between lines without a hyphen) (D2.1,4).

```
                        MASTER
                PROPOSAL DATA-ENTRY MAT

   1. FY __          2. PROPOSAL REF. _____
   3. PROPOSAL NUMBER _____
   4,5. PRINCIPAL INVESTIGATORS    __(20 characters)___
                                   _____
   6. DEPARTMENT CODE __
      (1=ADM, 2=AE, 3=AM, 4=CHE, 5=CSE,
       6=EE, 7=IE, 8=ME, 9=TE, 10=ENG,
       11=OAS, 12=EXT, 13=WRR)
   7-10. PROJECT TITLE    _____(40 characters)_____
                          _____
                          _____
                          _____
   11,12. SPONSOR _(15 characters)_
                  _____
   13-15. DOLLAR CONTRIBUTION:
               SPONSOR      $_____
               OAS          $_____
          SHARED DIRECT     $_____
   16-17. DURATION   mm-dd-yy  to  mm-dd-yy
   18. DATE MAILED TO VP-RES  mm-dd-yy
   19. DATE RECEIVED FROM VP-RES  mm-dd-yy
   20. NEW (N) or MODIFICATION (M)? _
       _
   21. ACCEPTANCE:  ACCEPTED (A), PENDING (P), REJECTED (R)_
   22. DEPT. MAILING (1=AE, 2=AM,    3=CHE, 4=CE,
                      6=EE, 7=IE, 8=ME, 9=TE, 10=ENG,
                      11=OAS, 12=EXT, 13=WRR)    __
```

Figure 7 Proposal Data-entry Mat

  6. Department code.   Enter one of the departmental codes given
     (D2.1,5).
7-10. Project title.   The project title is limited to 160 characters,
     and it will be output on four separate lines. Enter it exactly
     as it will be output (D2.1,6).
11,12. Sponsor.   The project sponsor is limited to 30 characters, and
     it will be output over two lines. Enter it exactly as it will be
     output (D2.1,7).
13-15. Dollar contribution.   Enter the contribution from the sponsor,
     OAS, and the amount of shared-direct overhead, as indicated
     (D2.1,8-10).
16-17. Duration.   Two options exist for recording duration. If the ex-
     act start and completion dates are known they may be entered
     directly. If they are not known but the number of months is

known, then enter the number of months as an unlabeled number. For example, if the duration is known to be 1 year, 6 months, enter 18. The system will interpret null entries in the remaining columns of this field to be indicative that the number of months duration is to be stored, as opposed to actual dates (D2.1,11).

18. Date mailed to VP-Research.   Enter the date mailed if the proposal has been already mailed. If not, enter null for this item (D2.1,12).
19. Date received from VP-Research.   Enter the date received if the proposal has been received; otherwise, enter null for this item (D2.1,13).
20. New or modification.   Enter N for new, M for modification. The default is N, as underlined on the mat. That is, a null entry will enter N (D2.1,14).
21. Acceptance.   Enter A for accepted, P for pending, or R for rejected. The default is P, as underlined on the mat (D2.1,15).
22. Dept. mailing.   The department that has assumed responsibility for mailing the proposal to the sponsor is entered, using the departmental codes given in Item 6 (D2.1,16).

When the mat given above is completed, the user will be queried to verify the screen. If an error is detected, any of the items may be changed. If not, the proposal record will be written to the proposal file (D2.1). Corrections or updates after this point will require the use of Option 3 of the PCC supervisory menu (SM). The user will be queried at this point to either enter another new proposal or return to the PCC supervisory menu.

2.2.2 Create New Contract

This entry on the PCC SM indicates that a new contract is to be created, i.e., a new record will be created in the contract file (D2.2). Only OAS logon codes will be able to enter new contracts. When selected, the user will be given two options by the following query:

```
DO YOU WISH TO CONVERT AN EXISTING
PROPOSAL (Y/N)? => _
```

If the response to this query is yes (Y), an existing proposal will be used to generate the major part of the contract information. At the same time, that proposal record will be updated to a status of "Accepted" in the proposal file. In this case the next query will be:

```
ENTER PROPOSAL REFERENCE NUMBER (FYXXXXXX)
(ENTER ? FOR HELP) => FY_____
```

```
                            MASTER
                   CONTRACT DATA-ENTRY MAT

     1. FY __              2. CONTRACT REF. _____
     3. ACCOUNT NUMBER _-_____
     4. CONTRACT NAME ___(20 characters)__
     5-6. PRINCIPAL INVESTIGATORS    __(20 characters)___
                                     -------------------
     7. DEPARTMENT CODE __
        (1=ADM, 2=AE, 3=AM, 4=CHE, 5=CSE,
         6=EE, 7=IE, 8=ME, 9=TE, 10=ENG,
         11=OAS, 12=EXT, 13=WRR)
     8-11. PROJECT TITLE  _____(40 characters)_____
                          --------------------------------------
                          --------------------------------------
                          --------------------------------------
     12-13. SPONSOR _(15 characters)_
                    -------------------
     14-16. CURRENT DOLLAR         17-19. PAST CONTRIBUTION:
            CONTRIBUTION:
               SPONSOR      $_____         SPONSOR      $_____
                   OAS      $_____             OAS      $_____
         SHARED INDIRECT $_____    SHARED INDIRECT $_____
     20-21. DURATION   mm-dd-yy  to  mm-dd-yy
     22. PROPOSAL REFERENCE NO. FY_____
     23. PROPOSAL NUMBER ____(12)____
     24. ADMINISTRATIVE DIGEST NUMBER _____
     25. ADMINISTRATIVE DIGEST DATE mm-dd-yy
     26. NEW (N) OR MODIFICATION (M) _
     27. RESEARCH ACTIVITIES REPORT (Y/N)_
```

Figure 8 Contract Data-entry Mat

The system will expect a fiscal year followed by the six-digit reference number. A question mark entered at this point will provide the user with a cross-reference list of reference numbers by proposal number. Once a legitimate proposal reference number is entered, the user will be presented with the data-entry mat given in Figure 8. The following contract items of Figure 8 will be copied from the proposal as defaults on this mat if a proposal is being converted to a contract: 5-6, 7, 8-11, 12-13, 14-19, 20-21, 22, and 23. The remaining items must be completed at this time.

In the event that the query:

```
DO YOU WISH TO CONVERT AN EXISTING
PROPOSAL (Y/N)? => _
```

is answered with no (N), the mat given in Figure 8 will appear immediately without any of the items copied from the proposal file. The following describes the various items of Figure 8.

1. FY. Fiscal year. The default for this item is output by the system from the general parameter file (D4.3). It is necessary to provide a unique contract reference number, which is a concatenation of the FY and the contract reference number, Item 2. If the user wants to elect the next fiscal year (FY), then enter this as the first entry on the mat; otherwise, depress ENTER to retain the default value (D2.2,1).

2. Contract reference. This is a sequential-transaction reference number which starts with 1 for the first contract of any given fiscal year. This is not supplied by the user (protected field). It is generated by the system and maintained in the general parameter file, D4.3 (D2.2,2).

3. Account number. The account number assigned by the accounting department. The system will check for a valid account number and it will not allow data entry to proceed unless a valid account number is entered (D2.2,3).

4. Contract name (alias project number). The contract name assigned to the contract (D2.2,4).

5-6. Principal investigators. Space limitation on output restricts this field to two entries of 20 characters each. Enter the principal investigator on these two lines exactly as they should be output (e.g., do not break a single word between two lines without hyphen) (D2.2,5).

7. Department code. Enter one of the departmental codes given (D2.2,6).

8-11. Project title. The project title is limited to 160 characters, and it will be output on four separate lines. Enter it exactly as it will be output (D2.2,7).

12-13. Sponsor. The project sponsor is limited to 30 characters, and it will be output over two lines. Enter it exactly as it will be output (D2.2,8).

14-16. Current dollar contribution. Enter the contribution from the sponsor, OAS, and the amount of shared, indirect overhead, as indicated (D2.2,9-11).

17-19. Past contribution. Enter any past contributions which have been made on this contract by the sponsor, OAS, and the amount of shared, indirect overhead (D2.2,12-14).

20-21. Duration. Enter the start and the completion dates for the contract (D2.2,15).

22. Proposal reference number. Enter the reference number of the proposal for this contract, including the fiscal year as the first two digits (D2.2,16).

23. Proposal number. Enter the OAS-assigned proposal number for this contract (D2.2,17).
24. Administrative digest number. Enter the reference number from the administrative digest when it is received back from administrative review (D2.2,18).
25. Administrative digest date. Enter the date from the administrative digest (D2.2,19).
26. New (N) or modification (M). If this is an extension or modification of an existing contract, enter M. If it is a new contract, enter N (D2.2,20).
27. Research activities report. If this contract is to be included in the research activities report listing, as well as the percent support listing, enter Y (D2.2,21).

Upon completion of the data entry for a new contract, the user is queried to verify the screen, and opportunity for correction is provided. When no further corrections are required, the contract information is written to the contract file (D2.2). Corrections or updates after this point will require the use of Option 4 of the PCC supervisory menu (SM). The user will be queried at this point to either enter another new contract or return to the PCC SM.

2.2.3 Modify Proposal

This option is selected when the user wishes to modify information for a given proposal. The following query will result:

```
ENTER PROPOSAL REFERENCE NUMBER (FYXXXXXX)
(ENTER ? FOR HELP) => FY_____
```

If ? is entered, a cross-reference listing of the proposal reference numbers and their corresponding proposal numbers and project numbers will be given, followed by the option to obtain this list on hard copy. When a legitimate proposal reference number is entered, the information for that proposal will be presented in the form of the completed data-entry mat of Figure 7. Any modifications required can then be made.

2.2.4 Modify Contract

This option is selected when the user wishes to modify information contained in a contract record. The following query will result:

```
ENTER CONTRACT REFERENCE NUMBER (FYXXXXXX)
(ENTER ? FOR HELP) => FY_____
```

If ? is entered, a cross-reference listing of the contract reference number, account numbers, and contract name will be given, followed by the option to obtain this listing on hard copy. When a legitimate contract reference number is entered, the information for that contract will be presented in the form of the completed data-entry mat of Figure 8. Any modifications required can then be made.

2.2.5 Generate (PCCM) Reports

This option indicates that the user wishes to generate one or more of the PCCM reports. It is important that all available updates to contracts and proposals be made prior to report generation.

The possible PCCM reports will appear on a menu similar to that given in Figure 9. Depending upon the report selected, the user will be queried for the information necessary to generate the report. The following subsections detail these queries for each report. Note that since options exist on most reports for either terminal or hard copy output, the queries for this will not be repetitively documented below. Also, due to screen-width constraints on the CRT terminals, certain outputs will require the user to specify those data items to be viewed. This will be done by optional scrolling right and left, as well as down the page. For examples of these reports, see Section 2.2 of the MASTER Maintenance Manual.

```
                         MASTER
            PROPOSAL AND CONTRACT CONTROL MODULE
                PCC REPORT-GENERATOR MENU

          1 - EXTRAMURAL PROPOSALS AND NEW FUNDING
              (EPNF)
          2 - SUMMARY OF ACTIVE PROJECTS (SAP)
          3 - RESEARCH ACTIVITIES REPORT (RAR)
          4 - PERCENT SUPPORT LIST (PSL)
          5 - PROPOSAL STATUS REPORT (PSR)
          6 - CONTRACT STATUS REPORT (CSR)
          7 - RETURN TO PCC SUPERVISORY MENU

          ENTER SELECTION =>
```

Figure 9 PCCM Report-generator Menu

2.2.5.1 Extramural Proposals and New Funding (EPNF) Queries

No queries are required for this report.

2.2.5.2 Summary of Active Projects (SAP) Queries

The following will appear after option 2 is selected from the PCCM RG Menu:

```
ENTER DEPARTMENT CODE
(ENTER ? FOR HELP) => ___
```

A three-character department code will be entered, or the word ALL will indicate that all departments are to be considered. If ? is entered, a listing of the departmental codes will be presented on the screen. A legitimate department code entry will lead to the following query:

```
ENTER CONTRACT REFERENCE NUMBER
or
ENTER ACCOUNT NUMBER
or
ENTER "RETURN" FOR ALL ACCOUNTS => FY _____
```

If a particular active contract is to be reviewed, it may be selected either by contract reference number (assigned by the system at the time of contract record creation), or by account number. If a listing of all accounts is desired, then neither will be entered.

Following this the query will appear:

```
DO YOU WANT ONLY OVERSPENT CONTRACTS
LISTED (Y/N)? => _
```

Once one of these options is chosen, the appropriate SAP report(s) will be generated.

2.2.5.3 Research Activities Report (RAR) Queries

No queries are required for this report.

2.2.5.4 Percent Support List (PSL) Queries

No queries are required for this report.

2.2.5.5 Proposal Status Report (PSR) Queries

The following query will appear after option 5 is selected from the PCCM RG Menu:

```
ENTER DEPARTMENT CODE
(ENTER ? FOR HELP) => ___
```

This query is identical to that discussed in Section 2.2.5.2. It is followed by:

```
ENTER PROPOSAL REFERENCE NUMBER
or
ENTER PROPOSAL NUMBER
or
ENTER "RETURN" FOR ALL
PROPOSALS => _____
```

If a particular proposal is to be reviewed, it may be selected either by proposal reference number (assigned by the system at the time of proposal record creation), or by proposal number (assigned by OAS). If a listing of all accounts is desired, then no numbers will be entered.

2.2.5.6 Contract Status Report (CSR) Queries

The CSR queries are identical to those given immediately above for PSR, with the exception that contract reference number and account number replace proposal reference number and proposal number, respectively.

2.2.5.7 Return to PCC Supervisory Menu

This option has no queries; it merely returns control to the PCC supervisory menu (Figure 6).

2.2.6 Delete Proposal

When a proposal is no longer to be maintained in the proposal file it may be deleted. At this point it will be written to the proposal purge buffer file, which may be backed up and saved off line. The following query will result from Option 6 of the PCC SM:

```
ENTER PROPOSAL REFERENCE NUMBER OF
PROPOSAL TO BE DELETED => _____
CONFIRM DELETE REQUEST BY ENTERING THE
OAS PROPOSAL NUMBER => _____
```

If these two are inconsistent (i.e., they do not refer to the same pro-posal), no action will be taken. If the Proposal Purge Buffer File is close to being full, a message to that effect will be given, and instructions will be provided for backing it up on tape prior to its deletion.

2.2.7 Delete Contract

Considerations for contract deletion are identical to those for proposal deletion with the exception that the contract reference number and the account number are used for identification and verification of the con-tract to be deleted.

2.2.8 Projection File Updates

This update must be performed monthly in order to reflect the most current contract and proposal activity before any proposals and/or con-tracts are deleted. The system will automatically generate a monthly reminder for this and other periodical user requirements. This re-minder will come up on the CRT with any OAS logon code until it has been cleared for that month. Further, all available contract and pro-posal information should be entered prior to performing this update. The following query will follow the entry of Option 8 from the PCC SM:

```
ENTER THE MONTH JUST COMPLETED FROM THE
FOLLOWING LIST:
1  - OCT        2 - NOV        3 - DEC        4 - JAN
5  - FEB        6 - MAR        7 - APR        8 - MAY
9  - JUN       10 - JUL       11 - AUG       12 - SEP
ENTER => __
```

The system will compare this date with its internal calendar and print a warning if there is an inconsistency. This will be followed by:

```
DO YOU WISH TO UPDATE THE CURRENT RATIOS
AND VALUE IN THE PROJECTION FILE FOR THE
MONTH OF XXX (Y/N)? _
```

If N is entered, the system will return to the PCC SM. If Y is entered, the system will go through the specified month's records and obtain an update for each item within the projection file (D2.3). These updates will be listed on the terminal and the printer for verification. Then the following query will be output:

```
DO YOU WISH THESE UPDATES TO BE WRITTEN TO
THE PROJECTION FILE CURRENT RATIOS (Y/N)? _
```

If Y they will be written; N will return control to the PCC SM.

If the current ratios are updated, the user will be given an opportunity to update the values in the projection file which are used for estimating. Generally, this will only need to be done once per year. The following query will result:

```
IS IT THE END OF THE FISCAL YEAR (Y/N)? _
```

If N is entered, control is returned to the PCC SM. If Y is entered, the following query will result:

```
DO YOU WISH TO UPDATE THE PROJECTION
PARAMETERS (Y/N)? _
```

If N, control is returned to the PCC SM. Otherwise the following query will result:

```
WHAT WEIGHT DO YOU WISH TO GIVE TO
THE CURRENT YEAR'S FIGURES?
ENTER A NUMBER BETWEEN 0 (for no weight)
and 100 (for 100% replacement) => _
```

This number will be reflected for verification. Then the updates will be made. A printout of the old and new figures will be provided, and provision will be made to manually change any of the figures in this file.

2.2.9 Return to Master Supervisory Menu

This option enables the system to return to the supervisory menu (Figure 1), from which any of the other MASTER modules can be accessed.

2.3 Extension Evaluation and Control

When the EEC module is selected from the MASTER supervisory menu (Figure 1), a second menu will appear as given in Figure 10. Only OAS logon codes will be able to access EEC functions. Each of these options will be discussed in separate subsections below. Note that these subsections are numbered corresponding to the EEC supervisory menu option (example: Subsection 2.3.5 is for menu option 5).

2.3.1 Create New Program Budget

This entry on the EEC supervisory menu indicates to the system that a new program budget is to be created, i.e., a new record will be created in the extension file (D3.1). When selected, the mat given in Figure 11

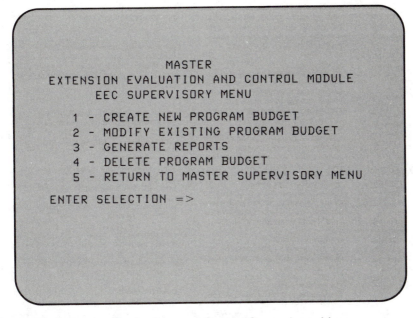

```
                    MASTER
    EXTENSION EVALUATION AND CONTROL MODULE
          EEC SUPERVISORY MENU

        1 - CREATE NEW PROGRAM BUDGET
        2 - MODIFY EXISTING PROGRAM BUDGET
        3 - GENERATE REPORTS
        4 - DELETE PROGRAM BUDGET
        5 - RETURN TO MASTER SUPERVISORY MENU

    ENTER SELECTION =>
```

Figure 10 Extension Evaluation and Control Supervisory Menu

will appear. Extension personnel will use the same procedure in completing the mat as they now use in their manual system. When creating a new budget record, all unknown values or values not yet available will be omitted by entering a RETURN only.

After completing the entries required, the user will be queried to make any modifications to the entries. When all items are verified and correct, the user will so indicate, and the record will be written to the file. Corrections or updates after this point will require the use of Option 2 of the EEC SM. The user will be queried at this point to either enter another new budget or return to the EEC SM.

2.3.2 Modify Existing Program Budget

This option is selected when the user wishes to modify information contained in an existing program budget (i.e., created as discussed in Section 2.3.1). The following query will result:

```
ENTER PROGRAM REFERENCE NUMBER
(ENTER ? FOR HELP) => _____
```

If ? is entered, a cross-reference listing of the program reference numbers by program description will be listed, with an option to obtain the same on hard copy. When a legitimate program reference number is entered, the information for that program budget will be presented in

```
                              MASTER
                   EEC BUDGET DATA-ENTRY MAT

    REF NUM _____                 DATE OF PROGRAM  MM-DD-YY
    DESCRIPTION __(20 characters)_____ TOTAL ATTENDING _____
    CHARGE/PERSON ____   EST. INCOME _____ ACT. INCOME _____

                               COST

       Direct Costs:                 Estimated         Actual
       Payments to Univ. Fac./Staff  _____         _____
       Faculty/Staff Travel          _____         _____
       Honoraria & Speaker's Fees    _____         _____
       Planning/Advisory Comm.       _____         _____
       Printing                      _____         _____
       Postage                       _____         _____
       Advertising                   _____         _____
       Telephone                     _____         _____
       Rentals (housing)             _____         _____
       Meals: Participants           _____         _____
              Program Personnel      _____         _____
              Breaks                 _____         _____
       Supplies: Office              _____         _____
                 Instructional       _____         _____
       Other: Transportation         _____         _____
              Miscellaneous          _____         _____
       Administrative Costs:
         Clerical, Secretarial Services _____      _____
         Coordination, Supervision   _____         _____
         General Administration      _____         _____
       Contributed Costs             _____         _____
```

Figure 11 EEC Budget Data-entry Mat

the form of the completed data-entry mat of Figure 11. Any modifications required can be made in identical fashion as the original data entry.

2.3.3 Generate Reports

This option indicates that the user wishes to generate one or more of the EEC reports. It is important that all available updates be made prior to selecting this option. The possible EEC reports will appear on a menu as given in Figure 12. Depending upon the report selected, the user will be queried for the information necessary to generate the report desired. The following subsections will detail these queries for each report. Note that since options exist on most reports for either terminal or hard copy output, the queries for this will not be repetitively documented below. For example reports, see Section 2.3 of the MASTER Maintenance Manual.

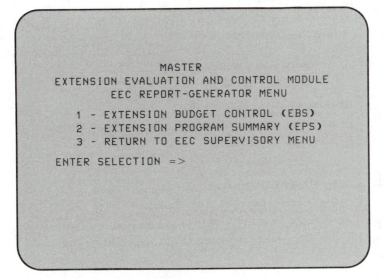

Figure 12 EECM Report-generator Menu

2.3.3.1 Extension Budget Control (EBC) Report

This report presents a formatted listing of the data in one or more of the program budgets formerly entered. The system will need to know at this point which of the program budgets to output. The user may request one, several, or all. In response to the following query:

```
ENTER REFERENCE NUMBER(S) OF PROGRAMS
(ENTER ? FOR HELP) =>
```

any one of the following entries may be made:

1. A single reference number. This is the most typical response; the output for that program will be generated immediately.
2. Several reference numbers separated by commas. All programs listed will be output.
3. Two reference numbers separated by a dash. This defines a range of reference numbers; all corresponding programs identified will be output.
4. The word ALL. This would cause all program reports to be generated.
5. ? This will present the options given above, and optionally output a listing of reference numbers and their corresponding descriptions.

Once a given set of reports is generated the system will return control to the EEC SM.

2.3.3.2 Extension Project Summary

This is a summary of several user-selected programs. The process of program selection is identical to that given in Section 2.3.3.1. The output will be a summation of those programs selected.

2.3.3.3 Return to EEC Supervisory Menu

This option has no queries; it merely returns control to the EEC supervisory menu (Figure 10).

2.3.4 Delete Program Budget

When a program budget is no longer to be maintained in the extension file it may be deleted. At this point, a hard copy report will be generated so that the information can be saved off line. The following query will result from this option:

```
ENTER PROGRAM REFERENCE NUMBER OF PROGRAM TO
BE DELETED => _____
```

The system will then display the description of this program on the screen for verification. The following will then appear:

```
IS THIS THE CORRECT PROGRAM FOR
DELETION (Y/N)? _
```

A Y will cause the deletion to take place. The entry of N will cause the system to return to the EEC supervisory menu.

2.3.5 Return to Master Supervisory Menu

This option enables the system to return to the supervisory menu (Figure 1), from which any of the other MASTER modules can be accessed.

Part 3 Technical Operations

The purpose of this section is to provide a separate check list of activities to enable users to easily access the MASTER system. The LOGON procedure is as follows:

1. Turn terminal switch to the ON position (if terminal is not left on).
2. Put terminal in ONLINE mode by depressing the control (CNTL) and the LOCAL keys simultaneously.
3. Depress the ENTER key; system should request logon codes as follows: ENTER SIGN-ON
4. Enter the following:

   $LOG.N accno userid password

   (Account numbers, userids and passwords will be assigned to all eligible MASTER users.)
5. The system will now be under MASTER control, and it will request a MASTER access code. Type access code followed by ENTER. At this point the MASTER system monitor will be in effect, and the menu in Figure 13 will appear.

The logoff procedure is as follows:

1. From any of the MASTER menus, return to the MASTER supervisory menu (Figure 1).
2. Enter Option 5 to get to the system menu (Figure 13).
3. Enter Option 2 to get to the operating system.
4. Enter OFF to logoff system; clear screen.

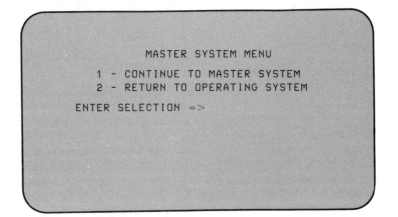

Figure 13 System Menu

MANAGEMENT ANALYSIS
TO SUPPORT TRANSACTIONS
IN EDUCATION AND RESEARCH

(M A S T E R)

Maintenance Manual

Designed and Developed by:
Scientific Computer Applications, Inc.
P.O. Box 1391
Auburn, Alabama 36830

MASTER MAINTENANCE MANUAL

Table of Contents

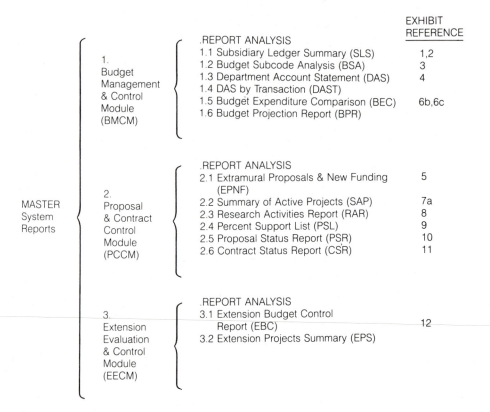

Figure 1.1 Analysis of MASTER Report Generation Components

1.0 Introduction

Management Analysis to Support Transactions in Education and Research (MASTER) is a system designed to provide management control to the University Office of Administrative Services (OAS). For an overview of the features of MASTER, see the MASTER System Overview and User Guide, Part 1. User requirements are given in Parts 2 and 3 of that volume. This manual is directed toward the software developers, and, ultimately, toward those who will maintain MASTER. It consists of three parts: (1) Output Report Specifications, (2) File Designs, and (3) Process Specifications. These define the outputs, inputs, and procedures for processing the data. This manual assumes that the reader is familiar with the MASTER System Overview and User Guide (O & UG), in that very little of the material contained therein is repeated here and many references are made to the O & UG.

Figure 1.1 demonstrates the analysis of the total reporting portion of MASTER into its constituent parts. From the MASTER supervisory menu, the user can select any of the three modules, each of which leads to a separate report generator (RG) menu. These are as follow:

1. Budget Management and Control Module (BMCM),
2. Proposal and Contract Control Module (PCCM), and
3. Extension Evaluation and Control Module (EECM).

Note that the column at the right of Figure 1.1 contains the reference to the report by external exhibit number. These exhibits are kept in a separate volume; however, most of these have been copied into their respective subsections below. Each of the modules are subdivided according to the listed reports as given. These are detailed in the subsections below according to the report number assigned in Figure 1.1, i.e., report X.Y is documented in Section 2.X.Y.

2.0 Output Report Specifications

This section presents the specifications of the reports required by the MASTER system. The reports are grouped together according to their general function for purposes of modular system design and development. This breakdown is presented in Figure 1.1.

The output specifications for each of the reports listed in Figure 1.1 are given in the subsections below. Each subsection will be prefaced by a short section describing the overall objective and target of the report. This will be followed by a description of the report variables.

2.1 Budget Management and Control Module (BMCM) Reports

2.1.1 Subsidiary Ledger Summary (SLS)

SLS has the following general characteristics:

Purpose—to provide department heads and OAS personnel with a quick means of reviewing all accounts for budgetary control.

Availability—on CRT or hard copy to each department head upon request (for that department); or a summary of all accounts for OAS review and control.

Brief description—this report presents a summary for each account which includes expenditures, encumbrances, and balance available.

Sort/Order—order by account number within each department.

All information for the SLS report will be obtained from the file as given at the end of the item description (in parentheses). The following variables are output (see Output 1.1):

```
                    UNIVERSITY OFFICE OF ADMINISTRATIVE SERVICES (1)
        08/02/84 (2)
                         MASTER SUBSIDIARY LEDGER SUMMARY (3)
        15:57:15 (5)                                      REPORT PAGE 1 (4)
        PGM = SLS (7)                                     USER ID IE35 (6)
                        SUMMARY OF EXPENDITURES & FUNDS AVAILABLE
             INDUSTRIAL ENGINEERING (10)
                                    (13)       (14)        (15)         (16)
          (11)          (12)       REVISED    CURRENT     CURRENT       FISCAL
        ACCOUNT      DESCRIPTION    BUDGET     MONTH       QUARTER    YEAR-TO-DATE
        -------    -------------   -------    -------     -------     ------------
        1-16280    INDUSTRIAL ENG.   5,010-     21.00-      21.00-       136.00-
                   LEDGER 1 TOTAL (21)  5,010-  21.00-      21.00-       136.00-

        2-10856    INDUSTRIAL ENG. 543,918  24,359.17   24,359.17     511,169.14
        2-12856    IE ENG EXP STAT  13,256   2,466.60    2,466.60       9,171.02
        2-12863    COST SHR 420393   5,040      39.98-      39.98-        503.92
        2-12867    COST SHR 420408  18,166   1,147.32    1,147.32       4,820.43
        2-12869    COST SHR 420421  16,080   1,303.19-   1,303.19-     11,288.90
        2-12878    IOAS82N1MORRISSE 18,964   1,845.68    1,845.68       9,017.82
        2-12879    IOAS-84-C-PARK   25,075   5,064.68    5,064.68      14,576.30
        2-12880    IOAS-84-C-BULFIN 26,710   1,520.99    1,520.99      18,312.57
        2-12898    COST SHR 420467  13,980     495.00      495.00         495.00
                   LEDGER 2 TOTAL (21) 681,189 35,556.27 35,556.27     579,355.10

        4-20247     AFOSR-79-0016   54,148
        4-20372    DAAH01-80-C-1480 37,434                                 70.10
        4-20393    USFS-19-80-407   21,240     413.00      413.00       6,705.45
        4-20397    OHTS810501A31501  4,996
                   LEDGER 4 TOTAL (21) 117,818 413.00      413.00       6,775.55
             INDUSTRIAL ENG. TOTAL 1,184,717 71,810.45  71,810.45     714,441.08
```

Output 1.1 Screen 1

1. Major heading (hard copy only).   This will be constant for all printed reports.
2. Computer date (hard copy only).   The date stored as the current date within the computer.
3. Report title.
4. Report page (hard copy only).
5. Time of day (hard copy only).   The time stored as the current time within the computer.
6. User identification number (hard copy only).   The user identification number of the user requesting the hard copy report.
7. Program code.   The reference code of the program which produces this report.
8. Reserved.
9. Reserved.

```
              UNIVERSITY OFFICE OF ADMINISTRATIVE SERVICES (1)
                            08/02/84 (2)
                  MASTER SUBSIDIARY LEDGER SUMMARY (3)
     15:57:15 (5)                                    REPORT PAGE 1 (4)
     PGM = SLS (7)                                    USER ID 48.1 (6)
                  SUMMARY OF EXPENDITURES & FUNDS AVAILABLE
                       INDUSTRIAL ENGINEERING (10)
                                    (18)         (18a)
                         (17)                     TOTAL     (19)        (20)
             (11)      TOTAL-TO-   CURRENT-MONTH  ENCUM-    BALANCE     EXPIRE
            ACCOUNT      DATE      ENCUMBRANCES   BRANCES   AVAILABLE    DATE
            -------    ---------   ------------   -------   ---------   -------
            1-16280       136.00-                            4,874-
            1-TOTAL       136.00-                            4,874-

            2-10856    511,169.14     850.32     1,673.19   31,075
            2-12856      9,171.02                    .45     4,084
            2-12863      4,995.99                               44    05/31/84
            2-12867     18,051.15                              114    06/30/84
            2-12869     14,284.14                            1,795    04/30/84
            2-12878      9,017.82     212.10      744.50     9,201    09/30/84
            2-12879     14,576.30                           10,498    09/30/84
            2-12880     18,312.57                            8,397    09/30/84
            2-12898        495.00                           13,485    06/30/85
            2-TOTAL    600,073.13   1,062.42     2,418.14   78,693

            4-20247     54,147.70                                     09/30/82
            4-20372     37,433.64                                     09/30/83
            4-20393     20,864.12                              375    05/31/84
            4-20397      4,995.91                                     09/30/83
            4-TOTAL    117,441.37                              375

            IE TOTAL   717,378.50   1,062.42     2,418.14   74,194
               (22)
```

Output 1.1 Screen 2

10. Department name. Determined by user identification number (see Item 6), specified at LOGON (D4.1,2).
11. Account number. Report will have one line of information for each account within the department, six digits, first two separated by dash (D1.1,2).
12. Description. Contract name (D2.2,4).
13. Revised budget. The budget as it exists after the last update; original budget plus or minus any budget changes made during the year (D1.1,9,16,23,...summed).
14. Current month expenditures. Total expenses reported for month. This will be automatically reset to zero on the first of each month (D1.1,10,17,24,...summed).
15. Current quarter expenditures. Total expenses to date for current quarter. This will be automatically reset to zero on the

first day of the quarter (D1.1,11,18,25,...summed). Quarters are:

$$10/1 - 12/31$$
$$1/1 - 3/31$$
$$4/1 - 6/30$$
$$7/1 - 9/30$$

16. Year-to-date expenditures. Total expenses for fiscal year (10/1 - 9/30) to date (D1.1,12,19,26,...summed).

17. Total-to-date expenditures. Total expenses since beginning of project to the current date. This amount would be different from year-to-date only on grants and contracts which run for more than one year and also for any accounts which have a "grant or project" year different from the University fiscal year (D1.1,13,20,27,...summed).

18. Current month encumbrances. Total amount of unpaid purchase orders incurred during the current month. This will be automatically reset to zero on the first of each month (D1.1,14,21,28,...summed).

18a. Total encumbrances. Total amount of unpaid purchase orders to date (D1.1,15,21,28,...summed).

19. Balance available. Total balance-to-date less encumbrances (calculated, Item 13 - Item 17 - Item 18a).

20. Expire date. The expiration date for contracts and grants (D2.2,15).

21. Ledger totals. Column totals for each subclassification as defined by the first digit of the account code; see Item 11 (calculated).

22. Grand totals. Column totals for all account codes (calculated).

2.1.2 Budget Subcode Analysis (BSA)

BSA has the following general characteristics:

Purpose—to provide a breakdown of each department by subcode, and to summarize all departments by subcode; further, for each subcode, to provide a breakdown by State, OAS, and extramural funds.

Availability—same as SLS (Section 2.1.1) but on hard copy only.

Brief description—this report summarizes the amount of funds obligated and the amount expended according to funding sources. It does this for any department or all departments.

Sort/Order—arrange by subcode.

```
                                          (12)
                                       YEAR-TO-DATE
  (10)                              --------------------------------------
  SUB-                (11)          (13)        (14)       (15)
  CODE      DESCRIPTION             INSTRUCTION  OAS     COST SHARING
  100       SALARIES
  140       WAGES
  150       EMPLOYEE BENEFITS
  200       SERVICES PURCHASED
  600       STUDENT AID
  700       EQUIPMENT PURCHASED
  800       PLANT FUND
  900       TRANSFERS AND OTHER


  -----------------------------------------------------------------------
        (16)          (17)          (18)          (19)         (20)

    EXTRAMURAL   GIFTS/GRANTS   EXTENSION   GRANT-IN-AID    TOTAL




                              (21)
                        TOTAL TO DATE
  ------------------------ (Repeat Items 10-20) -----------------------

                              (22)
                        BALANCE AVAILABLE
  ------------------------ (Repeat Items 10-20) -----------------------
            (23)
            TOTALS
```

Output 1.2

The following variables are output (see Output 1.2):

1–9. Same as Items 1-9, SLS (Section 2.1.1).
  10. Subcode.   Major budget control subcodes. The following sub-
      codes will be reported:

Subcode	Description	Includes
100	Salaries	100–139
140	Wages	140–149
150	Benefits	150–199
200	Service and Product	200–499
500	Purchases	500–599
600	Student Aid	600–699

Subcode	Description	Includes
700	Equipment	700–799
800	Plant Fund	800–899
900	Transfers	900–999

11. Description. Descriptive name for budget control subcode (D4.2,2).
12. Year-to-date. In this section, Items 13-20 will be listed for the fiscal year to date (D1.1,12,19,26,...).
13. Instruction. This will include all accounts with fund classification 1 (D1.1,5).
14. OAS. This will include all accounts with fund classification 2 (D1.1,5).
15. Cost sharing. This will include all accounts designated as fund classification 3 (D1.1,5).
16. Extramural. This will include all accounts with fund classifications 4, 5, or 6 (D1.1,5).
17. Gifts. This will include all accounts with fund classification 7 (D1.1,5).
18. Extension. This will include all accounts with fund classification 8 (D1.1,5).
19. Grant-in-aid. This will include all accounts with fund classification 9 (D1.1,5).
20. Total. This will include all accounts (calculated).
21. Total-to-date. Items 13–20 will be listed for the total expenditure to date under this major heading (D1.1,13,20,27,...).
22. Balance available. Items 13–20 will be listed for the balance available; calculated from original budget values (D1.1,9,16,23,...) minus the total-to-date items, under Item 21.
23. Totals. Column totals for all columns (calculated).

2.1.3 Departmental Account Statement (DAS)

DAS has the following general characteristics:

Purpose—to provide specific information on the status of each account by subcode.

Availability—same as SLS (Section 2.1.1).

Brief description—report available for each account which shows the revised budget, the unexpended budget balance, and the balance available by budget control code. Expenditures are shown totaled by subcode for the current month, the current quarter, and to-date. Also included is a detailed list of the status of all purchase orders as of the end of the current month.

Sort/Order—the first part of the report will be ordered by subcode; the "Open Encumbrances Status" portion will be ordered by date within each account.

The following variables are output (see Output 1.3):

1–9. Same as Items 1-9, SLS (Section 2.1.1).
  10. Expiration date (if grant or contract) (D2.2,15).
  11. Person directly responsible for the account (D2.2,5).
  12. Account name. Short name used to designate accounts (aliases - project number, description) (D2.2,4).
  13. Account number. The number of the account of interest, specified by the user.
  14. Subcode. Detail subcode or budget control code as given in Item 10, BSA, Section 2.1.2 (D4.2,1).

```
                   MASTER DEPARTMENTAL ACCOUNT SUMMARY

 08/02/84 (2)

    DEPARTMENTAL/ACCOUNT SUMMARY REPORT IN WHOLE DOLLARS FOR 11/80 (8)

 02:16:44                                              REPORT PAGE 1
 PGM = DAS                                             USER ID  IE35
     (10)               (11)            (12)                (13)
 EXP. DATE 09/83    DR. BRUSH     IHTSS810501A31501     4-20439


 (14)                       (14b)    (14c)    (14d)    (14e)
 SUB-                       REVISED  THIS     EXPENDED  YEAR-
 CODE    DESCRIPTION        BUDGET   MONTH    THIS QTR  TO-DATE
 ----    -----------        -------  -----    --------  -------

 100     SALARIES            8,500     708     1,417     1,417

 140     WAGES              24,939   1,904     3,831     3,831

 150     EMPLOYEE BENEFITS   4,681     366       366       735

       **PERSONNEL COSTS    38,120   2,978     5,983     5,983

 200     PROD & SRV PCHSD   20,939   1,808     3,513     3,513

 900     TRANSFERS & OTHER  16,802

 980     INDIRECT COSTS      3,128   1,557     3,128     3,128

       *TOTAL MAINTENANCE   19,930   1,557     3,128     3,128

        ****TOTAL****       79,489   6,343    12,624    12,624
```

Output 1.3 Screen 1

```
                  MASTER DEPARTMENTAL ACCOUNT SUMMARY

08/02/84

        DEPARTMENTAL/ACCOUNT SUMMARY REPORT IN WHOLE DOLLARS FOR 11/80

02:16:44                                              REPORT PAGE 1
PGM = DAS                                             USER ID IE35
EXP. DATE 09/83     DR. BRUSH      IHTSS810501A31501      4-20439

                                 (15)        (15a)
(14)                     (14f)    ENCUMBRANCES        (16)    (17)   (18)
SUB-                     BUDGET   ------------        BALANCE    %      %
CODE       DESCRIPTION   BALANCE  THIS MONTH  TOTAL   AVAL     EXP    AVAL
----       -----------   -------  ----------  -----   -------  ----   ----
100    SALARIES            7,083                       7,083    17     83
140    WAGES              21,108                      21,108    15     85
150    EMPLOYEE BENEFITS   3,946                       3,946    16     84
     **PERSONNEL COSTS    32,137                      32,137    16     84

200    PROD & SRV PCHSD   17,426              445     16,981    17     81
900    TRANSFERS & OTHER  16,802                      16,802           100
980    INDIRECT COSTS                                          100
      *TOTAL MAINTENANCE  16,802                      16,802    16     84

    **EXPENSE TOTAL       66,865              445     66,420    16     84
    ****TOTAL****         66,865              445     66,420    16     84
```

Output 1.3 Screen 2

14a. Description.   Descriptive name for subcode or budget control subcode (D4.2,2).

14b. Revised budget.   The budget as it exists at the end of the reporting month (D1.1,9,16,23,...).

14c. Expended current month.   Total expenses for month being reported (D1.1,10,17,24,...).

14d. Expended current quarter.   Total expenses to date for current quarter (D1.1,11,18,25...).

14e. Expended-to-date.   Total expenses since beginning of project (D1.1,13).

14f. Budget balance.   Revised budget (Item 14b) minus the expended-to-date (Item 14e).

15. Current month encumbrances.   Those purchase orders opened during the current month (D1.1,14,21,28,...).

15a. Total encumbrances. Unpaid purchase orders to date (D1.1,15,22,29,...).

16. Balance available. Budget balance minus open encumbrances (calculated).

17. Percent expended. Percentage of revised budget which has been expended (calculated).

18. Percent available. Percentage of revised budget which is available for expenditure (calculated).

19–26. Open encumbrances. All transactions will be listed for purchase orders which have D1.2,11 = N on the last transaction for the purchase order.

19. Account. Account number and subcode which purchase order was issued on (D1.2,4,5).

20. Reserved.

21. PO number. Purchase order number (D1.2,7).

22. Date. Date that the purchase order was processed by OAS (D1.2,3).

23. Description. Vendor name (D1.2,9).

24. Original amount. Amount of the purchase order plus or minus any change orders which have been processed (sum of D1.2,10 where D1.2,6 = 2).

```
                          OPEN ENCUMBRANCE STATUS


                                                                    (24)
        (19)            (21)         (22)             (23)          ORIG
      ACCOUNT          PO NO.        DATE          DESCRIPTION      AMOUNT
    -----------       -------      --------     --------------------  --------
    2-10856-231        012501      01/17/83     METRO COPY SYSTEMS      44.66
    2-10856-231        102853      10/29/83     METRO OFFICE EQUIPME   390.11
    2-10856-231        204231      11/20/83     METRO OFFICE EQUIPME  4,000.00
    2-10856-242        120383      07/29/83     KEPPCO, INC.           250.00
    2-10856-242        205934      06/30/84     SCIENCE ASSOCIATES      95.00
    2-10856-375        220983      07/26/84     IEE CONFERENCE OFFIC    30.00
    2-10856-471        202132      10/26/83     MARKETLINE SYSTEMS     180.00
    2-10856-771        219061      06/28/84     RADIO SUPPLY           110.96

                                               *ACCOUNT TOTAL*      5,100.73
```

Output 1.3 Screen 3

25. Liquidating expenditure. Includes total of vouchers processed against the purchase order (sum of D1.2,10 where D1.2,6 = 3).

26. Current amount. Original amount (Item 24) minus the liquidating expenditures (Item 25).

2.1.4 DAS by Transaction (DAST)

DAST has the following general characteristics:

Purpose—to provide an on-line listing of all transactions entered into the MASTER System.

Availability—same as SLS (Section 2.1.1).

Brief description—this report is available for each account and lists all transactions which were processed against the account within a user-specified time span. Exception reports by transaction-type code will also be available.

```
        OPEN ENCUMBRANCE STATUS
        (19)                    (24)              (25)                (26)
                             ORIGINAL          LIQUIDATING          CURRENT
        ACCOUNT              AMOUNT            EXPENDITURE          AMOUNT
        ----------           --------          -----------         --------

     2-10856-231                44.66              0.00                0.00
     2-10856-231               390.11            304.67               86.44
     2-10856-231             4,000.00          2,717.77            1,282.23
     2-10856-242               250.00              0.00              250.00
     2-10856-242                95.00             95.00                0.00
     2-10856-375                30.00              0.00               30.00
     2-10856-471               180.00              0.00              180.00
     2-10856-771               110.96              0.00              110.96
    *ACCOUNT TOTAL*          5,100.73          3,117.44            1,673.19
```

Output 1.3 Screen 4

Sort/Order—the transactions are listed in chronological order within subcodes in numerical sequence.

The following variables are output (see Output 1.4):

1-14a. Same as Items 1–14a for DAS (Section 2.1.3).
15. Date. Date transaction entered into MASTER (D1.2,3).
16. Type code. A code which enables the identification of the type of transaction being processed. The following codes will apply (D1.2,6):
    1. Nonpurchase-order expenditure
    2. Purchase-order encumbrance (new/modifications)
    3. Purchase-order payment (partial/final)
    4. Budget modification
17. Reference number. Sequential reference number assigned by system for reference purposes (D1.2,2).
18. PO number. Purchase-order number for PO transactions; otherwise, blank (D1.2,8).
19. Budget entries. The amount of budget change transactions; Item 16, code 4 (D1.2,10).
20. Expenditures. Expenditure activities for the period given; Item 16, code 1 or 3 (D1.2,10).
21. Encumbrances. Indicates the amount of purchase-order transactions, and also indicates the amount of the liquidation of purchase orders due to other transactions; Item 16, code 1, 2, or 3 (D1.2,10).
22. Cleared business office. This field will be output either as Y (yes, to indicate that the transaction cleared the business office), or blank. Blank will indicate nonclearance or not applicable (D1.2,14).

2.1.5 Budget Expenditure Comparison (BEC)

BEC has the following general characteristics:

Purpose—to provide OAS personnel and department heads with the ability to compare budgeted with actual expenditures by source of funds.

Availability—same as SLS (Section 2.1.1), but on hard copy only.

Brief description—this report lists the amount of money budgeted by funding source in major column headings. Within each of these, budgeted expenditures are compared with actual expenditures and the ratio of actual-to-budgeted is given. This is also compared to a standard ratio for this time period, obtained from several previous years' history.

```
            MASTER DEPARTMENTAL ACCOUNTS SUMMARY BY TRANSACTION
11/30/80                                              REPORT PAGE  1
02:16:44                                                 USER ID IE35
PGM = DAST                                            ACCOUNT 5-35272
EXP. DATE 09/30/81      DR. BROWNE                    OHTS810501A31501


(14)                                (16)      (17)       (18)
SUB-              (14a)       (15)   TYPE      REF        PO
CODE           DESCRIPTION    DATE   CODE      NO.        NO.
----     --------------------  -----  ----    ------     ------
101      SALARIES             11/12    1      567443
103      WAGES                11/12    1      567444
151      SOCIAL SECURITY      11/12    1      567445
400      BUDGET REALLOCATION  11/04    4      567446
420      UNIVERSITY BOOKSTORE 11/04    1      567447
         REX STATIONERY/PRTNG 11/07    3      567448     000169
         MCALIECE PAPER CO    11/08    3      567449     001963
         MEDICRAFT            11/08    3      567450     002049
         ACE SCIENTIFIC SUPP  11/08    2      567451     003894
         SCIEN GLASS APPARAT  11/09    3      567452     000186
         ARTHUR H THOMAS CO   11/10    3      567453     000641
         WARMAN PREC PRODS    11/11    2      567454     004491
         HELLMA CELL INC.     11/11    2      567455     004761
         DOVE BUSINESS        11/11    3      567456
         EMCO                 11/11    3      567457     000921
         ANDOR CHEMICAL CORP  11/11    3      567458     000946
430      JANITORIAL           11/12    3      567459     000629
445      LEARNING RESOURCES   11/09    1      567460
445      LEARNING RESOURCES   11/10    2      567461     000952
465      ACME MEDICAL SUPPLIES 11/09   3      567462     000853
499      A&M HARDWARE         11/11    3      567463     999318
499      PHARMACIA FINE CHEM  11/12    3      567464     031029
900      BUDGET REALLOCATION  11/04    4      567465
980      INDIRECT COSTS       11/02    4      567466
```

Output 1.4 Screen 1

```
                 MASTER DEPARTMENTAL ACCOUNTS SUMMARY BY TRANSACTION

 11/30/80                                                    REPORT PAGE  1
 02:16:44                                                    USER ID  IE35
 PGM = DAST    (7)                                    (13) ACCOUNT 5-35272
 EXP. DATE 09/30/81              DR. BROWNE                OHTS810501A31501
```

(14) SUB-CODE	(14a) DESCRIPTION	(15) DATE	(18) PO NUMBER	(19) BUDGET ENTRIES	(20) EXPEN-DITURES	(21) ENCUM-BRANCES	(22) CLEARED BUS OFC
101	SALARIES	11/12			708.33		Y
103	WAGES	11/12			1,903.75		Y
151	SOCIAL SECURITY	11/12			365.96		Y
400	BUDGET REALLOCATION	11/04		68.00-			
420	UNIVERSITY BOOKSTORE	11/04			1.96		
	REX STATIONERY/PRTNG	11/07	000169		4.41		
	MEDICRAFT	11/08	002049		23.11		
	ACE SCIENTIFIC SUPP	11/08	003894		62.47	58.85-	
	SCIEN GLASS APPARAT	11/09	000186		56.36	54.00-	
	ARTHUR H THOMAS CO	11/10	000641		17.20		
	WARMAN PREC PRODS	11/11	004491			73.02-	
	HELLMA CELL INC.	11/11	004761			160.00	
	DOVE BUSINESS	11/11			8.86		
	EMCO	11/11	000921		38.97	38.97-	
	ANDOR CHEMICAL CORP	11/11	000946		7.92		
430	JANITORIAL	11/12	000629		14.08		
445	LEARNING RESOURCES	11/09			5.45		
	LEARNING RESOURCES	11/10	000952		746.55		
465	ACME MEDICAL SUPPLIES	11/09	000853		752.00		Y
499	A&M HARDWARE	11/11	999318			40.00	Y
	PHARMACIA FINE CHEM	11/12	031029			68.00	
900	BUDGET REALLOCATION	11/04		68.00			
980	INDIRECT COSTS	11/02			1,557.00		
	TOTAL			00.00	6,343.00	197.17	Y

Output 1.4 Screen 2

Sort/Order—rows are arranged alphabetically by department.

The following variables are output (as in Exhibits 6b and 6c):

1-8. Same as SLS Items 1-8 (Section 2.1.1).
9. Department.   Name of department (6 alphanumeric characters) (D4.1,2).
10. Instructional budget.   The beginning-of-year approved amount allocated for instruction. See Section 2.1.2, Item 13 for definition of instruction accounts; sum of D1.1,9,16,23,..., etc., for accounts in which D1.1,5 = 1.
11. Instruction actual.   The actual amount expended to date for instruction (D1.1,13,20,27,...).
12. Instruction ratio.   Item 11 divided by Item 10 (calculated).
13. Instruction percent deviation.   A standard ratio will be computed for each month based upon several prior years' data (D2.3,3). When BEC is run, the current data (Item 12) will be compared to the standard ratio interpolated to that date. The percent deviation will indicate to what extent the actual expenditure is an increase (+) or a decrease (−) over that expected from previous years.
14-17. Same as Items 10-13 but for OAS. See Section 2.1.2, Item 14 for definition of OAS account (D1.1,5 = 2).
18-21. Same as Items 10-13 but for cost sharing. See Section 2.1.2, Item 15 for definition of cost sharing accounts (D1.1,5 = 3).
22-25. Same as Items 10-13 but for extramural. See Section 2.1.2, Item 16 for the definition of extramural accounts (D1.1,5 = 4, 5 or 6).
26-29. Same as Items 10-13 but for totals (all accounts).

2.1.6 Budget Projection Report (BPR)

BPR has the following general characteristics:

Purpose—to provide OAS personnel, project managers, and department heads with the ability to project current spending throughout part or all of the remaining contract period.

Availability—same as SLS (Section 2.1.1), but on hard copy only.

Brief description—report available for each account which shows revised budget, unexpended budget balance, and expenditures of the current month and to date; also, a projection is provided to show what the total expenditure and remaining balance would be if the current month and the average monthly expenditures are continued for either a specified number of months or over the remaining contract period.

Sort/Order—for each account, the rows of the report will be ordered by subcode.

The following variables are output (see Output 1.3, Section 2.1.3, and Output 1.6):

1-14c. Same as DAS (Section 2.1.3).
15-16. Same as DAS Items 14e and 14f.
  16a. Expiration date.   Contract expiration date (D2.2,15).
  16b. Projection period.   The number of months that are being projected, specified by the user at report generation (query).
   17. Current month projected.   This heading implies that the subheadings refer to a projection of the current month's expenditure over future months of the contract (calculated).
   18. Expenditure.   A projection of the expenditure (calculated).

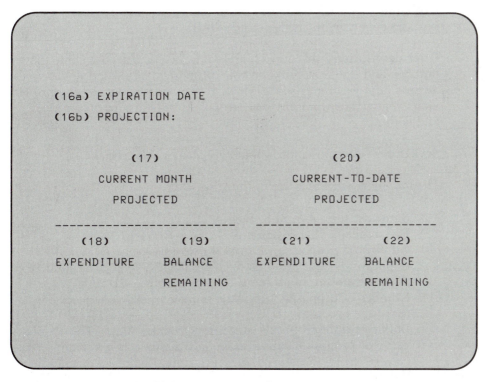

Output 1.6 Items for BPR in Addition to Output 1.3

19. Balance remaining. A projection of the balance remaining within each budget subcode (calculated).
20. Current-to-date projected. This heading implies that the sub-headings refer to a projection of the average monthly expenditure over future months of the contract.
21,22. Same as Items 18 and 19, but for heading 20 above, rather than heading 17.

2.2 Proposal and Contract Control Module (PCCM)

2.2.1 Extramural Proposals and New Funding (EPNF)

EPNF has the following general characteristics:

Purpose—to provide OAS personnel with a comparison of the proposals submitted and awards received against their counterparts for the previous years.

Availability—on CRT to OAS personnel; summary hard copy reports to departments as desired by OAS.

Brief description—this report presents, for each department, the number and dollar volume of both proposals submitted and new awards, and a comparison of these two of current versus the previous year. It also presents the proposals outstanding by number and dollars.

Sort/Order—rows arranged alphabetically by department.

The following variables are output (See Output 2.1):

1-8. Same as Items 1-8, SLS (Section 2.1.1).
9. Department.   Short name of department (D4.1,3).
10. Number of proposals submitted previous year. The number of proposals submitted as of this date last year, interpolated from the stored monthly number of proposals (D2.3,8).
11. Number of proposals submitted current year. Count of current year proposals (D2.1).
12. Dollar value of proposals submitted previous year.   The dollar value of proposals submitted as of this date last year, interpolated from the stored monthly dollar value of proposals (D2.3,9).
13. Dollar value of proposals submitted in current year.   Sum of D2.1, 8, 9, and 10 for current year proposals.

```
                       MASTER EXTRAMURAL PROPOSALS AND NEW FUNDING (3)
          (2)                                                                      (4)
          (5)                                                                      (6)
  PGM = EPNF (7)
               ANALYSIS OF RESEARCH PROPOSALS AND AWARDS AS OF MARCH 1984 (8)
                   PROPOSALS SUBMITTED                         NEW AWARDS
              ----------------------------------    ----------------------------------
                NUMBER           DOLLARS              NUMBER            DOLLARS
              ------------  --------------------    ------------  --------------------
              (10)  (11)                            (14)  (15)
      (9)      FY    FY      (12)        (13)        FY    FY       (16)        (17)
      DEPT    1983  1984     1983        1984       1983  1984      1983        1984
     ------   ----  ----   --------    --------     ----  ----    --------    --------

     DN-ENG     3     2      68,302      97,500       1     0      10,600            0

     AE        33    37   1,564,993   9,805,594      25    29   1,040,424    1,712,763

     CERE       9     9     538,078   2,508,053       1     1      64,731       39,243

     CHE       39    38   2,527,757   3,009,929      25    12   1,041,566      604,972

     CE        52    55   4,898,213   4,621,895      36    22   1,400,013      598,164

     EE        52    38  13,273,219   3,094,384      32    20   1,503,504    1,211,597

     ESM       14    16     824,690     610,392       2     8      64,942      215,101

     ISYE      19    32   1,232,273   3,360,289      18    21     748,316      787,458

     ME        16    24   1,311,762   1,254,844       9    13     428,774      648,814

     NE        56    84   3,283,041   3,496,351      48    81   1,070,281    1,453,152

     TE         9    16     729,954   1,218,705       6    11     362,878      374,295

     HS         3     5     222,137   3,085,073       1     2       5,500        3,996
              ---   ---   ----------  ----------     ---   ---   ----------   ----------
     TOTALS   306   356  30,474,419  36,163,009     204   220   7,741,529    7,649,555
     (20)



                   NOTE - COMPARISON OF EQUIVALENT NUMBER OF MONTHS
                          IN PREVIOUS AND CURRENT FISCAL YEARS
```

Output 2.1 Screen 1

14-17. Same as Items 10-13 but for new awards (count and sum of D2.2, 9, 10, and 11 for current year contracts).
  18. Number of proposals outstanding (D2.1,15).
  19. Dollar value of proposals outstanding (D2.1,8-10).
  20. Totals.   Totals for all columns (calculated).

```
                MASTER EXTRAMURAL PROPOSALS AND NEW FUNDING (3)
     (2)                                                                (4)
     (5)                                                                (6)
PGM = EPNF (7)
          ANALYSIS OF RESEARCH PROPOSALS AND AWARDS AS OF MARCH 1984 (8)
                 -------- PROPOSALS SUBMITTED --------   PROPOSALS OUTSTANDING
                                       DOLLARS           -----------------------
                 (10)      (11)
     (9)         FY        FY        (12)        (13)       (18)      (19)
     DEPT        1983      1984      1983        1984       NO.       DOLLARS
   ------        ----      ----    ----------  ----------   ---     ----------
   DN-ENG          3         2         68,302      97,500     1         32,500

   AE             33        37      1,564,993   9,805,594    14      6,980,454

   CERE            9         9        538,078   2,508,053     4      1,394,496

   CHE            39        38      2,527,757   3,009,929    32      2,807,678

   CE             52        55      4,898,213   4,621,895    45      4,806,761

   EE             52        38     13,273,219   3,094,384    23      3,373,092

   ESM            14        16        824,690     610,392     9        378,894

   ISYE           19        32      1,232,273   3,360,289    16      2,602,205

   ME             16        24      1,311,762   1,254,844    18        925,223

   NE             56        84      3,283,041   3,496,351     7        562,090

   TE              9        16        729,954   1,218,705     7        855,348

   HS              3         5        222,137   3,085,073     3      3,081,077
                 ---       ---     ----------  ----------   ---     ----------
   TOTALS        306       356     30,474,419  36,163,009   180     27,799,818
     (20)
```

```
               NOTE - COMPARISON OF EQUIVALENT NUMBER OF MONTHS
                      IN PREVIOUS AND CURRENT FISCAL YEARS
```

Output 2.1 Screen 2

2.2.2 Summary of Active Projects (SAP)

SAP has the following general characteristics:

Purpose—to provide OAS personnel and higher administrative officers a review of all projects, including duration and summary budgetary information.

Availability—to OAS personnel on hard copy upon demand for all projects; particular projects on demand via CRT.

Brief description—a listing of project number, principal investigator, account number, title, duration, total budget, expenses-to-date, and unobligated balance; additionally, as an option, a listing of only those projects for which the unobligated balance is negative.

Sort/Order—by account number within department (alphabetically ordered).

The following variables are output (as in Exhibit 7a):

1-8. Same as SLS Items 1-8 (Section 2.1.1).
9. Department.   The full name of the department responsible for the projects which follow (D4.1,2).
10. Project number.   20-alphanumeric-character code (D2.2,4).
11. Principal investigator.   40 alphanumeric characters, 20/line over 2 lines, for name or names of principal investigators (D2.2,5).
12. Account number.   Six digits numeric, first two digits separated by dash (D2.2,3).
13. Title.   160 alphanumeric characters, 40/line over four lines (D2.2,7).
14. Duration.   Two dates, mm/dd/yy, over two lines, with a dash following the first. These are the start and completion dates for the project (D2.2,15).
15. Amount.   The original budgeted amount for the project (sum of D1.1,9,16,23...).
16. Total expense.   Expenses incurred on project to date (sum of D1.1,13,20,27).
17. Unobligated balance.   Item 15 minus Item 16 (calculated).
18. Totals.   Totals for Items 15, 16, and 17 by department, at the end of each department's accounts (calculated).
19. Reserved.
20. Grand totals.   Same as Item 18 but summed for all departments (calculated).

2.2.3 Research Activities Report (RAR)

RAR has the following general characteristics:

Purpose—to provide the information for the OAS Research Activities Report such that it is formatted directly for typesetting.

Availability—on hard copy on an annual basis (or upon request) for OAS personnel.

Brief description—same basic information as SAP (Section 2.2.2), but restricted and reformatted for this Report.

Sort/Order—same as SAP (Section 2.2.2).

The following variables of the SAP (Section 2.2.2) will be output in the format given in Exhibit 8: 1-8, 9, 11, 13, 14, and 15. In addition, the following will be output: 21. Sponsor - 30 alphanumeric characters, 15/line over 2 lines (D2.2,8).

2.2.4 Percent Support List (PSL)

PSL has the following general characteristics:

Purpose—to provide the percent support from sponsors for the OAS Research Activities Annual Report.

Availability—same as RAR (Section 2.2.3).

Brief description—same basic structure as SAP (Section 2.2.2), but restricted and reformatted so that the percent of sponsor contribution is available.

Sort/Order—same as SAP (Section 2.2.2).

The following variables of the SAP (Section 2.2.2) will be output: 1-8, 9, 11, 13, 14. In addition, Item 21 of the RAR (Section 2.2.3) will be included along with the following:

22. Percent of sponsor support.   Rounded to the nearest whole percent (D2.2,9-11).

2.2.5 Proposal Status Report (PSR)

PSR has the following general characteristics:

Purpose—to provide OAS personnel with a worksheet to determine the status of any and/or all proposals.

Availability—same as SLS (Section 2.1.1), but on hard copy only.

Brief description—this report contains all necessary details for determining the status of any given proposal. Access to a particular proposal is by proposal number.

Sort/Order—chronological by date of entry.

The following variables are output (see Output 2.5):

1-8. Same as SLS Items 1-8 (Section 2.1.1).
9. Proposal number. Internal number assigned by OAS, 12 alphanumeric characters (D2.1,3).
10. Principal investigator. Same as SAP Item 11 (Section 2.2.2), but from proposal file (D2.1,4).
11. Department. Three-character departmental abbreviation (D2.1,5 => D4.1,1).
12. Title. Same as SAP Item 13 (Section 2.2.2), but from proposal file (D2.1,6).
13. Sponsor. Same as RAR Item 21 (Section 2.2.3), but from proposal file (D2.1,7).
14. Budgetary amounts as follows in 14a, b, and c:
14a. Total budget. Total amount independent of source (sum of D2.1,8-10).
14b. Extramural source. Dollar amount of budget from extramural source (D2.1,8).
14c. Cost sharing. Dollar amount of budget from internal source (D2.1,9).
15. Duration. Same as SAP Item 14 (Section 2.2.2), but from proposal file (D2.1,11).
16. Date mailed to VP for Research. mmddyy (D2.1,12).
17. Date received from VP for Research. mmddyy (D2.1,13).
18-19. New/Modification. Output NEW if a new proposal, otherwise output MOD (D2.1,14).
20-21. Accepted/Rejected. Output ACC if accepted, REJ if rejected, blank space if still pending (D2.1,15). NOTE: Items 18-21 may be placed on separate lines.
22. Department mailing. Three-character alphanumeric designation of department (or OAS) who mailed the proposal out (D2.1,16 => D4.1,1).

2.2.6 Contract Status Report (CSR)

CSR has the following general characteristics:

```
                    MASTER PROPOSAL STATUS REPORT  (3)

      (9)                  (10)
   PROPOSAL             PRINCIPAL                        (12)
    NUMBER             INVESTIGATOR                      TITLE
  -----------        ---------------        ------------------------------
  ER84-C-60          Philips/Gray           Analysis and Simulation of the
                       EE (11)                Marine Air Traffic Control and
                                              Landing System and Radar N.A.

  ER84-C-61            Smithe               Evaluation of the Effect of Heat Stress
                        IE                   on Worker Productivity for Selected
                                             Southern Forest Harvesting Tasks

  ER84-C-62            Yon                  Elastic Stability Research of Curved
                        ME                   Members

  ER84-C-63          Burk/Marstin           Development and Validation of a
                        AE                    Submissle Aerodynamics Prediction

  ER84-C-64          Dye/Maple              Improving the Efficiency, Safety, and
                        ME                    Utility of Woodburning Units
                                              New $105,673

  ER84-C-65            Tramey               Near-ground Tornado Wind Fields
                        CE

  ER84-C-66          Tarres/Vivel           Oil Recycling Proposal Development
                        CHE                   Project

  ER84-C-67            Browne               Planning and Evaluation-Upgrade
                        IE                    of the Local and Statewide Problem
                                              Identification and Evaluation
                                              Capability

  ER84-C-68          Vicker/Nagles          ADCC Operational Changes in CM/DCM
                        EE                    Environments

  ER84-C-69            Tatar                Investigation of Catalyst-support
                        CHE                   Interactions
```

Output 2.5 Screen 1

MASTER PROPOSAL STATUS REPORT (3)

(9) PROPOSAL NUMBER	(13) AGENCY	(14) AMOUNT	(15) DURATION FROM	TO	VP FOR RESEARCH MAILED TO	REC'D FROM	STATUS * * * * M N A R	DEPT MLD. W
ER84-C-60	Nat'l Fiber	48,497 151,865	12/10/82	12/31/84	5/05/84 (16)	5/05/84 (17)	X X	OAS
ER84-C-61	USDA FS	24,531 (14b) 13,780 (14c)	1 year		5/04/84	5/04/84	X X	IE
ER84-C-62	Struc- tural RSA CNCL	1,800	5/01/84	11/30/84	5/07/84			CE
ER84-C-63	US Army	649,016	7/01/84	9/30/86	5/07/84	5/07/84	X	AE
ER84-C-64	DOE	532,792	1 year		5/18/84	5/18/84	X	OAS
ER84-C-68	BMD Corp.	66,000	6 mos.		6/08/84	6/08/84	X	EE
ER84-C-69	Rsch Corp Atlnta	16,000	1 year		6/10/84	6/11/84	X	CHE

*LEGEND:
M = Modified   (18)
N = New   (19)
A = Accepted   (20)
R = Rejected   (21)
W = Who Mailed (22)

Output 2.5 Screen 2

Purpose—to provide OAS personnel with a worksheet to determine the status of any and/or all contracts.

Availability—same as SLS (Section 2.1.1).

Brief description—this report contains all of the necessary details for determining the status of any given contract.

Sort/Order—chronological by date of entry.
The following variables are output (see Output 2.6):

1-8. Same as SLS Items 1-8 (Section 2.1.1).
9. Contract name (alias project number). Same as SAP Item 10 (Section 2.2.2) (D2.2,4).
9a. Account number. Same as SAP Item 12 (Section 2.2.2) (D2.2,3).
10-15. Same as PSR Items 10-15 (Section 2.2.5).
16. Proposal number. Same as PSR Item 9 (Section 2.2.5) (D2.2,17).
17. Administrative digest number. Number from administrative digest (D2.2,18).
18. Administrative digest date. Date of administrative digest (D2.2,19).
19. New. 1 = yes; 2 = no, i.e., a continuation of an existing contract (D2.2,20). Conversion for output: 1 = >N for new, 2 = >E for existing.

2.3 Extension Evaluation and Control (EECM)

2.3.1 Extension Budget Control Report (EBC)

EBC has the following general characteristics:

Purpose—to provide a means for maintaining current budgetary information on extension activities by program.

Availability—on CRT or hard copy to OAS and extension personnel.

Brief description—this report includes the estimated and current actual costs for each program.

Sort/Order—none.

The following variables are output (see Output 3.1):

```
                   MASTER CONTRACT STATUS REPORT (3)

        (9)                 (10)
     CONTRACT            PRINCIPAL                          (12)
    NAME & NO.          INVESTIGATOR                        TITLE
    ----------       -----------------       ------------------------------------
    Chase MN         Placeke                 Phase and Compress Behavior in
    Placek              (11)                     Sedimentation and Thickening
    2-12894             CHE
      (9a)

    Air              Tarres/Guinn            The Role of Nonferrous Coal
    Products         Curtiss                    Minerals & By-products Metallic
    4-20315             CHE                      Wastes in Coal Liquefaction

    Cost Shr         Tarres/Guinn            See above
    4-20315          Curtiss
                        CHE

    NAS8-33886       Forster/Forzini         Rocket Motor Internal Ballistic
    4-20389             AE                      Performance Vibrations
                                                   N.A. $24,995

    Cost Shr         Forster/Forzini         See above
    4-20389             AE

    GB58277-         Gray/Nagler             Clutter Spectral Analysis and
      9107              EE                       Modeling Study
    4-20435

    US FS-19-        Smiths                  Assessment of the Physiological
      80-407            IE                      Studies of Selected Forest
    4-203939                                     Harvesting Activities in the
                                                 Southeastern United States

    Cost Shr         Smiths                  See above
    4-20393             IE
    2-12863

    DAAH01-          Williamson              Army Missle Command/Launcher
    82-P                AE                      Dynamics
    1729
    4-20457
```

Output 2.6 Screen 1

MASTER CONTRACT STATUS REPORT (3)

(9) CONTRACT NAME & NO.	(13) AGENCY	(14) AMOUNT	(15) DURATION	(16) PRO- POSAL NO.	ADMINISTRATIVE DIGEST (19)		N OR E
					NO.	DATE	
Chase MN Placeke 2-12894 (9a)	OAS	4,798	12/16/83- 9/30/84	n.a.	397-84 (17)	3/26/84 (18)	N
Air Products 4-20315	API	164,227	9/01/81- 5/31/84	n.a.	396-84	3/24/84	E
Cost Shr 4-20315	OAS	26,786	9/01/81- 5/31/85	n.a.	395-84	3/24/84	E
NAS8-33886 4-20389	NAS8	100,962	9/30/82- 5/31/85	n.a.	415-84	3/30/84	E
Cost Shr 4-20389	OAS	4,000	9/30/82- 3/31/85	n.a.	414-84	3/29/84	N
OSPFP-AU- WRC 8101 4-20421	OSPFP	67,000	4/14/83- 4/30/84	n.a.	429-84	4/02/84	N
GB58277- 9107 4-20435	Boeing	46,000	10/01/83- 9/30/84	n.a.	445-84	11/08/84	N
US FS-19- 82-407 4-203939	USFS	21,240	9/22/82- 5/31/84	n.a.	456-84	11/13/84	N
Cost Shr 4-20393	OAS	5,040	9/22/82- 5/31/84	n.a.	455-84	4/13/84	N
DAAH01- 84-P 1729 4-20457	U.S. Army	845,450	2/23/84-10/05/84	n.a.	436-84	4/14/84	E

Output 2.6 Screen 2

```
              UNIVERSITY CONTINUING EDUCATION ACTIVITY
      (9)  REFERENCE NUMBER _____  (9a) DATE OF PROGRAM: mm/dd/yy
      (9b) DESCRIPTION: _____

                        FEES AND INCOME
       INDIVIDUAL FEES                      TOTAL
      (10)              (11)            (12)              (13)
       No.              Amt.            Est.              Actual
       35       -       1150            40,250            46,000


                              COSTS
      DIRECT COSTS:                        ESTIMATED          ACTUAL
      (14) Payments to Fac. & Staff        8,000.00         7,650.00
      (15) Faculty/Staff Travel
      (16) Honoraria & Speakers' Fees      9,200.00         9,175.00
      (17) Planning/Advisory Comm.
      (18) Printing                        3,350.00         3,038.61
      (19) Postage                           100.00           100.00
      (20) Advertising                         0.00             0.00
      (21) Telephone                           0.00             0.00
      (22) Rentals                             0.00             0.00
      (23) Meals: Participants             1,350.00         1,399.78
           (24) Program Personnel
           (25) Breaks                       600.00           807.88
      (26) Supplies: Office                    0.00             0.00
      (27) Instructional Aids                375.00           352.00
      (28) Other: Rec. & Trans.              850.00           495.30
      (29) Miscellaneous                     400.00           293.13
      (30) Total Direct Costs             30,425.00        29,648.91

      ADMINISTRATIVE COSTS:
      (31) Clerical, Secretarial Srvcs.    1,100.00         1,100.00
      (32) Coordination, Supervision       3,300.00         3,300.00
      (33) General Administration          1,100.00         1,100.00
      (34) Total Administration            5,500.00         5,500.00

      (35) Contributed Costs                   0.00             0.00

      (56) Total Costs                    35,925.00        30,198.94
```

Output 3.1

1-8. Same as SLS Items 1-8 (Section 2.1.1).
  9. Reference number. A sequential number by which this project is referenced (D3.1,1).
9a. Date of program (D3.1,2).
9b. Description. Brief description of the program for reference purposes (D3.1,3).
10. Total number attending (D3.1,4).
11. Amount per individual attending (D3.1,5).
12. Estimated income (D3.1,6).
13. Actual income (D3.1,7).
14. Estimated payments to University Faculty/Staff (D3.1,8).
15. Estimated Faculty/Staff travel (D3.1,9).
16. Estimated honoraria and speakers' fees (D3.1,10).
17. Estimated Planning/Advisory Committee (D3.1,11).
18. Estimated printing (D3.1,12).
19. Estimated postage (D3.1,13).
20. Estimated advertising (D3.1,14).
21. Estimated telephone (D3.1,15).
22. Estimated rentals (housing) (D3.1,16).
23. Estimated meals: participants (D3.1,17).
24. Estimated meals: program personnel (D3.1,18).
25. Estimated meals: breaks (D3.1,19).
26. Estimated supplies: office (D3.1,20).
27. Estimated supplies: instructional (D3.1,21).
28. Estimated other: recreation and transportation (D3.1,22).
29. Estimated other: miscellaneous (D3.1,23).
30. Total direct costs (calculated, sum of Items 14-29).
31. Estimated clerical, secretarial services (D3.1,24).
32. Estimated coordination, supervision (D3.1,25).
33. Estimated general administration (D3.1,26).
34. Total administration (calculated, sum of Items 31-33).
35. Estimated contributed costs (D3.1,27).
36-55. Same as 14-34, but actual, rather than estimated; second column of costs in Output 3.1 (D3.1,28-47).
56. Total costs (calculated).

2.3.2 Extension Project Summary (EPS)

EPS has the following general characteristics:

Purpose—to provide a summary of EBC (Section 2.3.1) information for all projects to date.

Availability—same as EBC (Section 2.3.1).

Brief description—same as EBC (Section 2.3.1).

Sort/Order—none.

The variables output are identical to the EBC (Section 2.3.1). However, the variables will present the totals for all projects, or for a user-defined subset of projects.

3.0 File Design

MASTER files can be organized in the same way that the MASTER system was organized in Section 2.1. A separate section for each of the three modules will be given below. Recognize, however, that this organization does not preclude one module from utilizing or updating a file that is assigned to another. It is this common access to the various files that makes MASTER an integrated system. In addition to the three functional categories for the files, a fourth category, called "Parameter Files," has been established to handle those files which are not affected by day-to-day operations, such as category code labels and departmental abbreviations. Finally, a fifth category, called "Facilitation Files," includes those files established to facilitate programming and processing, such as cross-reference lists and temporary utility files.

Figure 3.1 presents an overall cross-reference of the files accessed to generate each of the MASTER reports. Note that the sections below are numbered according to the number of the file which each describes (example: Section 3.2.4 corresponds to File D2.4). All of the file layouts are given in their respective tables at the end of Section 3.

3.1 Budget Management and Control Module (BMCM) Files

Two primary files are required for the BMCM operation as follow:

1. Account file.   One record per account with account number as the primary key (D1.1).
2. Transaction file.   One record per transaction (e.g., expenditure), with a sequential transaction number as a primary key (D1.2). Each transaction will be linked to a unique account by the inclusion of the account number within each record.

A pointer from each account record (D1.1) to its associated contract record (D2.1) will be maintained to reduce time of file access. These files will be discussed in separate subsections below.

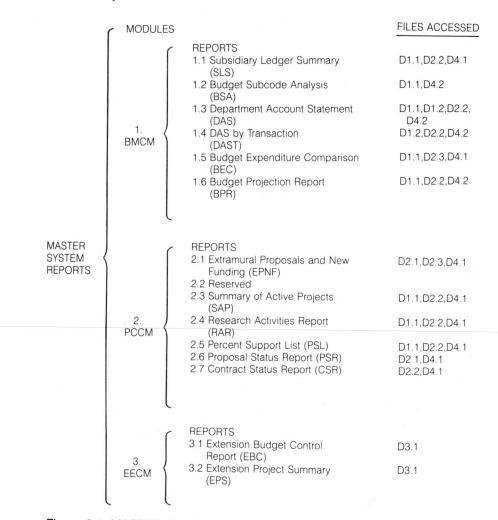

MODULES		FILES ACCESSED

MASTER SYSTEM REPORTS

1. BMCM

REPORTS
1.1 Subsidiary Ledger Summary (SLS) — D1.1,D2.2,D4.1
1.2 Budget Subcode Analysis (BSA) — D1.1,D4.2
1.3 Department Account Statement (DAS) — D1.1,D1.2,D2.2, D4.2
1.4 DAS by Transaction (DAST) — D1.2,D2.2,D4.2
1.5 Budget Expenditure Comparison (BEC) — D1.1,D2.3,D4.1
1.6 Budget Projection Report (BPR) — D1.1,D2.2,D4.2

2. PCCM

REPORTS
2.1 Extramural Proposals and New Funding (EPNF) — D2.1,D2.3,D4.1
2.2 Reserved
2.3 Summary of Active Projects (SAP) — D1.1,D2.2,D4.1
2.4 Research Activities Report (RAR) — D1.1,D2.2,D4.1
2.5 Percent Support List (PSL) — D1.1,D2.2,D4.1
2.6 Proposal Status Report (PSR) — D2.1,D4.1
2.7 Contract Status Report (CSR) — D2.2,D4.1

3. EECM

REPORTS
3.1 Extension Budget Control Report (EBC) — D3.1
3.2 Extension Project Summary (EPS) — D3.1

Figure 3.1 MASTER File Access Analysis

3.1.1 Account File

The layout for the account file is given in Table D1.1. An account record must be created before any transactions reflect activity within that account. Thus, the transaction programs must detect any attempt to enter a transaction on a nonexisting account number and prevent that from occurring prior to new account creation.

New accounts will be created by the new accounts/accounts update program. This program will create or update any of the parameters in that file. However, all updates of budgetary expenditures must be made by the transaction program. Thus, the user cannot arbitrarily update budgetary expenditures within the account file without the generation of a transaction record (D1.2). Whenever a transaction record is generated, the account file will be updated appropriately to reflect all transactions. Thus, this file can provide for on-line queries to provide most current account activity.

The BMC module also has an option to delete an account record from the account file. This will take place after all activity on that account is completed. A verification check for open encumbrances will be performed prior to this purge, and a hard copy of the account record will be generated prior to its deletion. In addition, a hard copy listing of all transactions involving that account will also be generated. Deleted account space on the file is reused before additional space for new accounts is taken at the bottom of the file.

Most of the data items in Table D1.1 are explicitly defined by the data entry. However, there are three fields used exclusively for data-structure purposes (see Figure 3.2 for structure diagram). These variables are referenced by their REF-NUM in Table D1.1 as follow:

1. Contract pointer.   This is a direct pointer to the physical location of the contract in file D2.2 which corresponds to the account record.
2. Next account.   This is a link which links the file records in order of account number (see Figure 3.2a).
3. Next-by-department.   This is a link which links the file records in order of account number within department. Note that when two records have identical departments in Figure 3.2b, the order of linkage is by account number.
4. First account transaction.   This is a pointer directly to the location in the transaction file (D1.2) which is the address of the transaction record from which all other transaction records for the account can be accessed by a traversal, using D1.2,15 as a link.

3.1.2 Transaction File

The layout for the transaction file (D1.2) is given in Table D1.2. Additions to this file will be made at the bottom in the chronological order in which transactions are made, unless previous deletions have left "holes" in the file in which case these holes will be filled first. The interactive transaction program will perform this file maintenance activity. Updates to the account file will be made simultaneously with the addition of a transaction.

The transaction file will be linked primarily by department, and secondarily on several other variables as shown in Figure 3.3, using the NEXT LINK field (D1.2,15). Thus, an ordered traversal through the linked list will produce a sequence of ordered subcodes within accounts

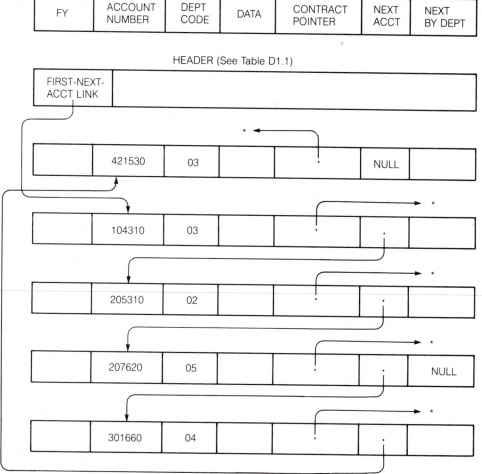

RECORD LAYOUT

FY	ACCOUNT NUMBER	DEPT CODE	DATA	CONTRACT POINTER	NEXT ACCT	NEXT BY DEPT

HEADER (See Table D1.1)

FIRST-NEXT-ACCT LINK	

	421530	03		.	NULL	

| | 104310 | 03 | | . | . | |

| | 205310 | 02 | | . | . | |

| | 207620 | 05 | | . | . | NULL |

| | 301660 | 04 | | . | . | |

* Corresponding contract record.

Figure 3.2a Account File Structure Diagram
(Contract Pointer and Next Account)

within department. This will be used mainly for presentation, since the account file will be updated on each addition to the transaction file. The transaction file will be linked according to the following arrangement: (1) department, (2) account, (3) account subcode, (4) purchase-order number, and (5) date entered. In this way, a complete chronological list of transactions may be generated for a given purchase order, a given account subcode for a given account, an entire account, or an entire department. The structure diagram for D1.2 is given in Figure 3.3. To get to the first transaction for a given account, use D1.1,68. When

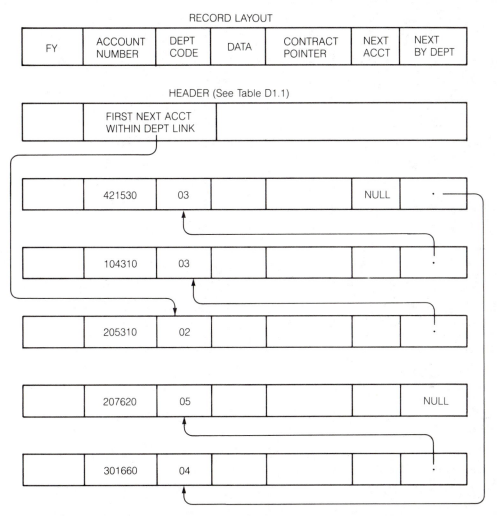

Figure 3.2b Account File Structure Diagram (Next by Department)

a subset of records, defined by an account, is purged from the transaction file, they will be listed to the line printer using the DASD report generator.

3.2 Proposal and Contract Control Module (PCCM) Files

Three primary files are required for PCCM operation, as follows:

1. Proposal file. One record per proposal arranged by OAS proposal number, with the proposal number as the primary key

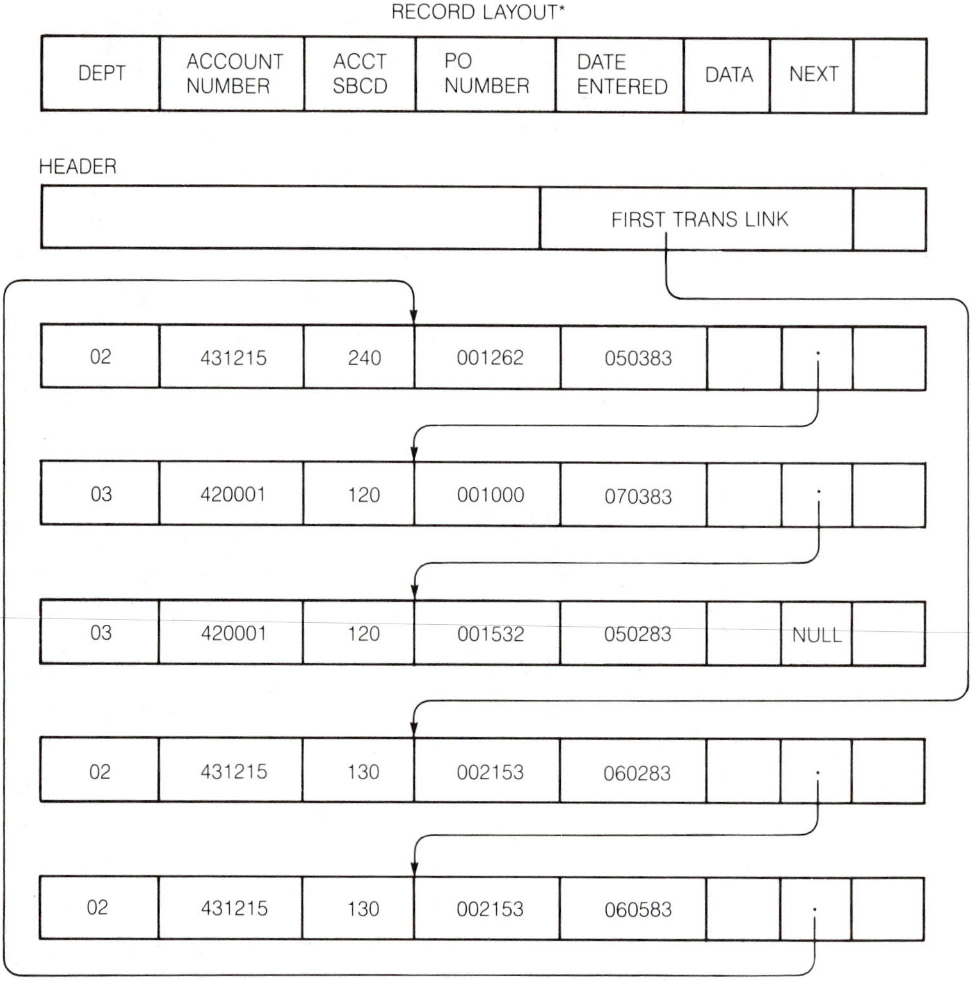

RECORD LAYOUT*

DEPT	ACCOUNT NUMBER	ACCT SBCD	PO NUMBER	DATE ENTERED	DATA	NEXT	

HEADER

		FIRST TRANS LINK	

02	431215	240	001262	050383		:	

03	420001	120	001000	070383		:	

03	420001	120	001532	050283		NULL	

02	431215	130	002153	060283		:	

02	431215	130	002153	060583		:	

* Variable ordering differs from D1.2 for clarity;
use Table D1.2 for layout.

Figure 3.3 Transaction File Structure Diagram

(D2.1). A record will be added to this file whenever a new proposal becomes active.

2. Contract file.   One record per contract added chronologically as proposals are accepted for funding (D2.2). All pertinent information in the proposal file will be written to the contract file when a new contract record is created. Additional elements of information will also be added, being input by the user. The primary key will be the contract account number.

3. Projection file.   One record per department, per year (D2.3). The

records will be arranged in chronological order, one set of departments per year, and in numerical order by department code. The most current (first) set of department records will hold a partial year of information, while the second set of department records will contain the most current full year of information.

3.2.1 Proposal File

The layout for the proposal file (D2.1) is given in Table D2.1. A new record will be created for each proposal generated. The NEXT LINK field (D2.1,17) within the record will link the records by proposal reference numbers (D2.1,2) within department (D2.1,5), to facilitate departmental listings. The structure diagram for this linkage is essentially identical to that diagrammed in Figure 3.2b with the exception that proposal reference number replaces account number. Records will be created, modified, or deleted by the data-entry program (P2.1).

3.2.2 Contract File

The layout for the contract file (D2.2) is given in Table D2.2. A new record will be created for each contract signed. Generally these will emanate from the proposal file; however, it will be permissible to enter new contracts for which there are no proposals. This same program will also enable changes to be made to any part of a contract record.

A link field within the contract record will link the records by a concatenation of fiscal year and contract reference number (D2.2,1 + 2) within department (D2.2,6), to facilitate departmental listings. The structure diagram for this linkage is essentially identical to that diagrammed in Figure 3.2a with the following exceptions: (1) the ACCOUNT POINTER replaces the CONTRACT POINTER, providing an immediate linkage back to the account record which corresponds to the contract; and (2) the NEXT LINK replaces the NEXT BY DEPARTMENT LINK, providing an ordered listing by account number within fiscal year, within department. All links and pointers, including the account-to-contract pointer (D1.1,65), are created by the contract creation/modification/deletion program (P2.2).

3.2.3 Projection File

The layout for the projection file (D2.3) is given in Table D2.3. This file is used exclusively for the budget expenditure comparison (BEC) report (see Section 2.1.5). It is updated by the exercise of Option 8 of the PCC supervisory menu (see O & UG Figure 6 and Section 2.2.8). The program specification, which defines all variables in the file, is PSWD 2.3.

3.3 Extension Evaluation and Control Module Files

There is one file used by EECM. Its layout is given in Table D3.1. An extension file record will be created when a new extension project is initiated. The extension entry program (P3.1) will be used to create, maintain, and delete records in this file. Hard-copy backup will mandatorily precede record deletion. Any item in the record may be changed by the update program (within P3.1).

3.4 Parameter Files

The following parameter files are required to enable the MASTER system to function as designed:

1. Departmental parameter file. One record per department (D4.1).
2. Account subcode parameter file. One record per account subcode (D4.2).
3. General parameter file. Contains constants for system operation (D4.3).
4. Access control file. Used to verify access to programs and files (D4.4).

3.4.1 Departmental Parameter File

The departmental parameter file is given in Table D4.1.

3.4.2 Account Subcode Parameter File

The account subcode parameter file is given in Table D4.2.

3.4.3 General Parameter File

The general parameter file contains information, such as the fiscal year, which is generally constant for a given period of system operation. This file is updated by the MASTER supervisory menu Option 4. The general parameter file layout is given in Table D4.3. It consists of two records, one for the current fiscal year and one for the next fiscal year. This enables the next year's proposals to be input prior to the end of the current year.

3.4.4 Access Control File

This file is used to verify whether users have clearance to utilize certain file maintenance programs. It is in the format given in Table D4.4. The header record to this file is in the following format:

Date of Last P1.3b (6 characters)	

This file is accessed through the MASTER SM, Option 4, parameter file maintenance. The header record is updated by P1.3b.

3.5 Facilitation Files

The following file is used to facilitate data retrieval: Proposal number cross-reference to proposal reference number file. This is documented in Section 3.5.1.

3.5.1 Proposal Number Cross-reference to Proposal Reference Number File

This file is in the following format:

D5.1

Proposal Number (12 characters)	Fiscal Year (2 char.)	Proposal Location (6 characters)

Sort: Ascending by Proposal Number
DSN = FAD51

Here the fiscal year and proposal number are identical to variables D2.1,1 and D2.1,3. The proposal location is a pointer to the actual, physical location of the corresponding record in D2.1. This file is maintained by the proposal creation, update, and delete program (P2.1).

Table D1.1 MASTER FILE LAYOUT FORM—D1.1

FILE TITLE: Account File							
COMMENTS: One record per account						PAGE 1 of 1	
						RECORD LENGTH: 672 bytes	
						DSN = FAD11	

SEQ	ST POS	END POS	FLD SIZ	FLD	REF NUM	FIELD DESCRIPTION	SOURCE
1	1	2	2	N		Fiscal Year	ADEM-1
2	3	8	6	N		Account Number	ADEM-2
3	9	14	6	N		Creation Date, mmddyy	ADEM-3
4	15	16	2	N		Departmental Code	ADEM-4
5	17	18	2	N		Fund Classification	ADEM-5
6	19	20	2	N		Overhead Rate	ADEM-6
7	21	29	9	N		Exception Subcodes	ADEM-7
8	30	31	2	N		Fringe Benefits Rate	ADEM-8
9	32	42	11	N		Revised Budget—Subcode 100	ADEM-9
10	43	53	11	N		Expended-this-Month SC 100	P1.2
11	54	64	11	N		Expended-this-Quarter SC 100	P1.2
12	65	75	11	N		Year-to-Date Exp. SC 100	P1.2
13	76	86	11	N		Total-to-Date Exp. SC 100	P1.2
14	87	97	11	N		Current Encumbrances SC 100	P1.2
15	98	108	77	N		Total Encumbrances SC 100	P1.2
-22	109	185	77	N		Seq 9-15 for SC 140	ADEM-10..
-29	186	262	77	N		Seq 9-15 for SC 150	ADEM-11..
-36	263	339	77	N		Seq 9-15 for SC 200	ADEM-12..
-43	340	416	77	N		Seq 9-15 for SC 600	ADEM-13..
-50	417	493	77	N		Seq 9-15 for SC 700	ADEM-14..
-57	494	570	77	N		Seq 9-15 for SC 800	ADEM-15..
-64	571	647	77	N		Seq 9-15 for SC 900	ADEM-16..
65	648	652	5	N	1	CONTRACT POINTER	P2.2.1
66	653	657	5	N	2	NEXT ACCOUNT	P1.1.1
67	658	662	5	N	3	NEXT BY DEPARTMENT	P1.1.1
68	663	667	5	N	4	FIRST ACCOUNT TRANSACTION	P1.2
69	668	672	5	AN		Reserved	

HEADER RECORD LAYOUT							
1	1	5	5	N		First Next-Account Link	P1.1
2	6	10	5	N		First Next-Account-within-	P1.1
						Department Link	
3	11	15	5	N		First Delete Link	P1.1
4	16	20	5	N.		Filesize (No. of active records)	P1.1
5	21	336	316	X		Filled with Blanks	P1.1

Table D1.2 MASTER FILE LAYOUT FORM—D1.2

FILE TITLE: Transaction File	PAGE 1 of 1
COMMENTS: One record per transaction	RECORD LENGTH: 92 bytes
	DSN = FAD12

SEQ	ST POS	END POS	FLD SIZ	FLD	REF NUM	FIELD DESCRIPTION	SOURCE
1	1	2	2	N		Fiscal Year	D4.3,1
2	3	8	6	N		Transaction Reference Number	D4.3,2
3	9	14	6	N		Date Entered (mmddyy)	TDEM-3
4	15	20	6	N		Account Number	TDEM-4
5	21	23	3	N		Account Subcode	TDEM-5
6	24	24	1	N		Type Code:	TDEM-6
						1. Non-PO Expenditure	
						2. PO Encumbrance	
						3. PO Expense	
						4. Budget Modifications	
7	25	30	6	N		Purchase-Order Number	TDEM-7
8	31	36	6	N		Purchase-Order Date (mmddyy)	TDEM-8
9	37	60	24	AN		Description	TDEM-9
10	61	71	11	N		(SXXXXXXX.XX)$ Amt. of Entry	TDEM-10
11	72	72	1	A		Final PO Transaction Indicator	TDEM-11
12	73	74	2	N		Department Code	TDEM-14
13	75	75	1	A		Exception Code	TDEM-12
14	76	76	1	A		University Clearance	TDEM-13
15	77	84	8	N		NEXT LINK	P1.2.7
16	85	92	8	N		Reserved	

HEADER RECORD LAYOUT							
1	1	5	5	N		First Transaction Link	P1.2
2	6	10	5	N		First Delete Link	P1.2
3	11	15	5	N		Filesize (No. of active records)	P1.2
4	16	92	77	X		Filled with Blanks	P1.2

Table D2.1 MASTER FILE LAYOUT FORM—D2.1

FILE TITLE: Proposal File   PAGE 1 of 1
COMMENTS: One record per proposal   RECORD LENGTH: 309 bytes
DSN = FAD21

SEQ	ST POS	END POS	FLD SIZ	FLD	REF NUM	FIELD DESCRIPTION	SOURCE
1	1	2	2	N		Fiscal Year	D4.3,1
2	3	8	6	N		Proposal Reference Number	D4.3,4
3	9	20	12	AN		Proposal Number	PDEM-3
4	21	60	40	A		Principal Investigators (20 ch split)	PDEM-4,5
5	61	62	2	N		Department Code	PDEM-6
6	63	222	160	AN		Project Title (40 ch split)	PDEM-7-10
7	223	252	30	AN		Sponsor (15 ch split)	PDEM-11, 12
8	253	259	7	N		Dollar Contribution of Sponsor	PDEM-13
9	260	266	7	N		Dollar Contribution, OAS	PDEM-14
10	267	273	7	N		Shared-Indirect Cost	PDEM-15
11	274	285	12	N		Duration (2 options)	PDEM-16, 17
12	286	291	6	N		Date Mailed VP-Res (mmddyy)	PDEM-18
13	292	297	6	N		Date Received VP-Res (mmddyy)	PDEM-19
14	298	298	1	N		New or Modification (1 = new, 2 = modification)	PDEM-20
15	299	299	1	N		Acceptance code (1 = accepted, 2 = pending, 3 = rejected)	PDEM-21
16	300	301	2	N		Department Mailing	PDEM-22
17	302	309	8	N	*	NEXT LINK	P2.1.1

*See Section 3.2.1 narrative.

HEADER RECORD LAYOUT							
1	1	5	5	N		First Proposal Link	P2.1
2	6	10	5	N		First Delete Link	P2.1
3	11	15	5	N		Filesize (No. of active records)	P2.1
4	16	309	294	X		Filled with Blanks	P2.1

Table D2.2 MASTER FILE LAYOUT FORM—D2.2

FILE TITLE: Contract File   PAGE 1 of 1
COMMENTS: One record per contract   RECORD LENGTH: 370 bytes
                                                  DSN = FAD22

SEQ	ST POS	END POS	FLD SIZ	FLD	REF NUM	FIELD DESCRIPTION	SOURCE
1	1	2	2	N		Fiscal Year	D4.3,1
2	3	8	6	N		Contract Reference Number	D4.3,5
3	9	14	6	N		Account Number	CDEM-3
4	15	34	20	AN		Contract Name (Project No.)	CDEM-4
5	35	74	40	A	#	Principal Investigators (20 ch split)	D2.1,4
6	75	76	2	N	#	Department Code	D2.1,5
7	77	236	160	AN	#	Project Title (40 ch split)	D2.1,6
8	237	266	30	AN	#	Sponsor (15 ch split)	D2.1,7
9	267	273	7	N	#	Current Contribution of Sponsor	D2.1,8
10	274	280	7	N	#	Current Contribution, OAS	D2.1,9
11	281	287	7	N	#	Current Shared-Indirect Costs	D2.1,10
12	288	294	7	N		Past Contribution of Sponsor	CDEM-17
13	295	301	7	N		Past Contribution, OAS	CDEM-18
14	302	308	7	N		Past Shared-Indirect Costs	CDEM-19
15	309	320	12	AN	#	Duration (mmddyy mmddyy)	D2.1,11
16	321	328	8	N	#	Proposal Reference Number	D2.1,1-2
17	329	340	12	AN		Proposal Number	CDEM-23
18	341	346	6	N		Admin. Dig. Number	CDEM-24
19	347	352	6	N		Admin. Dig. Date	CDEM-25
20	353	353	1	A		New Code (N = NEW, M = MOD)	CDEM-26
21	354	354	1	N		RAR Exception Code	CDEM-27
22	355	362	8	N	*	NEXT LINK	P2.2
23	363	370	8	N	*	ACCOUNT POINTER	P2.2

						HEADER RECORD LAYOUT	
1	1	5	5	N		First Contract Link	P2.2
2	6	10	5	N		First Delete Link	P2.2
3	11	15	5	N		Filesize (No. of active records)	P2.2
4	16	185	170	X		Filled with Blanks	P2.2

#These items are copied from D2.1 by P2.2; if no   *See Section 3.2.2 narrative.
corresponding proposal exists, then they are
entered directly from the CDEM.

Table D2.3 MASTER FILE LAYOUT FORM—D2.3

FILE TITLE: Projection File							

FILE TITLE: Projection File   PAGE 1 of 1
COMMENTS: One record per department,   RECORD LENGTH: 472 bytes
                    per year   DSN = FAD11

SEQ	ST POS	END POS	FLD SIZ	FLD	REF NUM	FIELD DESCRIPTION	SOURCE
1	1	2	2			FISCAL YEAR	P2.3
2	3	4	2			Departmental Code	P2.3
3	5	40	36			Instructional Ratios (12)	P2.3
						One ratio per month,	
						Three digits per ratio.	
4	41	76	36			OAS Ratios (12)	P2.3
5	77	112	36			Cost-Sharing Ratios (12)	P2.3
6	113	148	36			Extramural Ratios (12)	P2.3
7	149	184	36			Total Ratios (12)	P2.3
8	185	220	36			Proposals Submitted (12)	P2.3
						One per month,	
						Three digits each.	
9	221	328	108			Dollar Value, Proposals (12)	P2.3
10	329	364	36			New Awards (12)	P2.3
11	365	472	108			Dollar Value New Awards (12)	P2.3

Table **D3.1** MASTER FILE LAYOUT FORM—D3.1

FILE TITLE: Extension File						PAGE 1 of 1	
COMMENTS: One record per program						RECORD LENGTH: 511 bytes	
						DSN = FAD31	

SEQ	ST POS	END POS	FLD SIZ	FLD	REF NUM	FIELD DESCRIPTION	SOURCE
1	1	6	6	N		Reference Number	D4.3,6
2	7	12	6	N		Date of Program (mmddyy)	EBDEM
3	13	32	20	AN		Description	EBDEM
4	33	37	5	N		Total Number Attending	EBDEM
5	38	41	4	N		Amount per Individual	EBDEM
6	42	52	11	N		Estimated Income	EBDEM
7	53	63	11	N		Actual Income	EBDEM
8	64	74	11	N		Est. Payments—Fac/Staff	EBDEM
9	75	85	11	N		Est. Fac/Staff Travel	EBDEM
10	86	96	11	N		Est. Honoraria	EBDEM
11	97	107	11	N		Est. Planning/Advisory Com.	EBDEM
12	108	118	11	N		Est. Printing	EBDEM
13	119	129	11	N		Est. Postage	EBDEM
14	130	140	11	N		Est. Advertising	EBDEM
15	141	151	11	N		Est. Telephone	EBDEM
16	152	162	11	N		Est. Rentals (Housing)	EBDEM
17	163	173	11	N		Est. Meals: Participants	EBDEM
18	174	184	11	N		Est. Meals: Program Pers.	EBDEM
19	185	195	11	N		Est. Meals: Breaks	EBDEM
20	196	206	11	N		Est. Supplies: Office	EBDEM
21	207	217	11	N		Est. Supplies: Instructional	EBDEM
22	218	228	11	N		Est. Other: Rec. & Transp.	EBDEM
23	229	239	11	N		Est. Other: Misc.	EBDEM
24	240	250	11	N		Est. Clerical, Sec. Services	EBDEM
25	251	261	11	N		Est. Coor. Supervision	EBDEM
26	262	272	11	N		Est. General Admin.	EBDEM
27	273	283	11	N		Est. Contributed Costs	EBDEM
-47	284	503	220			Same as 8–27 but for actual rather than estimated.	EBDEM
48	504	511	8			NEXT LINK	

Table D4.1 MASTER FILE LAYOUT FORM—D4.1

FILE TITLE: Departmental Parameter File							PAGE 1 of 1
COMMENTS: One record Per Department						RECORD LENGTH: 65 bytes	
						DSN = FAD41	

SEQ	ST POS	END POS	FLD SIZ	FLD	REF NUM	FIELD DESCRIPTION	SOURCE
1	1	3	3	AN		Departmental Abbreviation	Direct
2	4	39	36	AN		Department Name	
3	40	45	6	AN		Short Departmental Name	Edit
4	46	65	20	AN		Department Head Name	

Table D4.2 MASTER FILE LAYOUT FORM—D4.2

FILE TITLE: Account Subcode Parameter File							PAGE 1 of 1
COMMENTS: One record per Account Subcode						RECORD LENGTH: 43 bytes	
						DSN = FAD42	

SEQ	ST POS	END POS	FLD SIZ	FLD	REF NUM	FIELD DESCRIPTION	SOURCE
1	1	3	3	N		Account Subcode	Direct
2	4	43	40	AN		Account Subcode Description	Edit

Table D4.3 MASTER FILE LAYOUT FORM—D4.3

FILE TITLE: General Parameter File							PAGE 1 of 1
COMMENTS: Two Records						RECORD LENGTH: 38 bytes	
						DSN = FAD43	

SEQ	ST POS	END POS	FLD SIZ	FLD	REF NUM	FIELD DESCRIPTION	SOURCE
1	1	2	2	N		FISCAL YEAR	P1.3B
2	3	8	6	N		Last Transaction Reference	P1.2
3	9	20	12	N		Reserved	
4	21	26	6	N		Last Proposal Reference	P2.1
5	27	32	6	N		Last Contract Reference	P2.2
6	33	38	6	N		Last Extension Reference	P2.3

Table D4.4 MASTER FILE LAYOUT FORM—D4.4

FILE TITLE: Access Control File							PAGE 1 of 1
COMMENTS: One record per logon code							RECORD LENGTH: 48 bytes DSN = FAD0

SEQ	ST POS	END POS	FLD SIZ	FLD	REF NUM	FIELD DESCRIPTION	SOURCE
1	1	5	5	X		Access Code	M4.1
2	6	25	20	AN		Name	M4.1
3	26	43	18	X		Time Last Logged On	P0.0
4	44	48	5	X		Password	M4.1

4.0 Process Specifications

Process specifications will be subdivided into two basic areas, both involving the synthesis of the total MASTER system. Section 4.1 presents the program-file interactions using data flow diagrams, Warnier flow diagrams, and the program-file cross-reference table. Section 4.2 gives the detailed program specifications.

4.1 Program-file Interactions

Two types of data flow diagrams will be given for the purposes of system specification. The first, given in Section 4.1.1, will be the standard data flow diagram with the following four symbols: (1) external entity (square), (2) process node (rounded rectangle), (3) data store (open rectangle), and (4) data flow (arrow). The second, given in Section 4.1.2, will be a standard Warnier diagram form. Further, Section 4.1.3 provides a file-program cross-reference table to facilitate program modification. These are adequately labeled to provide a table of contents to the more detailed components of the system. Specifically, more detailed information on data stores was given in Section 4.3, while more detailed specifications for the programs are given in Section 4.2.

4.1.1 Standard Data Flow Diagrams

The data flow diagrams will be organized according to the modular subdivisions of the system. The following is a table of contents of the standard data flow diagrams:

DFD1. Budget Management and Control—Inputs
DFD2. Budget Management and Control—Outputs
DFD3. Proposal and Contract Control—Inputs
DFD4. Proposal and Contract Control—Outputs
DFD5. Extension Evaluation and Control.

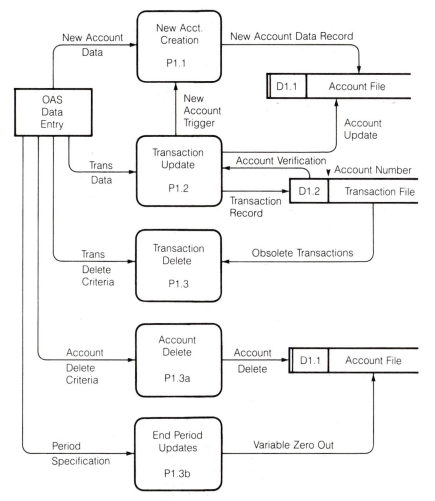

NOTE: Parameter Files Excluded.

DFD1 Budget Management and Control–Inputs

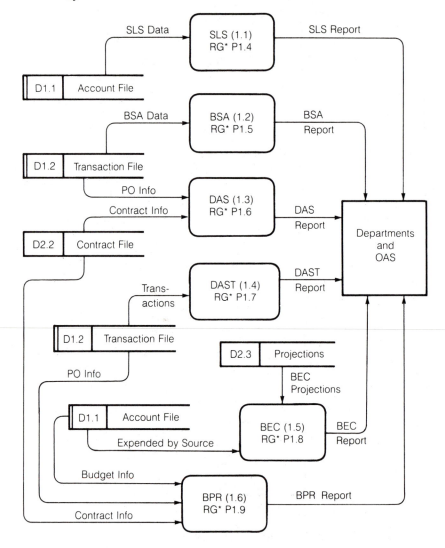

*RG = Report Generator.
NOTE: Parameter Files Excluded.

DFD2 Budget Management and Control–Outputs

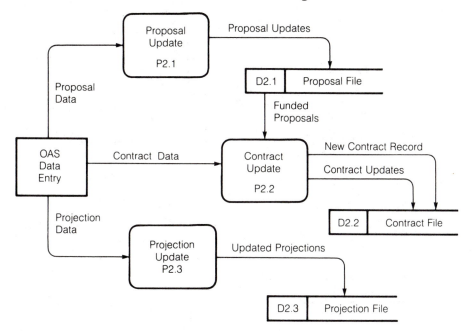

DFD3 Proposal and Contract Control–Inputs

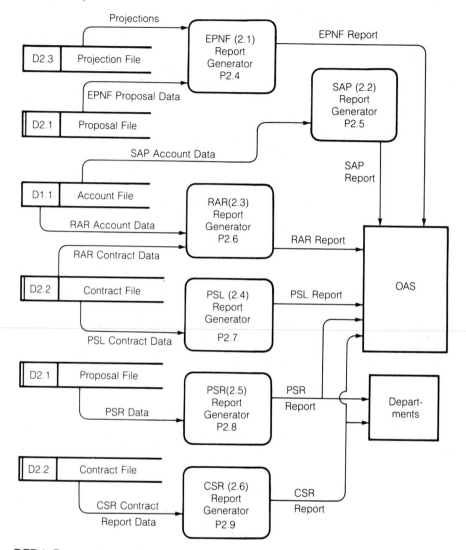

DFD4 Proposal and Contract Control–Outputs

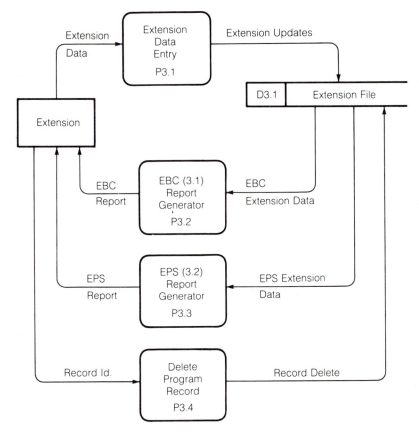

DFD5 Extension Evaluation and Control

4.1.2 Warnier Flow Diagrams

The Warnier flow diagrams are organized identically to the data flow diagrams. The external entities are depicted on the left as the receiver or supplier of information. The analysis proceeds to the program level for either data-entry or report-generator programs. Finally, the supplying or receiving file is given at the third level. Note that WFDx corresponds to DFDx.

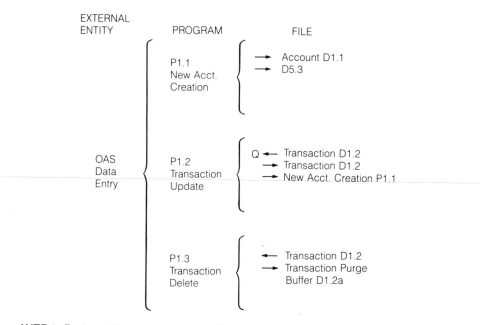

WFD1 Budget Management and Control–Inputs

```
EXTERNAL
ENTITY                PROGRAM                        FILE

                   ⎧  P1.4                    ◄—— Account D1.1
                   ⎪  SLS RG (1.1)       ⎰
                   ⎪  Creation           ⎱
                   ⎪
                   ⎪  P1.5                    ◄—— Transaction D1.2
                   ⎪  BSA RG (1.2)       ⎰
                   ⎪  Update             ⎱
                   ⎪
                   ⎪                     ⎰  ◄—— Transaction D1.2
                   ⎪  P1.6                  ◄—— Contract D2.2
                   ⎪  DAS RG (1.3)       ⎱
                   ⎪
  Departments   ⎨
  and OAS          ⎪  P1.7                  ◄—— Transaction D1.2
                   ⎪  DAST RG (1.4)      ⎱
                   ⎪
                   ⎪  P1.8                  ◄—— Account D1.1
                   ⎪  BEC RG (1.5)       ⎰
                   ⎪                     ⎱
                   ⎪
                   ⎪                     ⎰  ◄—— Account D1.1
                   ⎪  P1.9                  ◄—— Transaction D1.2
                   ⎩  BPR RG (1.6)       ⎱  ◄—— Contract D2.2
```

WFD2 Budget Management and Control–Outputs

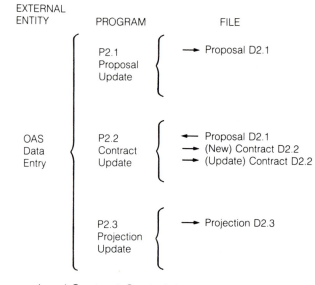

EXTERNAL
ENTITY   PROGRAM   FILE

P2.1   ⟶ Proposal D2.1
Proposal
Update

OAS   P2.2   ⟵ Proposal D2.1
Data   Contract   ⟶ (New) Contract D2.2
Entry   Update   ⟶ (Update) Contract D2.2

P2.3   ⟶ Projection D2.3
Projection
Update

WFD3 Proposal and Contract Control–Inputs

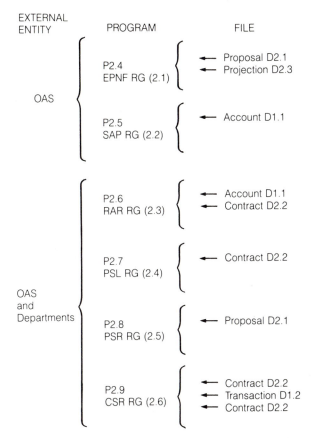

EXTERNAL
ENTITY PROGRAM FILE

OAS

P2.4
EPNF RG (2.1)
← Proposal D2.1
← Projection D2.3

P2.5
SAP RG (2.2)
← Account D1.1

OAS
and
Departments

P2.6
RAR RG (2.3)
← Account D1.1
← Contract D2.2

P2.7
PSL RG (2.4)
← Contract D2.2

P2.8
PSR RG (2.5)
← Proposal D2.1

P2.9
CSR RG (2.6)
← Contract D2.2
← Transaction D1.2
← Contract D2.2

WFD4 Proposal and Control–Outputs

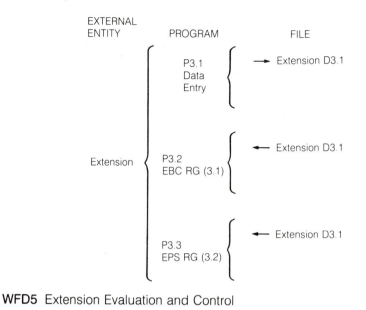

EXTERNAL
ENTITY　　　　PROGRAM　　　　　　FILE

P3.1
Data
Entry
　　　　　　　　　　　　→ Extension D3.1

Extension

P3.2
EBC RG (3.1)
　　　　　　　　　　　　← Extension D3.1

P3.3
EPS RG (3.2)
　　　　　　　　　　　　← Extension D3.1

WFD5　Extension Evaluation and Control

4.1.3 File-program Cross-reference

Table 4.1 presents the file-program cross-reference which should be used whenever program or file modifications are made to assure that system integrity is maintained. If a program does not appear in the listing, it has no file interaction.

Table 4.1   File-program Cross-reference

Program	Files									
	0.0	1.1	1.2	2.1	2.2	2.3	3.1	4.1	4.2	4.3
P0.0	I									
P1.1	B	B								I
P1.2		B	B						I	B
P1.3			B							
P1.4	I	I	I							
P1.5		I			I					
P1.6	I	I	I		I					
P1.7	I	I	I		I					
P1.8			I				I	I		
P1.9		I			I				I	
P2.1				B						B
P2.1A										
P2.2	B			I	B					B
P2.2A	I				B					
P2.3						B				
P2.4		I			I			I	I	
P2.5		I			I			I		
P2.6		I			I			I		
P2.7		I			I			I		
P2.8				I				I		
P2.9					I				I	
P3.1							B			
P3.2							I			
P3.3							I			
M1.0		C	C	C	C	C	C			
M1.1								C		
M1.2									C	

I = Input to program
O = Output from program
B = Both I and O
C = Creates
Mx.y = Maintenance program x.y

4.2 Program Specifications

Table 4.2 presents the program specifications in terms of inputs and outputs. This table, along with documentation given above, thoroughly specifies most of the simple programs. For example, P1.1 is the data-entry program for new account creation. The query and data-entry mat specifications are given in Section 2.1.1 of the Overview and User Guide. These, coupled with the output file specifications, given in Section 3.1.1 of the Design/Maintenance Manual above, are sufficient to initiate programming. To recap these specifications, and to further specify more sophisticated programs, a series of program specification Warnier diagrams (PSWDs) are presented in this section. Note that PSWD x.y corresponds to program Px.y.

Table 4.2 Key to Program Specifications

Program Number	Program Name	Inputs	Outputs
P1.1	New Acct. Creation	New acct. data (terminal) New acct. trigger (P1.2)	New acct. record (D1.1)
P1.2	Transaction Update	Transaction data (terminal) Acct. verification (D1.2)	Acct. update (D1.2) Transaction record (D1.2)
P1.3	Transaction Delete	Trans. delete criteria (terminal)	Deleted transactions (D1.2a) Obsolete transactions (D1.2)
P1.4	SLS RG*	SLS data (D1.1)	SLS Report
P1.5	BSA RG	BSA data (D1.2)	BSA Report
P1.6	DAS RG	PO Info (D1.2) Contract Info (D2.2)	DAS Report
P1.7	DAST RG	Transactions (D1.2)	DAST Report
P1.8	BEC RG	Expend by Source (D1.1)	BEC Report
P1.9	BPR RG	Budget Info (D1.1)	BPR Report
P2.1	Proposal Update	Proposal data (Terminal)	Proposal updates (D2.1)

Table 4.2   Key to Program Specifications (continued)

Program Number	Program Name	Inputs	Outputs
P2.2	Contract Update	Funded proposals (D2.1) Contract data (Terminal)	New contract records (D2.2) Contract updates (D2.2)
P2.3	Projection Update	Projection data (Terminal)	Updated Projections (D2.3)
P2.4	EPNF RG	Projections (D2.3) EPNF Proposal data (D2.1)	EPNF Report
P2.5	SAP RG	SAP Acct. Data (D1.1) SAP Contract Data (D2.2)	SAP Report
P2.6	RAR RG	RAR Acct. Data (D1.1) RAR Contract Data (D2.2)	RAR Report
P2.7	PSL RG	PSL Contract Data (D2.2)	PSL Report
P2.8	PSR RG	PSR Data (D2.1)	PSR Report
P2.9	CSR RG	CSR Contract Data (D2.2)	CSR Report
P3.1	Extension Data Entry	Extension Data (Terminal)	Extension Updates
P3.2	EBC RG	EBC Extension Data (D3.1)	EBC Report
P3.3	EPS RG	EPS Extension Data (D3.1)	EPS Report

*RG = Report Generator

.INPUT—User selection from supervisory
          menu—O&UG Part 1.

Perform Monthly Reminders—PSWD 0.1a

Chain to P0.1.1—PSWD 0.1.1
.cond—User selects 1
          Budget Management and Control Module
      (+)

Chain to P0.1.2—PSWD 0.1.2
.cond—User selects 2
          Proposal and Contract Control Module
      (+)

Chain to P0.1.3—PSWD 0.1.3
.cond—User selects 3
          Extension Evaluation and Control Module
      (+)

Chain to P0.1.4—PSWD 0.1.4·
.cond—User selects 4
          Parameter File Maintenance
      (+)

Logoff user
.cond—User selects 5
          Logoff/Terminate

.OUTPUT—Monthly reminders to CRT.

P0.1
System
Supervisory
Menu
Program
from
Logon
Procedures

PSWD 0.1 System Supervisory Menu

.INPUT—D4.4.

P01a
Perform
Monthly
Reminders
from
P0.1

Test
current
month
against
D4.4
header

New month—Perform Reminder

      (+)

Old month—EXIT

Reminder: THE END OF PERIOD UPDATE HAS
NOT BEEN PERFORMED FOR MONTH _ _
CLOSE OUT THAT MONTH USING BMC SM
Option 7 AS SOON AS POSSIBLE TO
KEEP MONTHLY STATISTICS TIMELY.

.OUTPUT—Monthly reminders to CRT.

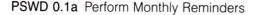

PSWD 0.1a Perform Monthly Reminders

.INPUT—User selection from BM and C
supervisory menu—O&UG 2.1.1.

Chain to P1.1
.cond—User selects 1
New Account Creation/Modification
(+)

Chain to P1.2 (Option: New)
.cond—User selects 2
Account Transaction
(+)

Chain to P1.2 (Option: Old)
.cond—User selects 3
Transaction Modification
(+)

P0.1.1
Budget
Mangement
and
Control
Supervisory
Menu
Program
from
P0.1

Chain to P0.1.1.1
.cond—User selects 4
Generate Reports
(+)

Chain to P1.3
.cond—User selects 5
Delect Transactions
(+)

Chain to P1.3a
.cond—User selects 6
Delete Account
(+)

Chain to P1.3b
.cond—User selects 7
Perform End Period Updates
(+)

Chain to P0.1
cond—User selects 8
Return to MASTER SM

OUTPUT—Option—Old, New.
(0,1)

PSWD 0.1.1 BM and C Supervisory Menu

.INPUT—User selection from BM and C
report-generator menu—O&UG 2.1.4.

Chain to P1.4
.cond—User selects 1
Subsidiary Ledger Summary
(+)

Chain to P1.5
.cond—User selects 2
Budget Subcode Analysis
(+)

Chain to P1.6
.cond—User selects 3
Department Account Statement
(+)

Chain to P1.7
.cond—User selects 4
DAS by Transaction
(+)

Chain to P1.8
.cond—User selects 5
Budget Expenditure Comparison
(+)

Chain to P1.9
.cond—User selects 6
Budget Projection Report
(+)

Chain to P0.1.1
.cond—User selects 7
BM and C Supervisory Menu

.OUTPUT—None.

P0.1.1.1
Budget
Management
and
Control
Report-
generator
Menu
Program
from
P0.1.1

PSWD 0.1.1.1 BM and C Report-generator Monitor

.INPUT—User selection from PC and C
        report generator menu—O&UG 2.2.

Chain to P2.1 (Option: New)
.cond—User selects 1
            Create New Proposal
        (+)

Chain to P2.2 (Option: New)
.cond—User selects 2
            Create New Contract
        (+)

Chain to P2.1 (Option: Old)
.cond—User selects 3
            Modify Proposal
        (+)

P0.1.2
Proposal
and
Contract
Control
Supervisory
Menu
Program
from
P0.1

Chain to P2.2 (Option: Old)
.cond—User selects 4
            Modify Contract
        (+)

Chain to P0.1.2.1
.cond—User selects 5
            Generate Reports
        (+)

Chain to P2.1 (Option: Del)
.cond—User selects 6
            Delete Proposal
        (+)

Chain to P2.2 (Option: Del)
.cond—User selects 7
            Delete Contract
        (+)

Chain to P2.3
.cond—User selects 8
            Projection Files Update
        (+)

Chain to P0.1
.cond—User selects 9
            Return to MASTER SM

.OUTPUT—Option—Old, New, Del.
            (0,1)

PSWD 0.1.2 PC and C Supervisory Menu

P0.1.2.1
Proposal
and
Contract
Control
Report-
generator
Menu
Program
from
P0.1.2

.INPUT—User selection from PC and C
           report-generator menu—O&UG 2.2.5.

Chain to P2.4
.cond—User selects 1
           Extramural Proposals and New Funding
      (+)

Chain to P2.5
.cond—User selects 3
           Summary of Active Reports
      (+)

Chain to P2.6
.cond—User selects 4
           Research Activities Report
      (+)

Chain to P2.7
.cond—User selects 5
           Percent Support List
      (+)

Chain to P2.8
.cond—User selects 6
           Proposal Status Report
      (+)

Chain to P2.9
.cond—User selects 7
           Contract Status Report
      (+)

Chain to P0.1.2
.cond—User selects 8

.OUTPUT—None.

PSWD 0.1.2.1 PC and C Report-generator Menu

P0.1.3
Extension
Evaluation
and
Control
Supervisory
Menu
Program
from
P0.1

.INPUT—User selection from EEC
      supervisory menu.

Chain to P3.1 (Option: New)
.cond—User selects 1
            Create New Program
      (+)

Chain to P3.1 (Option: Old)
.cond—User selects 2
            Modify Existing Program Budget
      (+)
Chain to P0.1.3.1
.cond—User selects 3
            Generate Reports
      (+)

Chain to P3.4
.cond—User selects 4
            Delete Program Budget
      (+)

Chain to P0.1
.cond—User selects 5
            Return to MASTER SM

.OUTPUT—Option—New, Old.
            (0,1)

PSWD 0.1.3 EEC Supervisory Menu

P0.1.3.1
Extension
Evaluation
and
Control
Report-
generator
Program
from
P0.1.3

.INPUT—User selection from EEC
      report-generator menu.

Chain to P3.2
.cond—User selects 1
            Extension Budget.Control
      (+)

Chain to P3.3
.cond—User selects 2
            Extension Project Summary Program
      (+)

Chain to P0.1.3
.cond—User selects 3
            Return to EEC SM

.OUTPUT—None.

PSWD 0.1.3.1 EEC Report-generator Menu

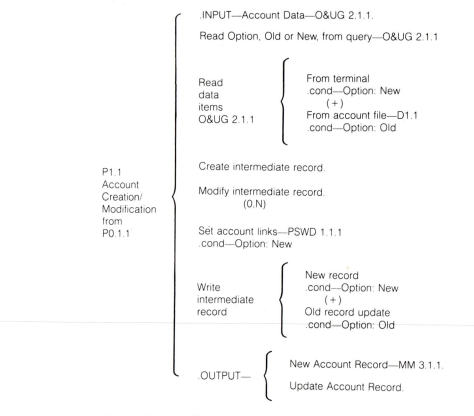

.INPUT—Account Data—O&UG 2.1.1.

Read Option, Old or New, from query—O&UG 2.1.1

Read
data
items
O&UG 2.1.1
{
From terminal
.cond—Option: New
(+)
From account file—D1.1
.cond—Option: Old
}

P1.1
Account
Creation/
Modification
from
P0.1.1

Create intermediate record.

Modify intermediate record.
(0,N)

Set account links—PSWD 1.1.1
.cond—Option: New

Write
intermediate
record
{
New record
.cond—Option: New
(+)
Old record update
.cond—Option: Old
}

.OUTPUT—
{
New Account Record—MM 3.1.1.

Update Account Record.
}

N = number of modifications required; allow reiteration
     and verification.

PSWD 1.1 Account Creation/Modification Program

P1.1.1
Set
Account
Links
from
P1.1

{

.INPUT—All information from P1.1.

Determine location of new record = LOC.

Read header of D1.1 and traverse file records using D1.1,66.

Find BEF66 = location of record with next lower value of D1.1,2.

Find AFT66 = location of record with next higher value of D1.1,2.

Set D1.1,66 of LOC record = AFT66.

Set D1.1,66 of BEF66 record = LOC.

Read header of D1.1 and traverse file records using D1.1,67.

Find BEF67 = location of record with next lower value of (D1.1, concatenated (4,2)).

Find AFT67 = location of record with next higher value of (D1.1, concatenated (4,2)).

Set D1.1,67 of LOC record = AFT67.

Set D1.1,67 of BEF67 record = LOC.

Set D1.1,65 = D1.1,68 = 0.

.OUTPUT—D1.1,65 through D1.1,68.

PSWD 1.1.1 Set Account Links

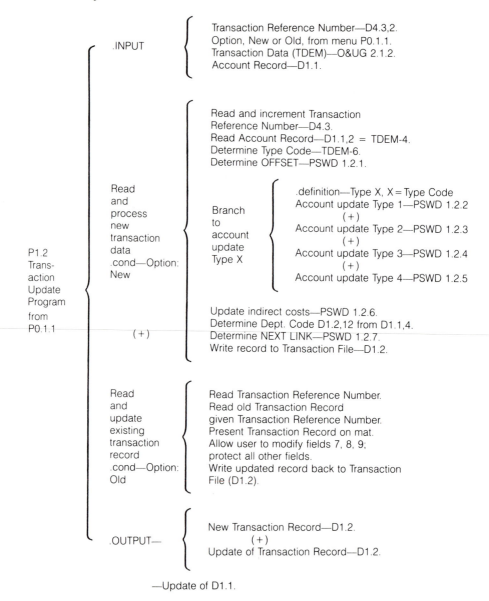

.INPUT {
Transaction Reference Number—D4.3,2.
Option, New or Old, from menu P0.1.1.
Transaction Data (TDEM)—O&UG 2.1.2.
Account Record—D1.1.
}

P1.2
Trans-
action
Update
Program
from
P0.1.1

Read
and
process
new
transaction
data
.cond—Option:
New {
Read and increment Transaction
Reference Number—D4.3.
Read Account Record—D1.1,2 = TDEM-4.
Determine Type Code—TDEM-6.
Determine OFFSET—PSWD 1.2.1.

Branch
to
account
update
Type X {
.definition—Type X, X = Type Code
Account update Type 1—PSWD 1.2.2
(+)
Account update Type 2—PSWD 1.2.3
(+)
Account update Type 3—PSWD 1.2.4
(+)
Account update Type 4—PSWD 1.2.5
}

(+)
Update indirect costs—PSWD 1.2.6.
Determine Dept. Code D1.2,12 from D1.1,4.
Determine NEXT LINK—PSWD 1.2.7.
Write record to Transaction File—D1.2.
}

Read
and
update
existing
transaction
record
.cond—Option:
Old {
Read Transaction Reference Number.
Read old Transaction Record
given Transaction Reference Number.
Present Transaction Record on mat.
Allow user to modify fields 7, 8, 9;
protect all other fields.
Write updated record back to Transaction
File (D1.2).
}

.OUTPUT— {
New Transaction Record—D1.2.
(+)
Update of Transaction Record—D1.2.
}

—Update of D1.1.

PSWD 1.2 Transaction Update Program

.INPUT—Account Subcode (TDEM-5).

Account Subcode	OFFSET (bytes)
100	0
140	77
150	154
200	231
600	308
700	385
800	462
900	539

P1.2.1
OFFSET
Determination
Subroutine
from
P1.2

Determine
OFFSET
from
Account
Subcode

Return OFFSET to main program—P1.2

.OUTPUT—OFFSET.

PSWD 1.2.1 OFFSET Determination Routine

.INPUT

Transaction Data-entry Mat (TDEM);
see O&UG 2.1.2.
Account Record—D1.1.
OFFSET.

.Description—Type Code 1 is for nonpurchase-
order expenditures.

P1.2.2
Account
Update
Type 1
.cond—
Type Code
= 1
from
P1.2

Add Dollar Amount,
TDEM-10, to D1.1
variables.

First byte position

$43 + OFFSET$

$54 + OFFSET$

$65 + OFFSET$

$76 + OFFSET$

Balance $= (D1.1,9) - (D1.1,13) - (D1.1,15)$

Determine if balance is negative and
print query for transaction verification.

If Balance is negative—EXIT.
.Desc—no change made to Account Record.

If Balance is positive, write updates
to D1.1.

.OUTPUT—Update to Account Record—D1.1.

PSWD 1.2.2 Account Update Type 1;
Nonpurchase-order Expenditure

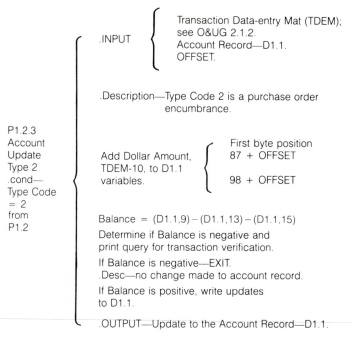

P1.2.3
Account
Update
Type 2
.cond—
Type Code
= 2
from
P1.2

.INPUT
{
Transaction Data-entry Mat (TDEM);
see O&UG 2.1.2.
Account Record—D1.1.
OFFSET.
}

.Description—Type Code 2 is a purchase order
encumbrance.

Add Dollar Amount,
TDEM-10, to D1.1
variables.
{
First byte position
87 + OFFSET

98 + OFFSET
}

Balance = (D1.1,9) − (D1.1,13) − (D1.1,15)

Determine if Balance is negative and
print query for transaction verification.

If Balance is negative—EXIT.
.Desc—no change made to account record.

If Balance is positive, write updates
to D1.1.

.OUTPUT—Update to the Account Record—D1.1.

PSWD 1.2.3 Account Update Type 2;
Purchase-order Encumbrance

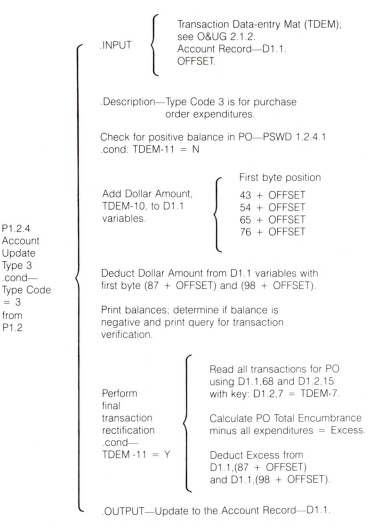

.INPUT
- Transaction Data-entry Mat (TDEM); see O&UG 2.1.2.
- Account Record—D1.1.
- OFFSET.

.Description—Type Code 3 is for purchase order expenditures.

Check for positive balance in PO—PSWD 1.2.4.1
.cond: TDEM-11 = N

Add Dollar Amount, TDEM-10, to D1.1 variables.

First byte position
43 + OFFSET
54 + OFFSET
65 + OFFSET
76 + OFFSET

P1.2.4
Account
Update
Type 3
.cond—
Type Code
= 3
from
P1.2

Deduct Dollar Amount from D1.1 variables with first byte (87 + OFFSET) and (98 + OFFSET).

Print balances; determine if balance is negative and print query for transaction verification.

Perform final transaction rectification .cond— TDEM -11 = Y

Read all transactions for PO using D1.1,68 and D1.2,15 with key: D1.2,7 = TDEM-7.

Calculate PO Total Encumbrance minus all expenditures = Excess.

Deduct Excess from D1.1,(87 + OFFSET) and D1.1,(98 + OFFSET).

.OUTPUT—Update to the Account Record—D1.1.

PSWD 1.2.4 Account Update Type 3; Purchase-order Expenditure

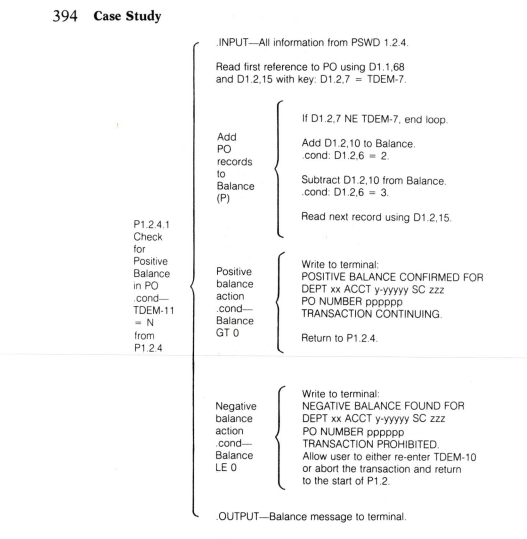

.INPUT—All information from PSWD 1.2.4.

Read first reference to PO using D1.1,68
and D1.2,15 with key: D1.2,7 = TDEM-7.

P1.2.4.1
Check
for
Positive
Balance
in PO
.cond—
TDEM-11
= N
from
P1.2.4

Add
PO
records
to
Balance
(P)

If D1.2,7 NE TDEM-7, end loop.

Add D1.2,10 to Balance.
.cond: D1.2,6 = 2.

Subtract D1.2,10 from Balance.
.cond: D1.2,6 = 3.

Read next record using D1.2,15.

Positive
balance
action
.cond—
Balance
GT 0

Write to terminal:
POSITIVE BALANCE CONFIRMED FOR
DEPT xx ACCT y-yyyyy SC zzz
PO NUMBER pppppp
TRANSACTION CONTINUING.

Return to P1.2.4.

Negative
balance
action
.cond—
Balance
LE 0

Write to terminal:
NEGATIVE BALANCE FOUND FOR
DEPT xx ACCT y-yyyyy SC zzz
PO NUMBER pppppp
TRANSACTION PROHIBITED.
Allow user to either re-enter TDEM-10
or abort the transaction and return
to the start of P1.2.

.OUTPUT—Balance message to terminal.

P = Number of records for PO.

PSWD 1.2.4.1 Check Purchase-order for Positive Balance

.INPUT {
   Transaction Data-entry Mat (TDEM);
   see O&UG 2.1.2.
   Account Record—D1.1.
   OFFSET.

P1.2.5
Account
Update
Type 4
.cond—
Type Code
= 4
from
P1.2

.Description—Type Code 4 is for
              budget modifications.

Change Dollar Amount, TDEM-10,
to D1.1 byte position (32 + OFFSET).

Test if amount remaining is negative;
if so, print error message and do
not make modification.

Print out new budget balance
for verification before saving.

.OUTPUT—Update to the Account Record
      in D1.1.

PSWD 1.2.5 Account Update Type 4; Budget Modification

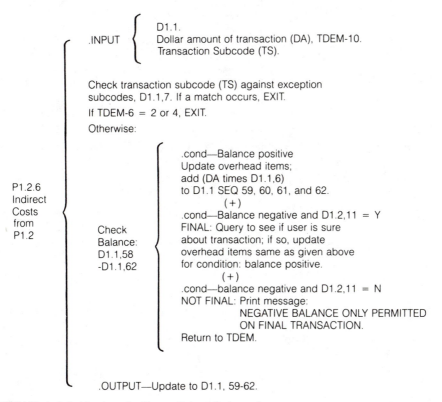

.INPUT {
   D1.1.
   Dollar amount of transaction (DA), TDEM-10.
   Transaction Subcode (TS).

Check transaction subcode (TS) against exception
subcodes, D1.1,7. If a match occurs, EXIT.

If TDEM-6 = 2 or 4, EXIT.

Otherwise:

P1.2.6
Indirect
Costs
from
P1.2

Check
Balance:
D1.1,58
-D1.1,62

.cond—Balance positive
Update overhead items;
add (DA times D1.1,6)
to D1.1 SEQ 59, 60, 61, and 62.
      (+)
.cond—Balance negative and D1.2,11 = Y
FINAL: Query to see if user is sure
about transaction; if so, update
overhead items same as given above
for condition: balance positive.
      (+)
.cond—balance negative and D1.2,11 = N
NOT FINAL: Print message:
        NEGATIVE BALANCE ONLY PERMITTED
        ON FINAL TRANSACTION.
Return to TDEM.

.OUTPUT—Update to D1.1, 59-62.

PSWD 1.2.6 Update Indirect Costs Subroutine

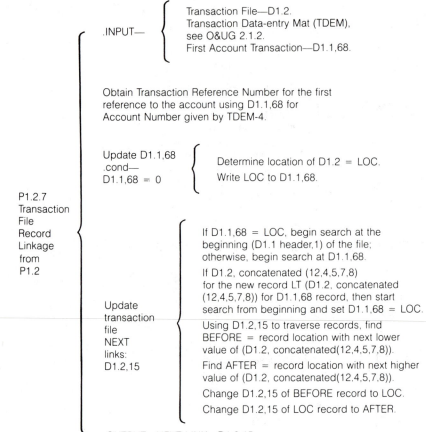

.INPUT—

Transaction File—D1.2.
Transaction Data-entry Mat (TDEM),
see O&UG 2.1.2.
First Account Transaction—D1.1,68.

Obtain Transaction Reference Number for the first
reference to the account using D1.1,68 for
Account Number given by TDEM-4.

Update D1.1,68
.cond—
D1.1,68 = 0

Determine location of D1.2 = LOC.
Write LOC to D1.1,68.

P1.2.7
Transaction
File
Record
Linkage
from
P1.2

Update
transaction
file
NEXT
links:
D1.2,15

If D1.1,68 = LOC, begin search at the
beginning (D1.1 header,1) of the file;
otherwise, begin search at D1.1,68.

If D1.2, concatenated (12,4,5,7,8)
for the new record LT (D1.2, concatenated
(12,4,5,7,8)) for D1.1,68 record, then start
search from beginning and set D1.1,68 = LOC.

Using D1.2,15 to traverse records, find
BEFORE = record location with next lower
value of (D1.2, concatenated(12,4,5,7,8)).

Find AFTER = record location with next higher
value of (D1.2, concatenated(12,4,5,7,8)).

Change D1.2,15 of BEFORE record to LOC.

Change D1.2,15 of LOC record to AFTER.

.OUTPUT—NEXT LINK—D1.2,15.

PSWD 1.2.7 Transaction File Linkage Routine

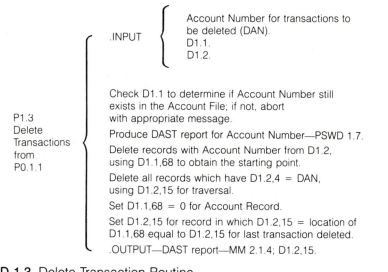

.INPUT
{
Account Number for transactions to
be deleted (DAN).
D1.1.
D1.2.
}

P1.3
Delete
Transactions
from
P0.1.1

Check D1.1 to determine if Account Number still
exists in the Account File; if not, abort
with appropriate message.

Produce DAST report for Account Number—PSWD 1.7.

Delete records with Account Number from D1.2,
using D1.1,68 to obtain the starting point.

Delete all records which have D1.2,4 = DAN,
using D1.2,15 for traversal.

Set D1.1,68 = 0 for Account Record.

Set D1.2,15 for record in which D1.2,15 = location of
D1.1,68 equal to D1.2,15 for last transaction deleted.

.OUTPUT—DAST report—MM 2.1.4; D1.2,15.

PSWD 1.3 Delete Transaction Routine

.INPUT { Account Number to be deleted (DAN). D1.1.

P1.3a
Delete
Account
from
P0.1.1

If (D1.1,65 NE 0) or (D1.1,68 NE 0), EXIT.

Produce DAS report for Account Number—PSWD 1.6.

Delete Account Record from D1.1 for which D1.1,2 = DAN.

Set D1.1,66 for the record in which D1.1,66 = location of account deleted equal to D1.1,66 of account deleted.

Set D1.1,67 for the record in which D1.1,67 = location of account deleted equal to D1.1,67 of account deleted.

.OUTPUT—DAS report—MM 2.1.3.

PSWD 1.3a Delete Account Routine

P1.3b
End-
Period
Updates
from
P0.1.1

.INPUT—Update Type = M, Q, Y.

Perform reminders—O&UG 2.1.7.

If user does not enter CONTINUE, EXIT.

Zero out Seq 11, 18, ... for all accounts.

Zero out Seq 15, 22, ... for all accounts.

Zero out Seq 12, 19, ... for all accounts.
.cond—Update Type Q or Y.

Zero out Seq 13, 20, ... for all accounts.
.cond—Update Type Y.

Zero out D4.3 and put next fiscal year in D4.3,1.
.cond—Update Type Y.

Put current date in header of D4.4.

.OUTPUT—D1.1 Update to all accounts.

PSWD 1.3b End-period Update Routine

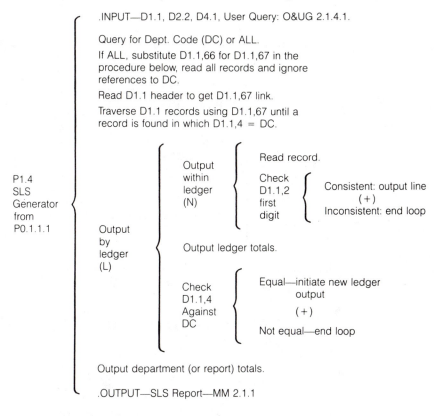

.INPUT—D1.1, D2.2, D4.1, User Query: O&UG 2.1.4.1.

Query for Dept. Code (DC) or ALL.

If ALL, substitute D1.1,66 for D1.1,67 in the procedure below, read all records and ignore references to DC.

Read D1.1 header to get D1.1,67 link.

Traverse D1.1 records using D1.1,67 until a record is found in which D1.1,4 = DC.

P1.4
SLS
Generator
from
P0.1.1.1

Output by ledger (L)

Output within ledger (N)

Read record.

Check D1.1,2 first digit

Consistent: output line (+)

Inconsistent: end loop

Output ledger totals.

Check D1.1,4 Against DC

Equal—initiate new ledger output (+)

Not equal—end loop

Output department (or report) totals.

.OUTPUT—SLS Report—MM 2.1.1

N = number of records for which digit D1.1,2 is consistent.
L = number of ledgers.

PSWD 1.4 Subsidiary Ledger Summary Report Generator

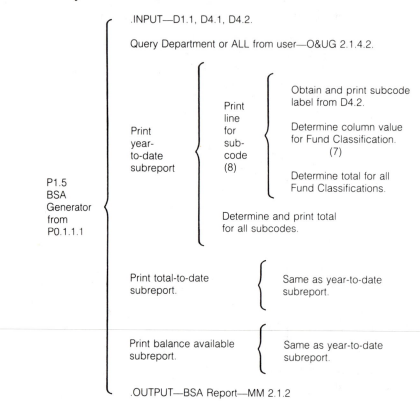

.INPUT—D1.1, D4.1, D4.2.

Query Department or ALL from user—O&UG 2.1.4.2.

P1.5
BSA
Generator
from
P0.1.1.1

Print
year-
to-date
subreport

Print
line
for
sub-
code
(8)

Obtain and print subcode
label from D4.2.

Determine column value
for Fund Classification.
(7)

Determine total for all
Fund Classifications.

Determine and print total
for all subcodes.

Print total-to-date
subreport.

Same as year-to-date
subreport.

Print balance available
subreport.

Same as year-to-date
subreport.

.OUTPUT—BSA Report—MM 2.1.2

PSWD 1.5 Budget Subcode Analysis (BSA) Report Generator

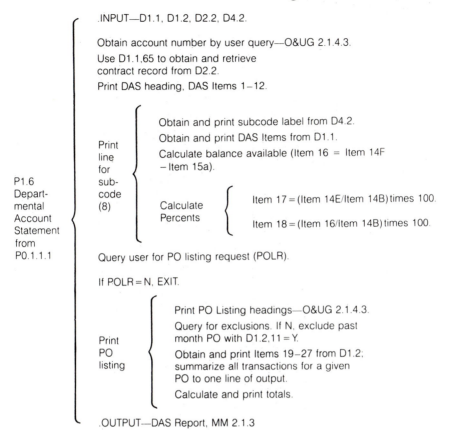

.INPUT—D1.1, D1.2, D2.2, D4.2.

Obtain account number by user query—O&UG 2.1.4.3.

Use D1.1,65 to obtain and retrieve
contract record from D2.2.

Print DAS heading, DAS Items 1–12.

P1.6
Depart-
mental
Account
Statement
from
P0.1.1.1

Print
line
for
sub-
code
(8)

Obtain and print subcode label from D4.2.

Obtain and print DAS Items from D1.1.

Calculate balance available (Item 16 = Item 14F
− Item 15a).

Calculate
Percents

Item 17 = (Item 14E/Item 14B) times 100.

Item 18 = (Item 16/Item 14B) times 100.

Query user for PO listing request (POLR).

If POLR = N, EXIT.

Print
PO
listing

Print PO Listing headings—O&UG 2.1.4.3.

Query for exclusions. If N, exclude past
month PO with D1.2,11 = Y.

Obtain and print Items 19–27 from D1.2;
summarize all transactions for a given
PO to one line of output.

Calculate and print totals.

.OUTPUT—DAS Report, MM 2.1.3

PSWD 1.6 Departmental Account Statement (DAS) RG

.INPUT—D1.1, D2.2, D4.2, D5.1, D5.5.

Obtain Account Number, Date Span, and Transaction Type (TT) by user query—O&UG 2.1.4.4.

Use D5.1 to obtain and retrieve contract record from D2.2.

Print DAST heading, Items 1–12.

Obtain First Transaction Reference Number from D5.5.

P1.7
Dept.
Account
Statement
by
Transaction
from
P0.1.1.1

Print
trans-
action
line
for
account
(N)
.cond—
D1.2,6
= TT

Read transaction record of Transaction Account.

.Initiate condition: D2.2,8 GE Date Span minimum.

.Terminate condition: D2.2,3 NE Account Number and D2.2,8 LE Date Span maximum.

Determine subcode.

Read subcode description from D4.2.

Assemble line output from D1.2 as indicated in MM 2.1.4.

Accumulate totals for Items 21, 22 and 23—PSWD 1.7.1

Print totals for Items 21, 22 and 23.

.OUTPUT—DAST Report, MM 2.1.4.

N = number of transactions of type TT for account.

PSWD 1.7 DAS by Transaction (DAST) RG

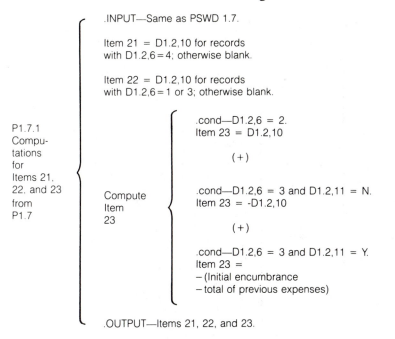

.INPUT—Same as PSWD 1.7.

Item 21 = D1.2,10 for records
with D1.2,6 = 4; otherwise blank.

Item 22 = D1.2,10 for records
with D1.2,6 = 1 or 3; otherwise blank.

P1.7.1
Compu-
tations
for
Items 21,
22, and 23
from
P1.7

Compute
Item
23

.cond—D1.2,6 = 2.
Item 23 = D1.2,10

(+)

.cond—D1.2,6 = 3 and D1.2,11 = N.
Item 23 = -D1.2,10

(+)

.cond—D1.2,6 = 3 and D1.2,11 = Y.
Item 23 =
− (Initial encumbrance
− total of previous expenses)

.OUTPUT—Items 21, 22, and 23.

PSWD 1.7.1 Computations for Items 21, 22, and 23

.INPUT—D1.2, D2.3, D4.1, Month.

Obtain department codes by user query;
see O&UG 2.1.4.5.

Use D4.1,3 to obtain short department name.

Print BEC heading, Items 1–9.

	D1.1,5	Fund Type	RUN
Determine fund type	1	Instruction	1
	2	OAS	2
	3	Cost Share	3
	4,5,6	Extramural	4

Perform calculations only for accounts with qualifying Fund Type.

Calculate Item (6 + RUN times 4):
sum (D1.1,10 + D1.1,17 + ...) for all qualifying accounts.

Calculate Item (7 + RUN times 4):
sum (D1.1,14 + D1.1,21 + ...) for all qualifying accounts.

Calculate Item (8 + RUN times 4):
Item (7 + RUN times 4)/Item (6 + RUN times 4).

Calculate Item
(9 + RUN times 4)   PSWD 1.8.1

Calculate and print section of report by Fund Type (4 RUNS) and department (12)

P1.8
BEC
RG
from
P0.1.1.1

Calculate and print section of report for totals;
sum all items and prorate where necessary.

.OUTPUT—BEC Report—MM 2.1.5.

PSWD 1.8 Budget Expenditure Comparison (BEC) RG

.INPUT—D2.3, Item 12, 16, 20, 24 from P1.8.

Read D2.3 for the department.

Item (I)	Comparison Field (X)
12	D2.3,2
16	D2.3,3
21	D2.3,4
24	D2.3,5
Totals	D2.3,6

Determine Comparison Field

P1.8.1
Percent
Deviation
Sub-
routine
from
P1.8

Compute percent = (I/X) times 100.

.OUTPUT—Percentages to P1.8.

PSWD 1.8.1 Percent Deviation Subroutine

.INPUT—D1.1, D2.2, D4.2.

Obtain Account Number and Months (MONTHS) from user
by query—O&UG 2.1.4.6.

Use D1.1,65 to obtain contract expiration date,
Item 16A, from D2.2,15.

Print BPR heading, Items 1–16a.

P1.9
BPR
RG
from
P0.1.1.1

Print
line
for
sub-
code
(8)

Obtain and print subcode label from D4.2.

Obtain and print Items 14b, 14c, 15, and 16
from D1.1.

Output projection period, Item 16b, MONTHS.

Item 18: project current month =
MONTHS times Item 14c.

Item 19: calculate projected balance
remaining =
Item 14b – Item 18.

Item 21: project current-to-date =
MONTHS times Item 15/(months into contract).

Item 22: calculate estimated balance
remaining =
Item 14b – Item 21.

Calculate and print section of report for totals;
sum all items and prorate where necessary.

.OUTPUT—BPR report—MM 2.1.6.

PSWD 1.9 Budget Projection Report (BPR) RG

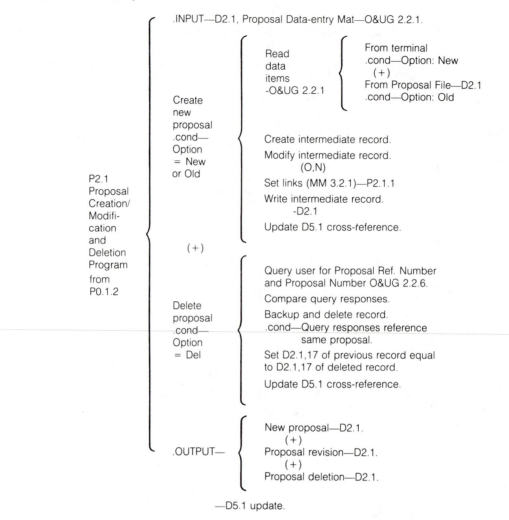

.INPUT—D2.1, Proposal Data-entry Mat—O&UG 2.2.1.

Read data items -O&UG 2.2.1
From terminal .cond—Option: New (+)
From Proposal File—D2.1 .cond—Option: Old

Create new proposal .cond— Option = New or Old

Create intermediate record.
Modify intermediate record. (O,N)
Set links (MM 3.2.1)—P2.1.1
Write intermediate record. -D2.1
Update D5.1 cross-reference.

P2.1 Proposal Creation/ Modifi- cation and Deletion Program from P0.1.2

(+)

Delete proposal .cond— Option = Del

Query user for Proposal Ref. Number and Proposal Number O&UG 2.2.6.
Compare query responses.
Backup and delete record. .cond—Query responses reference same proposal.
Set D2.1,17 of previous record equal to D2.1,17 of deleted record.
Update D5.1 cross-reference.

.OUTPUT—

New proposal—D2.1. (+)
Proposal revision—D2.1. (+)
Proposal deletion—D2.1.

—D5.1 update.

PSWD 2.1 Proposal Creation/Modification and Deletion Program

.INPUT—All information from P2.1.

Determine location of new record = LOC.

P2.1.1
Proposal
Linkage
Routine
from
P2.1

Read header of D2.1 and traverse file records using D2.1,17.

Find BEFORE = location of record with next lower value of (D2.1, concatenated (5,2)).

Find AFTER = location of record with next higher value of (D2.1, concatenated (5,2)).

Set D2.1,17 of LOC record = AFTER.

Set D2.1,17 of BEFORE record = LOC.

.OUTPUT—D2.1,17.

PSWD 2.1.1 Set Proposal Links

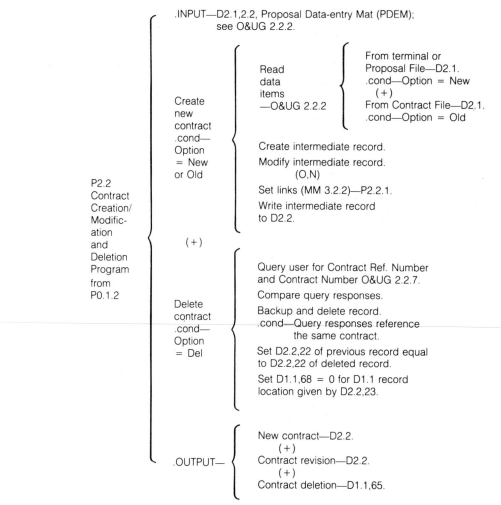

.INPUT—D2.1,2.2, Proposal Data-entry Mat (PDEM);
see O&UG 2.2.2.

P2.2
Contract
Creation/
Modific-
ation
and
Deletion
Program
from
P0.1.2

Create
new
contract
.cond—
Option
= New
or Old

Read
data
items
—O&UG 2.2.2

From terminal or
Proposal File—D2.1.
.cond—Option = New
(+)
From Contract File—D2.1.
.cond—Option = Old

Create intermediate record.

Modify intermediate record.
(O,N)

Set links (MM 3.2.2)—P2.2.1.

Write intermediate record
to D2.2.

(+)

Delete
contract
.cond—
Option
= Del

Query user for Contract Ref. Number
and Contract Number O&UG 2.2.7.

Compare query responses.

Backup and delete record.
.cond—Query responses reference
the same contract.

Set D2.2,22 of previous record equal
to D2.2,22 of deleted record.

Set D1.1,68 = 0 for D1.1 record
location given by D2.2,23.

.OUTPUT—

New contract—D2.2.
(+)
Contract revision—D2.2.
(+)
Contract deletion—D1.1,65.

PSWD 2.2 Contract Creation/Modification and Deletion Program

P2.2.1
Proposal
Linkage
Routine
from
P2.2

.INPUT—All information from P2.2.

Determine location of new record = LOC.

Read header of D2.2 and traverse file records using D2.2,22.

Find BEFORE = location of record with next lower value of (D2.2, concatenated (6,12)).

Find AFTER = location of record with next higher value of (D2.2, concatenated (6,12)).

Set D2.2,22 of LOC record = AFTER.

Set D2.2,22 of BEFORE record = LOC.

Set D2.2,23 = location of the record in D1.1 for which D1.1,2 = CDEM-3.

Set D1.1,65 in record with account number CDEM-3 equal to LOC.

.OUTPUT—D2.2,22; D2.2,23.

PSWD 2.2.1 Set Contract Links and Pointers

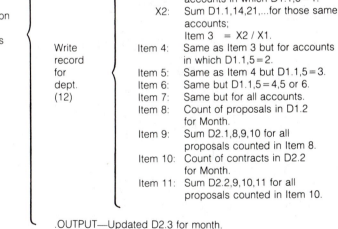

P2.3
Projection
File
Updates
from
P0.1.2

.INPUT—D1.1, D2.1, D2.2.

.Trigger—end of each month.

Determine Month from system.

Create new record.
.cond—Month = 1, October.

Print BPR heading, Items 1-16a.

Write record for dept. (12)

Item 1:   D4.3,1
Item 2:   D1.1,3
Item 3:
  X1:   Sum D1.1,10,17,...for all Dept. accounts in which D1.1,5 = 1.
  X2:   Sum D1.1,14,21,...for those same accounts;
         Item 3 = X2 / X1.
Item 4:   Same as Item 3 but for accounts in which D1.1,5 = 2.
Item 5:   Same as Item 4 but D1.1,5 = 3.
Item 6:   Same but D1.1,5 = 4,5 or 6.
Item 7:   Same but for all accounts.
Item 8:   Count of proposals in D1.2 for Month.
Item 9:   Sum D2.1,8,9,10 for all proposals counted in Item 8.
Item 10:   Count of contracts in D2.2 for Month.
Item 11:   Sum D2.2,9,10,11 for all proposals counted in Item 10.

.OUTPUT—Updated D2.3 for month.

PSWD 2.3 Projection File Updates

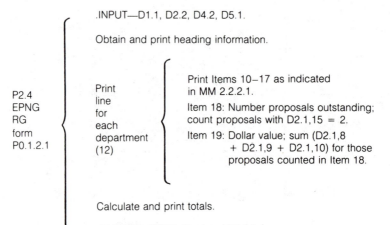

P2.4
EPNG
RG
form
P0.1.2.1

{
.INPUT—D1.1, D2.2, D4.2, D5.1.

Obtain and print heading information.

Print
line
for
each
department
(12)

{
Print Items 10–17 as indicated
in MM 2.2.2.1.

Item 18: Number proposals outstanding;
count proposals with D2.1,15 = 2.

Item 19: Dollar value; sum (D2.1,8
       + D2.1,9 + D2.1,10) for those
       proposals counted in Item 18.

Calculate and print totals.

.OUTPUT—EPNG Report—MM 2.2.1.
}

PSWD 2.4 Extramural Proposals and New Funding RG

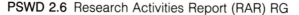

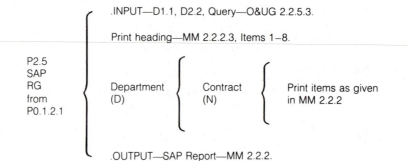

P2.5
SAP
RG
from
P0.1.2.1

{
.INPUT—D1.1, D2.2, Query—O&UG 2.2.5.3.

Print heading—MM 2.2.2.3, Items 1–8.

Department
(D)
{
Contract
(N)
{
Print items as given
in MM 2.2.2

.OUTPUT—SAP Report—MM 2.2.2.
}

D = number of departments.
N = number of contracts within each department.

PSWD 2.5 Summary of Active Projects (SAP) RG

P2.6
RAR
RG
P2.6
from
P0.1.2.1

{
.INPUT—D1.1, D2.2.

Same as PSWD 2.5 with
minor format changes
given in MM 2.2.3.

.OUTPUT—RAR Report—MM 2.2.3.
}

PSWD 2.6 Research Activities Report (RAR) RG

P2.7
PSL
RG
from
P0.1.2.1
{
.INPUT—D1.1, D2.2.

Same as PSWD 2.5 with
minor format changes
given in MM 2.2.4.

.OUTPUT—PSL Report—MM 2.2.4.
}

PSWD 2.7 Percent Support List (PSL) RG

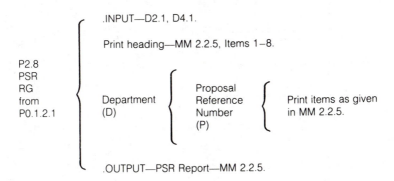

P2.8
PSR
RG
from
P0.1.2.1
{
.INPUT—D2.1, D4.1.

Print heading—MM 2.2.5, Items 1–8.

Department (D) { Proposal Reference Number (P) { Print items as given in MM 2.2.5.

.OUTPUT—PSR Report—MM 2.2.5.
}

D = number of departments.
P = number of proposals within each department.

PSWD 2.8 Proposal Status Report (PSR) RG

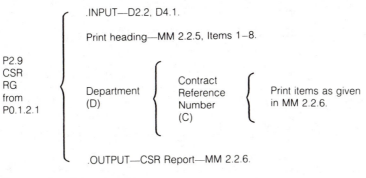

P2.9
CSR
RG
from
P0.1.2.1
{
.INPUT—D2.2, D4.1.

Print heading—MM 2.2.5, Items 1–8.

Department (D) { Contract Reference Number (C) { Print items as given in MM 2.2.6.

.OUTPUT—CSR Report—MM 2.2.6.
}

D = number of departments.
C = number of contracts within each department.

PSWD 2.9 Contract Status Report (CSR) RG

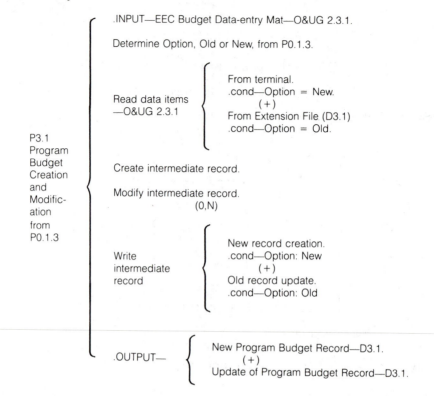

P3.1 Program Budget Creation and Modification from P0.1.3

.INPUT—EEC Budget Data-entry Mat—O&UG 2.3.1.

Determine Option, Old or New, from P0.1.3.

Read data items —O&UG 2.3.1

From terminal.
.cond—Option = New.
(+)
From Extension File (D3.1)
.cond—Option = Old.

Create intermediate record.

Modify intermediate record.
(0,N)

Write intermediate record

New record creation.
.cond—Option: New
(+)
Old record update.
.cond—Option: Old

.OUTPUT—

New Program Budget Record—D3.1.
(+)
Update of Program Budget Record—D3.1.

N = number of modifications required by user.

PSWD 3.1 Program Budget Creation/Modification Program

P3.2
EBC
Generator
from
P0.1.3.1

{

.INPUT—D3.1.

.Reference—MM 2.3.1 for source of all output variables.

Obtain program reference numbers by query—O&UG 2.3.3.1.

Print
EBC
for
program
(P)

{

Retrieve record from D3.1.

Produce output for program budget as described in MM 2.3.1.

.OUTPUT—EBC Report, MM 2.3.1.

P = number of programs selected.

PSWD 3.2 Extension Budget Control (EBC) Report Generator

P3.3
EPS
RG
from
P0.1.3.1

{

.INPUT—D3.1.

.Reference—MM 2.3.1 for source of all output variables.

Obtain program numbers—query, O&UG 2.3.3.2.

Read file and add variables to produce totals for user-defined set of programs.

Produce output described in MM 2.3.2.

.OUTPUT—EPS Report, MM 2.3.2.

PSWD 3.3 Extension Project Summary (EPS)

P3.4
Delete
Program
Budget
from
P0.1.3

{

.INPUT—D3.1.

Query user for Program Reference Number; see O&UG 2.3.4.

Produce EBS Report—P3.2.

Delete Program Budget Record from D3.1.

OUTPUT—EBS Report.

PSWD 3.4 Delete Program Budget

MANAGEMENT ANALYSIS
TO SUPPORT TRANSACTIONS
IN EDUCATION AND RESEARCH

(M A S T E R)

Commentary on the
MASTER System

Introduction

The purpose of this section is to set the stage for, and provide some explanation of, the case study example documentation presented immediately above. In reality, these documents, namely, the Overview, User Guide, and Design/Maintenance Manual, will not be accompanied by additional explanation. Rather, these two self-contained documents should thoroughly furnish programmers, users, and involved managers with all that they need to perform their required functions with regard to the software being developed.* It is for this reason that the documents have been presented above in their completed form without supplementary explanation. They provide a comprehensive example of the output of techniques and methodology presented in this book. It is recommended that the MASTER system documentation be reviewed, if not read in detail, before proceeding.

While in reality the completed documents should be adequate for software development and operation, the reader is not "in reality." There are many questions about the MASTER system which would be easy for anyone to answer who worked within the organization for which it was designed. Since most readers will not have this or similar experience, it is necessary to present some background such that many of these questions are answered. This will be accomplished by performing a general Warnier analysis of the affected management system (in the next subsection), as would be required of any analyst designing software for a new application.

In addition to a presentation of the background and environment of the case study, some explanation is required as to the reason for particular decisions which were made in the design of the MASTER system. Obviously, the explanation of every detail of MASTER cannot be included within the documentation itself without defeating its purpose. For example, the programmer does not need to know (and usually does not care) why three, rather than four, modules were chosen, or why the system was named MASTER. While such explanations and justifications are inappropriate within the documentation, they typify the interests of the designer. The thought processes involved are quite important to the novice designer in order to put an example into its proper perspective. For this reason, a final subsection will contain comments intended to reveal the rationale for the design decisions made in the MASTER system. Further, these will be cross-referenced back to those relevant sections of the book which deal with either the techniques or methodology for their resolution. These comments will be ordered by the specific sections in the MASTER documentation being discussed.

---

*The only exception might be exhibits which cannot be produced on a word processor, and thus will be bound separately.

Background and Analysis of the AMS

This case study involves the development of a management information system for a research and extension arm of a major university. Specifically, the software system will primarily serve the university's Office of Administrative Services (OAS) in providing information on research proposals, contracts, and extension (short course) activities. OAS provides coordination and other support services for the departments within the university. In addition to teaching and research, the faculty is involved in several hundred short courses per year under the heading of extension activities.

Preliminary interviews with the OAS staff led to a rough definition of problems encountered in administering the research and extension programs. These were documented by the systems analyst who was investigating the possibility of designing and developing this system on a consulting basis. The following contents of a memorandum written to the head of OAS by the analyst provides additional information for understanding the project background:

> As you requested, I have reviewed the informational needs expressed by you and your staff. The primary problem emphasized by almost everyone involved the lack of responsiveness in your manual accounting system. While University Computer Services is doing an adequate job in providing for the overall accounting needs of the various departments, two shortcomings have been identified with regard to their providing your informational needs: (1) because they serve the entire university, the delay time of from two to six weeks in obtaining their standardized reports makes them of little value for control purposes, and (2) the content of the accounting reports is not adequate for control, even if such were timely. It is anticipated that there will be a small amount of redundancy between the information to be entered for a management information system (MIS) and that currently being entered within the University Computer System. However, it will be much less than your current manual system, which is obviously necessary to maintain your management capabilities. While additional detailed study is required before recommendations can be made, it appears that the development of a computerized MIS might be able to replace the current manual system within OAS while providing a large number of additional capabilities.
>
> A second problem area was identified with regard to the originating and maintaining of information on proposals and contracts. A number of reports are currently being manually produced at regular intervals which could easily be mechanized. While these reports draw, in part, from the financial information, their primary source is unique to the files which are cur-

rently maintained for each contract and each proposal. Similarly, a third problem area was identified in the area of extension. Information is required on each short course which must be updated and consulted frequently in order to maintain control. While the current manual system is holding its own, it is evident that it will not be able to handle anticipated growth and additional report requirements.

Your decision to utilize the excess capacity on the minicomputer which was recently acquired by OAS should be easy to implement. That machine currently has far more capacity than should ever be required by the proposed MIS. As computer demands for research continue to increase, a point will be reached where computer time and space will be at a premium. However, at that point, additional capacity should be justified to handle both the research and the MIS requirements.

In order for us to develop the design for this software, it will be necessary to spend a number of days in detailed analysis of your current operation and future needs. We will document your current system and procedures in a Warnier outline format so that we can develop a complete understanding of needs. This will ultimately result in a documented set of system output report specifications. These will be presented to you and your staff before we proceed any further with the design.

Programming requirements will depend heavily upon the size of the design. As we proceed through the design capabilities, we will work very closely with you to assure that the design is cost-beneficial to you. We estimate that the design itself will take approximately 60 man-days to complete. If you are in favor of proceeding we will put together a complete plan for the design project and submit it for your approval.

With the approval of chargeable work, a design plan, nearly identical to that presented in Chapter 7, was created. The only modification was in the sharp curtailment in time requirements necessitated by the limited scope of this overall project. Time and space do not permit a thorough documentation of all actions taken during the 60-day design period. However, in order to present the environment in which the design documentation was developed, the transformed Warnier analysis is presented in Figure C.1. This is the output of Activity 2.1 (see Figure 7.7) entitled: Transform Warnier Analysis.

While taking several pages, Figure C.1 is greatly condensed and simplified. Its purpose here is to present the environment of the design documentation rather than to serve as a tool for determining system output requirements. In practice, the management system Warnier diagram (MSWD) would begin with the *existing* (usually manual) system. This step has been bypassed in favor of the presentation of the MSWD after transformation into the anticipated new system, a transformation which usually results in much valuable simplification.

Figure C.1 demonstrates how the overall missions of the management information system (MIS) are analyzed into projects, activities, decision-actions, and procedure steps. The analyst may take the liberty of omitting unnecessary detail, as in the analysis of the third mission where the entire project level was omitted. Note, however, that the objective of this analysis, i.e., to facilitate system output specification, should not be jeopardized either by omission or simplification.

Readers who have already reviewed the MASTER system documentation (as recommended), will note that the mission breakdown of OAS is used to define the modularity of the system. This is typically the case. Further, Table C.1 illustrates how the output reports, documented in the second section of the Design/Maintenance Manual, relate to the Data Retrieval Transactions (DRTs) of the MSWD. While it is easy for us to set up such a table at this point, its presentation tends to conceal the considerable effort involved in transforming the procedural requirements of the MSWD into a set of well-defined output reports. To those who have experienced this transformation we need say no more; to the novice we can only appeal to imagination. Output report specifications require considerable interaction with users, resulting in much reiteration until the finally developed output is approved.

Table C1.   Relationship Between DRTs and MASTER Outputs

Data Retrieval Transaction	Master Output Report	Maintenance Manual Section
DRT 1	Department Account Statement (DAS)	2.1.3
DRT 2	DAS by Transaction (DAST)	2.1.4
DRT 3	Subsidiary Ledger Summary (SLS)	2.1.1
DRT 4	(None—obtained directly from DST 4)	
DRT 5	Budget Subcode Analysis (BSA)	2.1.2
DRT 6	Budget Expenditure Comparison (BEC)	2.1.5
DRT 7	Budget Projection Report (BPR)	2.1.6
DRT 8	Extramural Proposals and New Funding (EPNF)	2.2.1
DRT 9	Proposal Status Report (PSR)	2.2.5
DRT 10	Contract Status Report (CSR)	2.2.6
DRT 11	Summary of Active Projects (SAP)	2.2.2
DRT 12	Research Activities Report (RAR)	2.2.3
DRT 13	Percent Support List (PSL)	2.2.4
DRT 14	Extension Budget Control Report (EBC)	2.3.1
DRT 15	Extension Projects Summary (EPS)	2.3.2

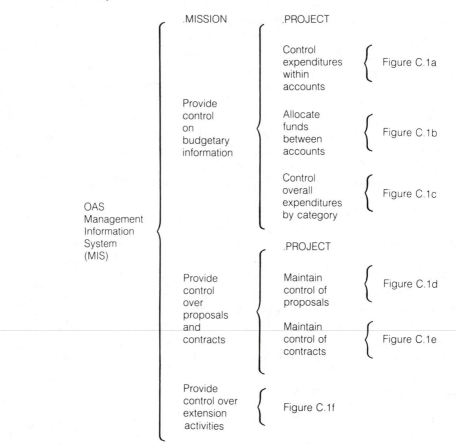

Figure C.1 MSWD of OAS MIS

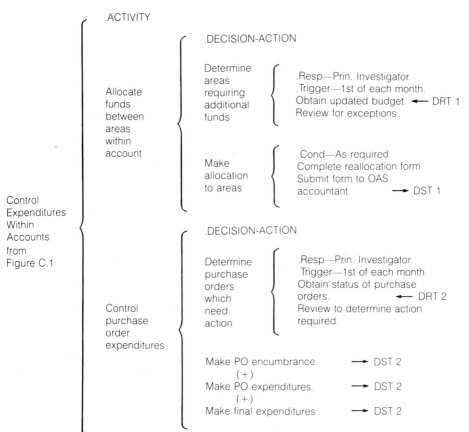

.ACTIVITY

.DECISION-ACTION

Allocate funds between areas within account

Determine areas requiring additional funds

.Resp—Prin. Investigator.
.Trigger—1st of each month.
Obtain updated budget. ◄— DRT 1
Review for exceptions.

Make allocation to areas

.Cond—As required.
Complete reallocation form.
Submit form to OAS
accountant. —► DST 1

.DECISION-ACTION

Control Expenditures Within Accounts from Figure C.1

Control purchase order expenditures

Determine purchase orders which need action

.Resp—Prin. Investigator.
.Trigger—1st of each month.
Obtain status of purchase
orders. ◄— DRT 2
Review to determine action
required.

Make PO encumbrance. —► DST 2
    (+)
Make PO expenditures. —► DST 2
    (+)
Make final expenditures. —► DST 2

Figure C.1a MSWD of OAS MIS (continued)

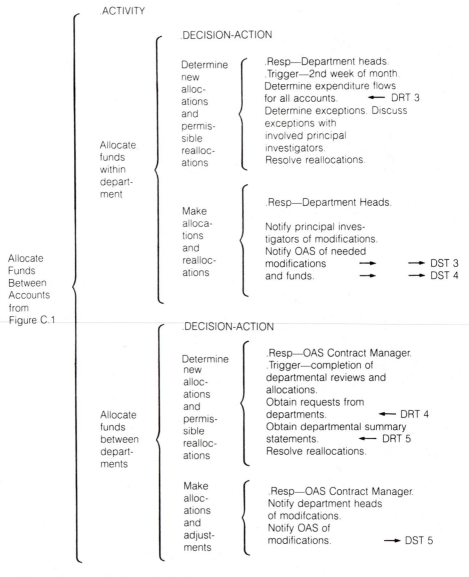

.ACTIVITY

.DECISION-ACTION

Allocate Funds Between Accounts from Figure C.1

Allocate funds within department

Determine new alloc-ations and permis-sible realloc-ations
.Resp—Department heads.
.Trigger—2nd week of month.
Determine expenditure flows for all accounts. ← DRT 3
Determine exceptions. Discuss exceptions with involved principal investigators.
Resolve reallocations.

Make alloca-tions and realloc-ations
.Resp—Department Heads.

Notify principal inves-tigators of modifications.
Notify OAS of needed modifications → → DST 3
and funds. → → DST 4

.DECISION-ACTION

Allocate funds between depart-ments

Determine new alloc-ations and permis-sible realloc-ations
.Resp—OAS Contract Manager.
.Trigger—completion of departmental reviews and allocations.
Obtain requests from departments. ← DRT 4
Obtain departmental summary statements. ← DRT 5
Resolve reallocations.

Make alloc-ations and adjust-ments
.Resp—OAS Contract Manager.
Notify department heads of modifcations.
Notify OAS of modifications. → DST 5

Figure C.1b MSWD of OAS MIS (continued)

.DECISION-ACTION

Control
Overall
Expenditures
by
Category
from
Figure C.1

Determine
status
of
expend-
itures
by
category

.Resp—Principal Investigator
and Department Head.
.Trigger—1st week each month.

Obtain projection on expenditures
by funding source.   ← DRT 6

Obtain projection on expenditures
by account subcode.   ← DRT 7

Determine all significant
variations from normal.

Take
corrective
action
on
exceptions

.Resp—Principal Investigator.

Increase expenditures in
categories which are under budget.
      (+)
Restrict expenditure in categories
which are over budget.
      (+)
Submit budget modifications
for changes in anticipated
cash flow.   → DST 6

Figure C.1c MSWD of OAS MIS (continued)

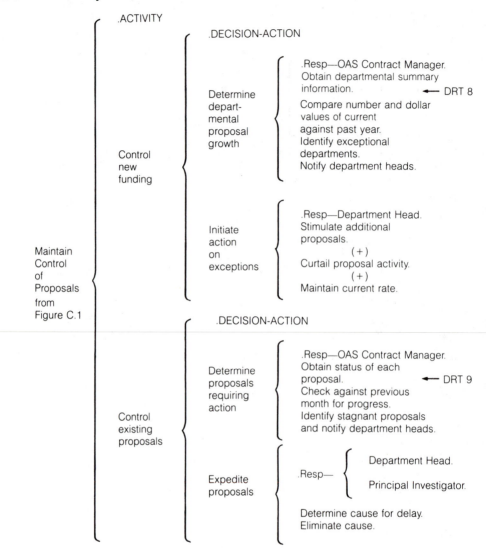

Figure C.1d MSWD of OAS MIS (continued)

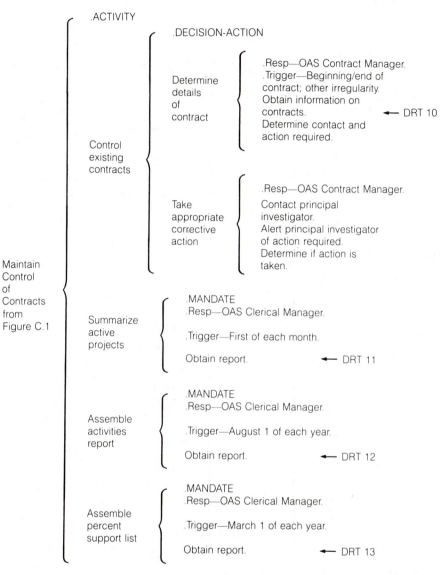

Figure C.1e MSWD of OAS MIS (continued)

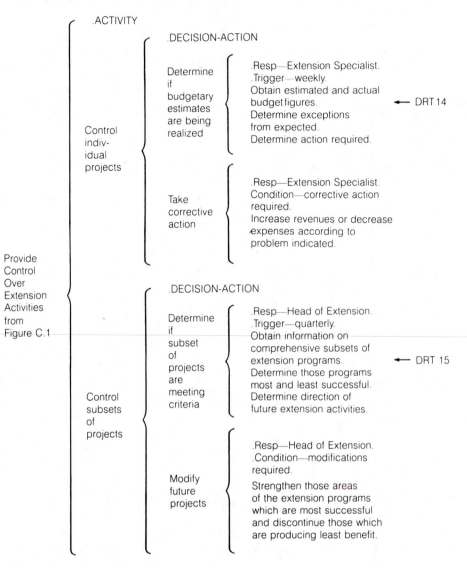

.ACTIVITY

Provide Control Over Extension Activities from Figure C.1

Control individual projects

.DECISION-ACTION

Determine if budgetary estimates are being realized
- .Resp—Extension Specialist.
- .Trigger—weekly.
- Obtain estimated and actual budget figures.
- Determine exceptions from expected.
- Determine action required.

← DRT 14

Take corrective action
- .Resp—Extension Specialist.
- Condition—corrective action required.
- Increase revenues or decrease expenses according to problem indicated.

Control subsets of projects

.DECISION-ACTION

Determine if subset of projects are meeting criteria
- .Resp—Head of Extension.
- .Trigger—quarterly.
- Obtain information on comprehensive subsets of extension programs.
- Determine those programs most and least successful.
- Determine direction of future extension activities.

← DRT 15

Modify future projects
- .Resp—Head of Extension.
- .Condition—modifications required.
- Strengthen those areas of the extension programs which are most successful and discontinue those which are producing least benefit.

Figure C.1f MSWD of OAS MIS (continued)

The analyst who finds that users have a good and precise idea of what they want is indeed blessed. In most cases, it will be the job of the analyst to "cut and try" the output reports and/or capabilities, using the methods discussed in Chapter 3.

The purpose of this subsection was to set the stage for the case study output documentation presented above. This was done in terms of the early analytical activity outputs. We have assumed that the analysis of the existing affected management system (AMS) has been completed. Figure C.1 presented the first activity of defining system outputs, i.e., that of transforming the AMS analysis to apply to the new proposed system. The results of the remaining activities are contained in the MASTER system documentation given above. The thought processes which led to these results are given in the next subsection, ordered according to the MASTER documentation for ease of reference.

Referenced Commentary

The purpose of this subsection is to provide additional information in a narrative format to clarify the thought processes in the development of the MASTER system documentation. It is important for the reader to realize that much of the documentation is not ordered chronologically according to development. In other words, some parts at the middle or near the end of the documentation were written before other parts nearer the beginning of it, as was brought out a number of times in the text. For this reason the commentary will generally not be ordered chronologically. For the chronological development methodology see Chapters 3, 5, and 6. As this commentary continues, subsections of these chapters will be cited to indicate the origin of each part of the documentation. This is also presented more concisely in Figure 7.5.

Each paragraph which follows will be preceded by a section reference from the MASTER documentation. The abbreviations O&UG (for Overview & User Guide) and MM (for the Design/Maintenance Manual) will be used to precede the individual section numbers. When there are several paragraphs referencing the same section, the paragraph reference will be followed by a letter.

Overview A. The most visible aspect of a system to the greatest number of people is its name. The naming of a system is essential; it cannot be discussed without a name. Therefore, if not assigned, it will acquire one, or perhaps many, pseudonames by default. Do not allow this to happen. The assignment of a clear name from the outset gives the system personality. While the choice of names can be quite difficult,

Table C.2   Examples of Good Acronyms

Acronym	Background Meaning	Initial and Longterm Impact
RAPID	Records Analysis for Problem Identification and Definition	System will perform its functions rapidly and efficiently.
AIM	Area Identification Module	Module aims to find the right area.
ACE	Accident Countermeasure Evaluation	System is number one.
CORRECT	Cost/benefit Optimization for the Reduction of Roadway Environment-Caused Tragedies	System is designed to correct accident problems.
CARE	Cities Accident RAPID Evaluation	Those who care will use this system.
DIR/ECT	Directory Encoding Control Technique	The system will get directly to the solution.
FARE	Functional Analysis for Research in Engineering	(?)
WORCS	Work Order Review and Control System	The system will work to provide the required information.

acronyms which impart an *immediate impact* and yet which have letters standing for meaningful descriptions are best. Soon everyone forgets the background meaning and the impact remains, providing a positive psychological feeling toward the system. Table C.2 presents some good examples of system names which illustrate this point. Bad names are acronyms which have no (or misleading) initial impact; are difficult, confusing, or embarrassing to pronounce; or have an ambiguous background meaning (in that order). FARE is only a fair name, mainly because its initial impact meaning is ambiguous. The rest of the acronyms in Table C.2 satisfy the criteria given above to varying extents.

Overview B. Once a name has been assigned to the system, by all means, use it. Encourage the users to be comfortable with reading it and using it in conversation. The same thing should be true of all acronyms used within the documentation. Certain terms will be understood by all readers, such as the term OAS in the first paragraph of the overview. However, always define an acronym upon first usage to avoid any misunderstandings.

Overview C. Note the heavy upper-management orientation of the Overview. No technical details are given except those of specific concern to them. Rather, emphasis is upon a review of the capabilities and features from the broadest perspective. The terms used are familiar to them, and it would be a laborious insult to their intelligence to over-define them. For example, "academic department heads" in the first sentence does not need the type of elaboration which was given in the introductory subsections above. Secondary targets of the Overview are the users, and to some extent, the programmers. It should provide them with the same introduction to the system, as well as furnishing them with the background for more detailed study.

Overview D. The first three paragraphs contain many general statements that could be made about most good user-friendly systems. To be more specific, the supervisory menu is used as a display within the documentation. This gives the reader something tangible from the system to identify the initial man-machine interface. Note that the ordering of the entries on this menu are in the importance of their respective features to higher management. The main objective of MASTER is embodied in the budget management and control feature, while the other objectives are secondary. Very little attention is given to parameter file maintenance, since it is of little concern to upper management.

Overview E. After presenting the features in terms of the supervisory menu, the Overview closes with some veiled motivational appeals for upper- and middle-management support. An alternative to this approach might be to include example outputs in lieu of the systematic menu-based review. This would be preferred if the Overview were to be used as a brochure for *outside* marketing. Actual example outputs, if they were understandable to the readers, might be more impressive than the narrative given in the MASTER Overview.

Overview F. The timing of Overview creation was not as well defined in the text as were the other parts of the documentation. An Overview should be generated as soon as required by upper management, possibly before the approval of the design project. A document generated this early will lack the specific details as given by this example. As the project continues through design and development, the Overview should evolve into a hard-hitting sales tool for the system. The more specifically that management needs can be addressed without going into boring detail, the more effective the Overview will be. The MASTER Overview given was the result of a number of modifications throughout design and development. It was maintained to provide a communication of overall system objectives when needed.

O&UG 2.0. This brief introduction provides the reader with the ordering of the manual to follow. This ordering is according to the supervisory menu already presented in the Overview, which is in itself a logically ordered presentation of overall system capabilities. The objective here is to make the manual easy to use both as an introduction to the system, and later, as a reference. All of the sections of Part 2 of the O&UG were written in their final form during Step 6 (Specify Input Requirements), very near the end of the design process.

O&UG 2.1. This introductory subsection shows that the entire Section 2.1 will be ordered according to the BMC supervisory menu given in Figure 2. The ordering of the discussion within each subsection is, with few exceptions, just as it will occur during system operation. Exceptions may include trivial and/or frequently recurring queries or other operations that do not bear mentioning on every occurrence.

O&UG 2.1.1A. Users may not care that there is a file called the account file, and most certainly they do not need to know that its reference number is D1.1. These are included in the first sentence mainly to help the programmers. The primary consideration is that these designations do not get in the way and prevent users from obtaining what they need from this part of the documentation. In this case they certainly do not cause user distraction, while they provide great aid to the programmers, who will use this documentation to aid in writing the programs. In particular, the specific wording of the menus, queries, and data-entry mats are found exclusively in this part of the documentation. The "D" numbers at the end of each data-entry mat item description are also included to aid the programmer.

O&UG 2.1.1B. Each item of the data-entry mat is discussed, whether or not the user is required to make an entry. This is done primarily to give the user an understanding of the protected fields, in this case fiscal year and date of entry. It is also done for completeness of reference to the programmers. For data items supplied by the system, both the source and the destination are referenced.

O&UG 2.1.1C. The design decision to combine account creation and modification within the same menu is largely a matter of personal preference. The alternative is to enlarge the menu, given in Figure 2, to 9 choices, with a separate option for account modification. If this were done it would not be necessary to determine by further query whether the user wanted a creation or modification. The combined menu option was chosen for three reasons: (1) the already large number of options on the menu, (2) the possibility of confusing account modification with account transaction or transaction modification, and (3) the fact that account modification is not a primary function of BMC, and therefore does not deserve BMC supervisory menu status. Given that this decision was made, the most efficient method is given (i.e., entry of RETURN in lieu of an account number) for determining if the user wants to create or modify an account. Note, also, the side benefit of reducing the instruction narrative, which results from combining the two functions into one menu option.

O&UG 2.1.1D. The last paragraph in Section 2.1.1 is a brief but adequate specification for performing modifications. This can be programmed in any one of a number of different ways, which do not require specification here. This allows the programmer to use the best or easiest method at hand. It is implied that the user will be given sufficient instructions on the screen at data-entry time such that additional documentation is not needed here.

O&UG 2.1.2. The User Guide is not intended to teach users about the OAS accounting system; therefore, it is assumed that the user understands the nature of OAS accounts and transactions.* Thus, there is very little that the user needs other than the details of data entry, such as special codes, and possibly some interpretation of abbreviated terms. In this subsection, there are two warnings given after data entry is completed. The first warns of a negative balance and gives the user the option to proceed or stop. Programming for the calculations required for this are not at all trivial, and a detailed program specification (PSWD 1.2.4.1) is required. The second is a notice to record the fiscal year and transaction reference number on the purchase order for an initial encumbrance transaction. This relates to the manual part of MASTER which requires such purchase orders to go through the university accounting system before they are returned with the purchase-order number assigned. When the purchase order is returned to OAS, they will update this first transaction to indicate the purchase-order number by obtaining it from the returned purchase order itself. If the users follow these directions for manually writing down the transaction number on the purchase order itself, this update will be quite easy. The reminder to do this is not just buried in the documentation; rather, it is put on the screen for each purchase-order encumbrance. Further, it requires a user action (namely, the depression of the ENTER key) before processing continues. Thus, every effort is made to assure that this required *manual* step is performed.

---

*This would be an invalid assumption if MASTER did not adopt the terminology commonly used for this application within OAS, or if MASTER were going to be marketed as a commercially available software package.

O&UG 2.1.3. The explanation as to the reasons that a transaction modification might be required is given. The details, however, are merely referenced, being the same as those given in the previous section. Heavy referencing is recommended so that when modifications are made, they will need to be made only once in the documentation.

O&UG 2.1.4. This introductory section for the report-generator menu option introduces the menu itself, as well as the report subsections to follow. It also describes, and should detail, if necessary, those queries common to all reports so that they will not have to be repeated in every subsection to follow. The detailed report specifications are referenced to the Maintenance Manual, where they are documented in detail. Actual example report exhibits could be integrated into these subsections if the designer feels that the user needs to see them. Since the users of MASTER were familiar with most of the reports already, and because all output reports are thoroughly documented in the Maintenance Manual, this was not done.

O&UG 2.1.4.1-7. The various options of the BMCM report-generator menu are presented in order according to user requirements. These queries are quite self-explanatory and it is doubtful that users will refer to these query definitions very often inasmuch as they are presented on the screen. However, as an initial aid to the programmer, these specifications are invaluable. Not only do they provide an ordered flow of the procedures, but they also specify the *wording* of the queries. Most programmers are delayed considerably if they have to worry about the wording (and sometimes the spelling) of queries. It is rare that the designer does not later modify most programmer-worded queries.

O&UG 2.1.5-6. The file maintenance ramifications of deleting records will be discussed with other file consideration in the discussion below. From the user's point of view it is critical that a deletion not take place which would inadvertently result in the loss of valuable data. We are assuming here that periodic (and frequent) backup is being performed at the *total systems* level. Good system management dictates that backup be done on the total system, rather than by every application. If the system backups are not deemed to be adequate, additional backup at the applications level will be required above that which is provided here. Given that the system backup is adequate, the objective here is to make the user need it as little as possible. Thus a series of warnings, reminders, and barriers are integrated into the software to try to prevent the loss of data. In this case an account's transactions cannot be deleted as long as the account exists. Faulty transactions can only be corrected by another transaction, thus providing an audit trail. When an account is deleted, a permanent hard-copy record is made of all available information, including its transactions. Once the account is deleted, the user might realize that a mistake was made. Restoration of the account may follow by using the system backup/restore utilities. This is a relatively easy step as opposed to restoring both the account and transactions. This is the reason for having separate delete steps, with the account deletion coming first.

O&UG 2.1.7. The performance of end-period updates could be accomplished by the system automatically at the beginning of each month. This capability was not elected for MASTER, however, because it would not allow for lateness in data entry or report generation. Rather, the system clock is used to generate monthly reminders that certain updates are required. These reminders continue until the particular update is performed, at which point they are "turned off." They are "turned on" once again when a new month appears on the clock.

O&UG 2.2. Proposals and contracts were set up in a separate module from accounts because of the difference in data-entry timing and report-generation characteristics. While this is fairly obvious once the design is specified, it required considerable analysis during the early analytical activities. Proposals and contracts could have been subdivided into two modules. This was not done because of the similarity between the data requirements associated with each. An OAS project starts out as a proposal. Only about a third of the proposals get funded. Once a proposal is accepted for funding (either internally or externally), an account number is assigned and an account record is established. Shortly thereafter, the pertinent proposal information is converted to contract information along with any additional information. This introductory section would review this process in detail if it were not commonly known within OAS. Since this is not required, the introductory paragraph is much the same as that for Section 2.1.

O&UG 2.2.1. This is a data-entry specification much like Sections 2.1.1 and 2.1.2. In some cases here, several numbered items are discussed together. This is done for several possible reasons, among them: (1) to impress upon the user that there is a separation between two very similar data fields, such as in Items 4 and 5; (2) to eliminate redundancy in the descriptions, as in the case of Items 13-15; and (3) to maintain a one-to-one correspondence of the items on the mat and the items in the file.

O&UG 2.2.2. It was decided to put the two data-entry options first. An alternative logical grouping might be: all proposal operations first, followed by all contract operations, followed by the report generator. In any case, the menu options reflect the natural flow of the process. The option to copy pertinent proposal information establishes a linkage with the previous section, maintaining the reader's train of thought.

Whenever possible, information already stored within the system should be used for default, eliminating unnecessary effort in data entry. This example demonstrates this principal quite well. Some of the information on Figure 8 is not on Figure 7 and must be added at this time. Also, all data items can be modified if they are not totally correct. Even the fiscal year was made alterable, since OAS personnel indicated that they might want to enter some of next year's contracts before the actual beginning of the fiscal year (see file D4.3 for details).

O&UG 2.2.3–4. See comments on O&UG 2.1.3. Proposal and contract modifications were considered to be of sufficient user demand to give them PCC supervisory menu status.

O&UG 2.2.5. See comments on O&UG 2.1.4. Even though some of the reports have no queries documented, a separate section is maintained for each to keep the subsections numbered consistently with the menu.

O&UG 2.2.6–7. See comments on O&UG 2.1.5-6.

O&UG 2.2.8. See comments on O&UG 2.1.7. Here the queries are considerably more complex, since some on-line decision making is required of the user. Flexibility is maintained to enable the user to perform those updates at a convenient time.

O&UG 2.3. This third module is the simplest of the three modules to design and develop. It consists of a direct mechanization of the exact form used for controlling the expenses involved in short-course presentation. For this reason the data-entry requirements are not documented in detail. To facilitate programming, the file containing this information is arranged in the exact order and the variables are given the same names as in Figure 11.

O&UG Part 3. This short but significant part of the User Guide was given little consideration in the text. Most of the information is hardware-dependent, and therefore provided by the manufacturers. It includes those procedures necessary for getting to the system supervisory menu. Any precautions with regard to the use of the hardware, and avoidance of its misuse, should be included. If necessary, a checklist should be provided such that it can be copied and posted near each terminal.

MM 1.0. This introduction is intended to give the reader an understanding of the nature of the Design/Maintenance Manual as it fits together with the Overview and User Guide. It also reviews the modular breakdown of MASTER in terms of the output report requirements. It is noted that some of the exhibits are not bound in this volume. Typically, wide printouts will not be entered on the word processor, and these can be maintained in a separate exhibit file for ease of reference.* Figure 1.1, set up by standard Warnier conventions, forms a table of contents for the output report specifications. Note that this document is merely referenced as the Maintenance Manual (MM), since once the software is developed this is its primary function.

MM 2.0. Output report specifications are developed as one of the early design activities, namely, as the fourth activity (Document Output Reports) in the second step (Analyze the New System Output Requirements). Thus, this section of the Maintenance Manual is the first part of the system documentation to be written.

MM 2.1. If general introductory comments related to all of the BMCM reports were required, they could go under this general heading; in this case there were none.

MM 2.1.1. The format followed is that given in Section 3.2.4 of the text. The availability is on CRT or hard copy, necessitating two separate formats. Some of the hard copy information will not be output on the CRT. The screen outputs are given, with certain overlapping columns repeated. Each variable in the output is referenced to the variable list by the reference numbers (circled or in parentheses). Generally, the source of each of these variables is given in terms of its file and the sequential position within the file. Thus, the programmer has all that is needed to generate the output reports.

MM 2.1.2. Items 1-9 are referenced to save space and time. It should be obvious that Output 1.2 is not an actual prototype. Item 16 carries on across the page, while Items 21 and 22 replace Item 12 to provide additional pages. It was felt that the MASTER programmers would have no problem understanding this, although a more explicit prototype would certainly be in order.

MM 2.1.3. Item 12 includes some aliases. While not recommended, aliases are usually unavoidable because the users have several names for the same item. Therefore, select one name (the most descriptive) and use it consistently. Frequent referencing of the aliases with the primary name, as done here, will eliminate most problems. Also note the use of reference numbers 14a, 14b, etc. These were added between 14 and 15 after the initial numbering. Also, an item number with "Re-

---

*Some of the outputs have been excluded; those given are sufficient to exemplify this case study.

served" in place of a description occurs when a data item previously considered important is discarded, leaving a gap. As a rule, *do not re-number.* Because the number of modifications are quite large, an analyst could make a career of renumbering every time an item of data is inserted or deleted, since every reference to that data item will also require renumbering. If renumbering is thought essential for neatness, let it be done when the rest of the design documentation is totally completed.

MM 2.1.4–2.3.2. The reports all follow the same basic documentation pattern. The purpose of some of these reports might be questioned, since some are very close in content to others. The response must come from the user; it is difficult to question the need for computer-generated reports when upper management is currently requiring the same to be generated by hand. It is with very little difficulty that a program is written to generate a slight modification of an existing report. For example, the PSL report (MM 2.2.4) is only a slight variation of the SAP report (MM 2.2.2). However, it contains exactly the information required for the research activities annual report; therefore, the small additional effort required to produce it is justified. If there were a very large number of small variations, then a generalized report generator might be in order.

MM 3.0. This introductory section guides the reader to the part that is required for a given application. The Maintenance Manual is strictly a reference document; it is not anticipated that anyone would read it from start to finish except for review. These introductory sections facilitate the reference process by guiding the reader's future reference efforts. This third major section of the Maintenance Manual is the product of the five activities within step 4 (specify file layouts and structures). The five activities are: (1) specify file record layouts, (2) develop variable specifications, (3) develop organizational Warnier diagrams, (4) develop structure diagrams, and (5) assure file integrity. It should be clear that the various parts of this section do not evolve sequentially. At some point toward the end of the file design process, someone needs to review the documentation generated, organize it, and assure that it is complete and in order. Figure 3.1 of the MM is an excellent aid in this process.

MM 3.1. The introductory section for the module gives a general picture of all of the files in that module, and their interaction. Also, major interactions with other modules can be noted.

MM 3.1.1. The rationale for organizing variables into files was given early in Chapter 5. The account file is the central file for queries and report generation. There are three structural considerations which bear discussion. The first is the CONTRACT POINTER (D1.1,65), by which contract file (D2.1) information becomes immediately accessible to the BMC reports. The second is the NEXT-BY-DEPARTMENT link, which links the records in account order within departments; and a third is the NEXT-ACCOUNT link, which links the records in account order irrespective of departments. The decision to include these links was made after considerable reiteration, including some actual testing on the hardware. Because of the hardware and system software constraints, the large 671-byte record could be read as quickly as a smaller record. Thus, there was no advantage, and considerable disadvantage, in maintaining the links in a separate file. This will certainly not be true on all hardware.

MM 3.1.2. The transaction record is set up to contain all transaction information. It is linked as indicated to provide for the reports which require a compilation of purchase-order information within account.

MM 3.2. The proposal and contract files follow the same pattern for creation and structure as the files discussed in MM 3.1.1-2. The projection file records are not created by data entry. Rather, during monthly updates, data from other files are used to compile the projection information. Thus, a detailed program specification is required in order to define the variables within these records. It should be clear that the projection values could only be maintained in a separate file.

MM 3.3. Little documentation is required for the file in the EEC module.

MM 3.4. Parameter files are required mainly to keep track of things. They are called *parameter* files because they vary very little, or else, in a very systematic way, once they are created.

MM 3.5. Facilitation files are those created solely to facilitate the inner workings of the programs themselves. Any permanent file which is set up to speed file reading or writing can be classified as a facilitation file. Temporary files to this effect will be documented in their respective PSWD, rather than as facilitation files.

MM 4.1. The flow diagrams within Sections 4.1.1 and 4.1.2 will be completed during step 5, Solidify Flow Diagrams, as discussed in the early part of Chapter 6. These, in turn, lead to the generation of the File-program Cross-reference given in Table 4.1 in the Maintenance Manual.

MM 4.2. The program specifications are the final step in the design process. They are the central focus of communication to the programmer, since all other documentation is included by reference. For this reason, the following paragraphs have been developed to provide greater understanding of the development of the program specification Warnier diagrams. These will be referenced by PSWD numbers.

PSWD 0.1. Note first that all menu programs are numbered with process number 0, and they do not appear in the system data-flow diagram (see text Section 6.3.1). This first one is the system supervisory menu program, which is also used to generate "monthly reminders" prior to actually generating the options of the menu. The options are related by an exclusive OR, since only one can be selected. Within each option the program and corresponding PSWD, which generally have the same numeric value, are referenced. Under this action designation is the condition by user response and user intent. All of the options lead to further menus, as indicated by the respective PSWD numbers. Input is referenced by the documentation of the menu, and output is strictly the monthly reminders, since transfer of control is not considered to be a program output data flow.

PSWD 0.1a. The appendage of a letter to a PSWD number is not the usual way of referencing a subordinate PSWD, and this is only done a few times in this case study. There are two reasons that such nomenclature might be adopted: (1) to avoid extensive renumbering, or (2) to reference a minor subroutine that will not fit on the same page as the calling routine. In this case, the second reason was in effect. The monthly reminder is spelled out in this procedure since it is not elsewhere documented in detail. It is only written out on the screen when the month changes, as detected by a comparison of the month generated by the system clock against the month in the D4.4 header which is updated whenever the monthly updates are performed.

PSWD 0.1.1. This is the first option of P0.1. With the exception of the fourth option, which is a report-generator menu, all of the options chain to programs referenced in the data-flow diagram. Note that P1.2 is invoked by both menu selections 2 and 3; however, this menu program will pass a different value of the variable OPTION depending upon the selection which will control the execution of P1.2. Note the ordering of the PSWDs. After P0.1, P0.1a is cleared first, followed by P0.1.1; however, P0.1.1.1 will be cleared prior to going back to the supervisory menu. Thus, all menu programs are presented first, followed by the nonmenu programs in their numerical order. Since the appendage of a number or letter onto a PSWD number is only done to illustrate that it is a subroutine, all such PSWDs are cleared before proceeding to the next sequentially numbered PSWD.

PSWD 0.1.1.1. To illustrate what was stated immediately above, P0.1.1.1 is considered a subroutine of P0.1.1, and therefore it is placed after P0.1.1 and before P0.1.2. This is the BM & C report generator and all programs invoked generate reports. However, the output from this program is stated to be "none," since this menu program is strictly for control transfer.

PSWD 0.1.2–PSWD 0.1.3.1. These are the menu program specifications for the other two modules in the MASTER system. The considerations are similar to those given for the first module above.

PSWD 1.1. This is the first nonmenu program specification. Consider first the input and output, which are referenced to the documentation. Since the Overview and User Guide (O&UG) provide a complete specification of queries and data-entry mats, this need not be repeated. Further, the file specifications and the source definition for each of the variables in the file layout in the Maintenance Manual define the output. In this case the transition is quite simple (direct moves), with the exception of D1.1,66 and D1.1,67, which are link definitions that are handled by PSWD 1.1.1. The remaining parts of the procedure must be written to maintain the distinction between the handling of a new-record creation as opposed to an old-record modification. The procedure clearly gives the user a chance to reiterate the data entry for review and correction.

PSWD 1.1.1. In Table D2.1 there is a reference to the narrative definitions of D1.1,66 and D1.1,67. Further, MM 3.1.1 contains structure diagrams which define this linkage. These, along with PSWD 1.1.1, might seem to be somewhat of an overkill as far as specification is concerned, since these are obviously redundant. They are justified, however, since different types of specification appeal to different programmers. The program specification is not attempting to dictate the coding technique which is at the discretion of the programmer. Therefore, there is an advantage to giving the programmer the concept of what is to occur, through narrative and structure diagrams, as well as the PSWD. Thus, in our judgment, the redundancy is justified. While this PSWD does define and use variables in a quasi-code type of specification, the programmer should be instructed during the walkthrough to use his/her judgment in bringing this linkage into existence. Finally, note that D1.1,65 and D1.1,68 are set equal to zero at this point since their value determination will be handled by other programs.

PSWD 1.2. This program, along with its subroutines (P1.2.1-P1.2.7), forms the most complicated procedure within the MASTER system. For this reason we will attempt to simulate a structured walkthrough of this as would be required prior to programming. This is given by the following:

a. Note first the inputs: (1) D4.3,2, which gives the last transaction reference so that the next number in the sequence can be assigned to a new transaction; (2) the value of OPTION (see P0.1.1), which will define whether the procedure is to be applied to a new transaction, or to update a transaction already entered; (3) the transaction data-entry mat; and (4) the affected account record.

b. Next note the outputs: (1) either a new transaction record or an update of the transaction record, and (2) an update of D1.1.

c. There are two major divisions in this procedure, so the query to determine the value of OPTION will be made first and the appropriate major procedure will be executed. Hence, the exclusive OR.

d. Let us first consider the situation when OPTION = NEW. This will require that D4.3,2 be read and incremented (it is inferred that this incremented value will be assigned to the new transaction and written back to D4.3,2).

e. The next step is to read the account record for the account number given in TDEM-4. After this, the type code is defined as the entry at TDEM-6.

f. The next step is to determine the "offset." We will cover this in PSWD 1.2.1. In the structured walkthrough it is essential that the entire major procedure be traversed before going into details on the subroutines. Thus, at this point in the walkthrough, the programmers and designers should not be concerned with understanding or explaining how the offset is determined. They might just mention that it will be used to determine the specific variable locations for updating the account record.

g. The next step is described as "Branch-to-account update," which is essentially what the calling program does. The particular routine will depend upon the type code determined from data entry as stated in Item e above. Note that a generic definition is given, rather than a highly repetitious condition on each subroutine, mainly to save space. It is inferred that these subroutines will update the account file.

h. A separate PSWD is required for updating indirect costs, since this is a rather involved procedure. This routine is performed next.

i. One variable which was not defined by the TDEM was the department code. This needs to be written in D1.2,12 since it will be used for linkage. It is obtained from D1.1,4.

j. At this point the linkage routine can be called to establish D1.2,15.

k. Finally, the assembled record is written to D1.2. The OPTION = OLD procedure is just a file update routine, and we will not elaborate on it. Further walkthroughs are required on each of the subroutines of P1.2 referenced above. The description for each of the respective PSWDs will provide the information generally covered in such walkthroughs.

PSWD 1.2.1. We have defined the term *offset* to mean the number of bytes from some fixed reference point in the account record at which a variable value is to be written or updated. A simple decision Warnier diagram (DWD) is used to define the offset as a function of the subcode.

PSWD 1.2.2. Based upon the offset, the record is tentatively updated. Then a balance is computed, and the record is updated if there is a positive balance. If not, the user is allowed to re-enter the data, but no updates are made.

PSWD 1.2.3. This is the third of seven subprocedures invoked by PSWD 1.2. PSWD 1.2.2 through 1.2.5 deal with each of the four types of transactions, respectively. This PSWD is for nonpurchase-order updates. The particular updates are changed, but the remainder of the procedure is identical to PSWD 1.2.2.

PSWD 1.2.4. The purchase-order expenditure transaction is, by far, the most complicated. The procedure is conditioned upon whether or not the user indicates that this is a final transaction on the purchase order (TDEM-11 = Y or N). If N, there must be a check for a positive balance in the purchase order, a procedure which is documented in P1.2.4.1 (this will be considered below). Then, assuming that control is returned to this point, the D1.1 file is tentatively updated. Balances are now output to the screen, followed by the same verification step as applied in the two previous types of transactions. Following this, a final transaction rectification step is performed, which compensates in differences between the encumbrance and the actual expenditures.

PSWD 1.2.4.1. This is the subroutine to check for a positive balance in a purchase order. Due to the nature of the transaction file, all transactions which relate to the purchase order must be read to determine this balance. This is no problem since the file structure was designed to link together all transaction records related to a given purchase order. The "Compute PO Balance" reads all records for the specified purchase order and determines the value of the balance. Based upon whether this value is positive or negative, one of the next two major procedure steps is executed. In the case of a positive balance, control is returned to P1.2.4, and it continues with the next statement there. In the case of a negative balance the user is given the opportunity to either re-enter the dollar value of the transaction (TDEM-10), or to abort the transaction and start the P1.2 process all over.

PSWD 1.2.5. This is the final PSWD for the four types of transactions. Budget modification is one of the simplest transaction types.

PSWD 1.2.6. Indirect costs are updated as a fixed percent of expenditures. Provision exists to exclude up to three transaction subcodes from indirect costs (see NADEM-7). These are checked first and if a match is found, no action is taken. Also, transaction types 2 and 4 (PO encumbrance and budget modifications) are also excluded. Finally, the balance is checked and, depending upon whether this is a final transaction or not, the appropriate action is taken.

PSWD 1.2.7. The final subroutine of PSWD 1.2 is the transaction file linkage routine. Now that the updates to D1.1 have been verified and made as a result of the transaction, the transaction record itself can be written to D1.2. Prior to this, however, it is necessary to set the linkage of both the D1.1 and D1.2 records. The first step in this procedure is to obtain the first reference pointer from D1.1,68 in order to have a starting point for searching the D1.2 file. If the value of this pointer is zero, this indicates that the account has no prior transactions, and, therefore, the new transaction record will be the first transaction for that account and its location can thus be written to D1.1,68. (Note that these exceptional cases, called boundary conditions, account for over half of the routine specification.) There are two boundary conditions that are considered within the "update transaction file Next Links" subroutine. The first is a consequence of the one discussed above, in which case the search for the D1.2,15 linkage must start at the very "beginning" of the file as opposed to the D1.1,68 position. The second boundary condition needing consideration is where the current record being written should be the D1.1,68 location (i.e., it has key fields "lower" than the current first record). This is also handled by the search being initiated from the beginning of the file, although other methods are available for performing this. The typical case (not a boundary condition) is processed by the last four steps in the subprocedure, where the links within the D1.2 records are set so that retrieval by purchase order is facilitated.

PSWD 1.3. A check is first performed to be sure that the account file exists. For security, the DAST report is generated using PSWD 1.7. Then the records are deleted, starting at the first account transaction for the account (D1.1,68) and proceeding through the linked list, using D1.2,15 until all records for the account are found and deleted. Then D1.1,68 is set equal to zero for control within PSWD 1.3a. Finally, the records are relinked around the deleted records.

PSWD 1.3a. Before an account record is deleted, it is required that both the transaction records and the contract record be deleted. Note that this is in the reverse order in which the records are created. Since the deletion of these respective records will result in a zero written to D1.1,65 and D1.1,68, these are checked before proceeding. For security reasons, the DAS report is produced before record deletion. Following this the record is deleted and a relinking of the remaining records takes place.

PSWD 1.3b. At the end of each month, quarter, and year, after all relevant reports have been generated, it is necessary to zero out the registers which contain the totals for those variables which are accumulated on a periodic basis. MASTER puts it in the control of the user to produce all of the required reports, and to run the end-of-period updates. The first step in this procedure is to perform the reminders as documented in O&UG 2.1.7. Given that the user wants to continue, the appropriate variables are zeroed out. Following this, D4.3 is prepared for a new year's record. Then the current data is placed in the D4.4 header. This file is used during logon procedures, and the data comparison enables P0.1a to remind the user if the end-of-period updates are required.

PSWD 1.4. This is the first of the budget management and control report generators. Report-generator PSWDs are somewhat different from those primarily used for file update. The query input is referenced to the User Guide, and the output is referenced to the Maintenance Manual, leaving only the specifications of the program loops required by the report. This is done conveniently by the Warnier frequency qualifiers. Note that, following the first specification (for the query), the second specification modifies the procedure as a function of the query response. The next steps are to read the header and begin the traversal of the D1.1,67 link. Once the department code is found, outputs are generated within ledgers, and a summary is generated for the department. If ALL departments are specified, reference to the DC is ignored and all records are output by account number by using the D1.1,66 link. Since the first digit of the account number is, by definition, the ledger number, this assures that the output is by ledger.

PSWD 1.5. Since there are few computations involved in this report, the program specification concentrates upon the loops required. Specific file variable assignments to generate the output variables are given in the output specifications.

PSWD 1.6. This report-generator program specification is more involved since it includes the definition of calculated variables.

PSWD 1.7. It might be noted at this point that without the output report specification in the Maintenance Manual, these program specifications are rather meaningless. This is because every effort has been made to minimize the redundancy in specification. Thus, the two must be considered together to obtain the full picture. In this case the exception reporting capabilities, first by transaction type and secondly by date, are handled easily by Warnier techniques.

PSWD 1.7.1. These are additional computations required within PSWD 1.7.

PSWD 1.8. A simple decision Warnier diagram (DWD) is integrated to define the variable RUN, which is used to compute the item numbers for the output report.

PSWD 1.8.1. A second DWD is broken out on a separate page, since it could not fit within PSWD 1.8.

PSWD 1.9. This is the final report-generator PSWD, which follows the same pattern as established above.

PSWD 2.1 This is similar in most respects to PSWD 1.1, with the exception that the deletion routine is included. The OPTION parameter is obtained from the menu program P0.1.2.

PSWD 2.1.1. The key fields for establishing the linkage in D2.1 are field 5 and 2, concatenated in that order. The terminology used here for specifying this as a variable value is "(D2.1, concatenated (5,2))."

PSWD 2.2–2.2.1. These two procedures were developed easily by copying PSWD 2.1–2.1.1 and making modifications. Great care must be taken in such modification since it is quite easy to neglect details peculiar to the new application. This is a newly introduced danger in using the capabilities of modern word processors.

PSWD 2.3. File D2.3 is unique in that its variables are not obtained from a data-entry mat; rather, they are calculated from other information stored in the system. For this reason the PSWD must specify each variable of the file as a function of other variables defined within the system. Note that the Warnier frequency qualifier indicates that the process is repeated for each department, which corresponds to the layout of D2.3 records.

PSWD 2.4–2.9. These report generators are basically direct outputs from the files. Because of the complete file descriptions and the complete output descriptions, both already documented, these specifications can and should be very brief. The Warnier frequency qualifiers aid greatly in specifying the ordering and arrangement of the report.

PSWD 3.1. This is the first PSWD for module 3. Note the similarity to PSWD 1.1. Once one data-entry PSWD is established, it may be used as a pattern for the others.

PSWD 3.2–3.3. The two report generators for module 3 are quite simple, heavily referencing their input and output documentation.

PSWD 3.4. The delete-record routine for module 3 is quite simple, since the records in file D3.1 are not linked. It might be noted, however, that D3.1,48 provides space in the file for linking. This was included in anticipation of a decision in the future for processing that requires linkage. Because of the small number of records (200–300 per year maximum), this is not a large amount of wasted space.

Index